Riding the Tiger

Tiger conservation in human-dominated landscapes

Beauty, grace and power m[...] most loved animals, yet it i[s...] have been its downfall. Poaching for skins and body parts, loss of habitat and prey and conflicts between people and wild tigers have caused catastrophic declines in tiger numbers throughout their range. If wild tigers are to survive through the next century, we must act now. *Riding the Tiger* is a comprehensive, scientific and eminently readable account of the problems and possible solutions of securing a future for wild tigers. Lavishly illustrated in full colour, it is written by leading conservationists working throughout Asia. It is a vital information resource for tiger conservationists in the field, necessary reading for serious students of carnivore conservation and conservation biologists in general, and an accessible overview of tiger conservation for general readers.

JOHN SEIDENSTICKER pioneered the use of radio-telemetry to study the mountain lion in North America. He was co-leader of the team that captured and radio-tracked the first wild tiger in Nepal. He co-authored *The Javan Tiger and the Meru-Betiri Reserve: A Plan for Management* and *Sunderbans Wildlife Management Plan: Conservation in the Bangladesh Coastal Zone*. He is a conservation biologist and curator of mammals at the Smithsonian Institution's National Zoological Park, and affiliate professor of biology at George Mason University, Virginia. He is author or editor of more than 120 articles and books including the widely acclaimed *Great Cats* (1992), *Dangerous Animals* (1995) and *Tigers* (1996). Dr Seidensticker also serves as Chairman of the Save the Tiger Fund.

SARAH CHRISTIE has coordinated the zoo breeding programmes for the Amur and Sumatran tigers in Europe for the last six years, and maintains a network that links zoos with conservation efforts for wild tigers. She is Conservation Programmes Coordinator at London Zoo, Zoological Society of London, and is a member of IUCN's Cat and Conservation Breeding Specialist Groups.

PETER JACKSON has been Chairman of the Cat Specialist Group, World Conservation Union (IUCN) since 1983. He serves on the Save the Tiger Fund Council and is a member of the Asian Elephant specialist group. His previous books include *Endangered Species: Tigers* (1990), *Endangered Species: Elephants* (1990) and *Wild Cats: Status Survey and Conservation Action Plan* (1996).

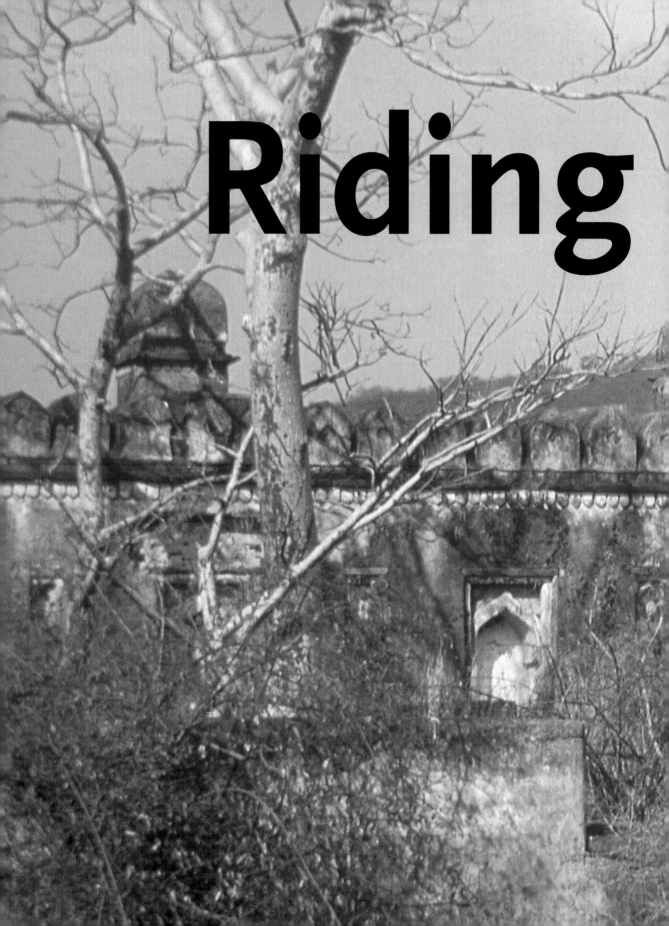

the Tiger

Tiger conservation in human-dominated landscapes

**Edited by
John Seidensticker,
Sarah Christie and Peter Jackson**

THE ZOOLOGICAL
SOCIETY OF LONDON

CAMBRIDGE
UNIVERSITY PRESS

PUBLISHED BY THE PRESS SYNDICATE OF THE UNIVERSITY OF CAMBRIDGE
The Pitt Building, Trumpington Street, Cambridge, United Kingdom

CAMBRIDGE UNIVERSITY PRESS
The Edinburgh Building, Cambridge CB2 2RU, UK http://www.cup.am.ac.uk
40 West 20th Street, New York, NY 10011–4211, USA http://www.cup.org
10 Stamford Road, Oakleigh, Melbourne 3166, Australia

First published 1999

Printed in the United Kingdom at the University Press, Cambridge

Typeset in FF Scala 9/13 pts. in QuarkXpress™ [HM]

A catalogue record for this book is available from the British Library

Library of Congress cataloguing in publication data

Riding the Tiger: Tiger conservation in human-dominated landscapes /
editors, John Seidensticker, Sarah Christie and Peter Jackson.
    p.   cm.
ISBN 0 521 64057 1. – ISBN 0 521 64835 1 (pbk.)
1. Tigers. 2. Wildlife conservation. I. Seidensticker, John.
II. Christie, Sarah, 1957– . III. Jackson, Peter, 1926–
QL737.C23R536 1998
333.95'97516 – dc21   98–24734 CIP

ISBN 0 521 64057 1 hardback
ISBN 0 521 64835 1 paperback

Title page illustration:
Tiger walking on the wall of the thousand-year-old abandoned fortress in Ranthambhore
National Park, India. Control of this fortress was once a key to controlling the surrounding
region. Today it provides a rare opportunity for tigers to replace people in the landscape.
Over most of the tiger's once vast range, people now predominate at the tiger's expense.

Contents

Contributors

Sean C. Ahearn, Department of Geology and Geography, Hunter College, City University of New York, NY 10021, USA.

Anne-Marie Alden, Tiger Information Center, c/o Minnesota Zoo, 13000 Zoo Boulevard, Apple Valley, MN 55124, USA.

Bastoni, Sumatran Tiger Project, PO Box 190, Metro, Lampung 34101, Sumatra, Indonesia.

Dorene Bolze, Wildlife Conservation Society, International Programs, 2300 Southern Boulevard, Bronx, New York, NY 10460–1099, USA.

Marnie Bookbinder, World Wildlife Fund-US, Conservation Science Program, 1250 24th Street NW, Washington DC 20037, USA.

Gerald Brady, Potter Park Zoo, 1301 S. Pennsylvania Avenue, Lansing, MI 48912, USA.

Richard Burge, Director General, Zoological Society of London, Regent's Park, London NW1 4RY, UK.

Mary Carrington, SAIC Frederick, National Cancer Institute, Frederick, MD 21702–1201, USA.

K. M. Chinnappa, PO: Kakur, Srimangala, Kodagu District, Karnataka-571 217, India.

Sarah Christie, London Zoo, Zoological Society of London, Regent's Park, London NW1 4RY, UK.

Raghunandan S. Chundawat, Wildlife Institute of India, PO Box 18, Chandrabani, Dehra Dun 248 001, India.

Kathy Conforti, Department of fisheries and Wildlife, 200 Hodson Hall, 1980 Folwell Avenue, University of Minnesota, St. Paul, MN 55108, USA.

Melissa Connor, Wildlife Conservation Society, International Programs, 2300 Southern Boulevard, Bronx, New York, NY 10460–1099, USA.

Eric Dinerstein, World Wildlife Fund-US, Conservation Science Program, 1250 24th Street NW, Washington DC 20037, USA.

Yuri M. Dunishenko, Far Eastern Branch, All Russia Institute of Hunting and Wildlife Management, Khabarovsk, Russia.

Karin Vaud Eliot, Global Survival Network, 1348 'T' Street, NW, Washington DC 20009, USA.

Neil Franklin, Sumatran Tiger Project, PO Box 190, Metro, Lampung 34101, Sumatra, Indonesia, and University of York, Heslington, York Yo1 5DD, UK.

Steven Russell Galster, Global Survival Network, 1348 'T' Street, NW, Washington DC 20009, USA.

Neel Gogate, Wildlife Institute of India, PO Box 18, Chandrabani, Dehra Dun 248 001, India.

Prashant Hedao, World Wildlife Fund-US, Conservation Science Program, 1250 24th Street NW, Washington DC 20037, USA.

Ginette Hemley, World Wildlife Fund-US, 1250 24th Street NW, Washington DC 20037, USA.

Maurice G. Hornocker, Hornocker Wildlife Institute, PO Box 3246, Moscow, Idaho 83843, USA.

Peter Jackson, Cat Specialist Group, World Conservation Union (IUCN) 1172 Bougy, Switzerland.

A. J. T. Johnsingh, Wildlife Institute of India, PO Box 18, Chandrabani, Dehra Dun 248 001, India.

Warren Johnson, Laboratory of Genomic Diversity, National Cancer Institute, Frederick, MD 21702–1201, USA.

Leslie Johnston, 1946 Calvart St., NW #1, Washington, DC 20009, USA.

Anup Joshi, Department of Fisheries and Wildlife, 200 Hodson Hall, University of Minnesota, St. Paul, MN 55108, USA.

Budsabong Kanchanasaka, Wildlife Technical Division, Royal Forest Department, Paholyothin Road, Chatuchak, Bangkok 10900, Thailand.

K. Ullas Karanth, Wildlife Conservation Society, International Programs, 2300 Southern Boulevard, Bronx, New York, NY 10460–1099, USA, and 403 Seebo Apartments, 26–2, Aga Abbas Ali Road, Bangalore 560 042, India.

Bijaya Kattel, Department of National Parks and Wildlife Conservation, Babar Mahal, Kathmandu, Nepal.

Andrew C. Kitchener, Department of Geology and Zoology, Royal Museum of Scotland, Chambers Street, Edinburgh EH1 1JF, Scotland.

Ashok Kumar, Wildlife Protection Society of India, Thapar House, 124 Janpath, New Delhi 110 001, India.

Kathy MacKinnon, The World Bank Group, 1818 H Street NW, Washington DC 20433, USA.

Jansen Manansang, Taman Safari Indonesia, Jalan Raya Puncak, Cisarua, Bogor, Indonesia.

Thomas Mathew, World Wildlife Fund-US, 1250 24th Street NW, Washington DC 20037, USA.

Charles McDougal, Tiger Tops, PO Box 242, Kathmandu, Nepal.

Marilyn Menotti-Raymond, Laboratory of Genomic Diversity, National Cancer Institute, Frederick, MD 21702–1201, USA.

Troy W. Merrill, Hornocker Wildlife Institute, PO Box 3246, Moscow, Idaho 83843, USA.

Judy A. Mills, TRAFFIC East Asia, Room 1701, Double Building, 22 Stanley Street, Central, Hong Kong.

Dale G. Miquelle, Hornocker Wildlife Institute, PO Box 3246, Moscow, Idaho 83843, USA.

Hemanta Mishra, Global Environment Facility Secretariat, The World Bank Group, 1818 H Street, NW, Washington, D.C. 20433, USA.

Sriyanie Miththapala, National Zoological Park, Smithsonian Institution, Washington DC 20008, USA.

Jessica Mott, World Bank Group, 1818 H Street Northwest, Washington DC 20433, USA.

Alexander E. Myslenkov, Sikhote-Alin Biosphere Reserve, Terney, Primorski Krai, Russia 692 150.

Philip Nyhus, Sumatran Tiger Project, c/o Minnesota Zoo, 13000 Zoo Boulevard, Apple Valley, MN 55124, USA, and University of Wisconsin, Madison, WI 53706, USA.

Stephen J. O'Brien, Laboratory of Genomic Diversity, National Cancer Institute, Frederick, MD 21702–1201, USA.

David Olson, World Wildlife Fund-US, Conservation Science Program,1250 24th Street NW, Washington DC 20037, USA.

Jill Pecon-Slattery, Laboratory of Genomic Diversity, National Cancer Institute, Frederick, MD 21702–1201, USA.

Dimitriy G. Pikunov, Institute of Geology, Far Eastern Branch, Russian Academy of Sciences, Vladivostok, Russia.

Howard B. Quigley, Hornocker Wildlife Institute, PO Box 3246, Moscow, Idaho 83843, USA.

Alan Rabinowitz, Wildlife Conservation Society, International Programs, 2300 Southern Boulevard, Bronx, New York, NY 10460–1099, USA.

Arup Rajuria, King Mahendra Trust for Nature Conservation, Nepal Conservation Research and Training Centre, P.O. Bachhauli, Sauraha, Chitwan District, Nepal.

Arun Rijal, King Mahendra Trust for Nature Conservation, Nepal Conservation Research and Training Centre, P.O. Bachhauli, Sauraha, Chitwan District, Nepal.

John G. Robinson, Wildlife Conservation Society, International Programs, 2300 Southern Boulevard, Bronx, New York, NY 10460–1099, USA.

Bittu Sahgal, Editor of Sanctuary Asia. 602 Maker Chambers V, Nariman Point, Bombay 400 021, India.

Bart Schleyer, Hornocker Wildlife Institute, PO Box 3246, Moscow, Idaho 83843, USA.

John Seidensticker, National Zoological Park, Smithsonian Institution, Washington DC 20008, USA.

Huang Shi-Qiang, Beijing Zoo, Beijing, People's Republic of China.

Saksit Simcharoen, Wildlife Technical Division, Royal Forest Department, Paholyothin Road, Chatuchak, Bangkok 10900, Thailand.

Dwiatmo Siswomartono, Directorate General of Forest Protection and Nature Conservation, Department Kehutanan Lt. 7/Blk. VII, Jalan Gatot Subroto, Jakarta 10270, Indonesia.

Evgeny N. Smirnov, Sikhote-Alin Biosphere Reserve, Primorski Krai, Russia 690 150.

James L. David Smith, Dept of Fisheries and Wildlife, 200 Hodson Hall, 1980 Folwell Avenue, St Paul, MN 55108, USA.

Sriyanto, Sumatran Tiger Project, PO Box 190, Metro, Lampung 34101, Sumatra, Indonesia.

J. Claiborne Stephens, Laboratory of Genomic Diversity, National Cancer Institute, Frederick, MD 21702–1201, USA.

Bradley M. Stith, Department of Wildlife Ecology and Conservation, University of Florida, Gainesville, FL 32611, USA.

Sumianto, Sumatran Tiger Project, PO Box 190, Metro, Lampung 34101, Sumatra, Indonesia.

Fiona Sunquist,125 Mason Road, Melrose, FL 32666, USA.

Mel Sunquist, 118 Newins–Zeigler Hall, Florida Museum of Natural History, University of Florida, Gainesville, FL 32611, USA.

Sompon Tanhan, Forest Engineering Division, Royal Forest Department, Paholyothin Road, Chatuchak, Bangkok 10900, Thailand.

Valmik Thapar, 19 Kautilya Marg, Chanakyapuri, New Delhi 110021, India.

Janet Tilson, Tiger Information Center, c/o Minnesota Zoo, 13000 Zoo Boulevard, Apple Valley, MN 55124, USA.

Ronald Tilson, Minnesota Zoo, 13000 Zoo Boulevard, Apple Valley, MN 55124, USA.

Schwann Tunhikorn, Wildlife Research Division, Royal Forest Department, Paholyothin Road, Chatuchak, Bangkok 10900, Thailand.

Pan Wenshi, Department of Environmental Biology and Ecology, College of Life Science, Beijing University, Beijing 100 871, China.

Joelle Wentzel, SAIC Frederick, National Cancer Institute, Frederick, MD 21702–1201, USA, and University of Pretoria, Department of Zoology and Entymology, Pretoria, South Africa 00002.

Eric D. Wikramanayake, Conservation Science Fellow, World Wildlife Fund-US, Conservation Science Program,1250 24th Street NW, Washington DC 20037, USA, and Conservation Scientist, Worldwide Fund for Nature – Indochina Programme, 116 Yet Kieu St., Hanoi, Vietnam.

Belinda Wright, Wildlife Protection Society of India, Thapar House, 124 Janpath, New Delhi 110 001, India.

Naoya Yuhki, Laboratory of Genomic Diversity, National Cancer Institute, Frederick, MD 21702–1201, USA.

Lu Zhi, Department of Environmental Biology and Ecology, College of Life Sciences, Beijing University, Beijing 100 871, China.

Picture credits

Photo editor: Sarah Christie
Photo captions: Sarah Christie and John Seidensticker

Foreword

Richard Burge
Director General, Zoological Society of London

Looking down at me in my office in ZSL is a painting by Joseph Wolf, a man whom Landseer called the greatest wildlife artist of his age. The painting was completed in 1850, and shows a tiger creeping through the jungle, while above him a troop of macaques screams defiance and fear. It is an extraordinary work of art. The tiger is almost invisible in the foliage, dense and dripping with moisture in a way familiar to all of us who have had the privilege of working in tropical forests. It is the tiger in its home; a fearsome predator at one with its world.

That painting is an important statement of the value with which we endow this extraordinary animal, and the responsibility we feel for its environment. The tiger is more than the charismatic predator: it is a keystone species in its environment. By saving the tiger in the world, we save complex ecosystems and habitats that would otherwise be destroyed in the relentless march of human need and, all too often, greed.

For all conservation zoos around the world, *in situ* conservation is our prime target and an article of faith. The full force of strong and convincing science, the skills harnessed within the zoological community, and the motivation that seeing living animals can give to people all around the world are the ways in which we at the Zoological Society of London (ZSL) have contributed, and will continue to contribute, to tiger conservation.

The Tigers 2000 meeting at which the information in this book was first presented was another major step in ZSL's ongoing commitment to tigers. This was the first time in a decade that 'tiger people' had come together to share information and ideas, and it occurred at a critical juncture in the international effort to save the species. Activity since the meeting has demonstrated the power and effectiveness of bringing minds together and moving forward, in step, with a common purpose. There were some important adjustments made to this route at the London meeting. The habitat assessment framework serves well as an example; this work, which examined all remaining tiger habitats and graded them for conservation importance, is of crucial importance in showing clearly where the priority areas are and in providing range states with the information they need to develop common and inclusive strategies for conservation action. It also provides a useful model for future action in all areas of tiger conservation; such holistic approaches will be of vital importance in pulling together innovative measures such as those described in these pages into integrated, co-operative regional and international initiatives across the tiger's range.

Tigers 2000 also saw the launch of 21st Century Tiger, a new fund-raising partnership between ZSL, Tusk Force and Global Tiger Patrol. We were delighted to welcome to the meeting Ian Upson, Managing Director of Esso UK plc, who introduced the partnership and Esso's role in it as founding sponsor; and John Gummer, then UK Secretary of State for the Environment, who both welcomed this initiative and announced a donation of £50 000 from the UK government.

ZSL is immensely grateful to a number of organisations for supporting the meeting and its participants. In particular we thank Esso UK plc, which funded the Tigers 2000 symposium, British Airways for providing flights for many speakers, CVI Ltd for printing services, and the Save the Tiger Fund (a joint project of the US National Fish and Wildlife Foundation and the Exxon Corporation) for grants which have enabled ZSL to produce this book in full colour and distribute it to relevant scientists and conservationists around the world.

We would also like to thank the many Tigers

2000 participants whose advice and ideas were invaluable in planning the symposium. Sriyanie Miththapala, Georgina Mace, Paul Beier, Sarah Durant and Mike Bruford kindly reviewed various of the papers, Susan Lumpkin provided the book's title and the title page picture caption, and Jean McConville was invaluable in checking manuscripts throughout. Thanks are also due to the many photographers whose work features here; all the pictures were provided free or at substantial discounts.

ZSL hopes that this book will prove an important source of primary reference, of information, and of inspiration for all those committed to and active in tiger conservation.

Preface

John Seidensticker, Sarah Christie and Peter Jackson

More than a quarter of a century has passed since the tiger was first internationally recognised as endangered and soon to be extinct in the wild if the forces resulting in its decline continued unabated. Over the ensuing years, considerable resources have been invested in saving the tiger, with mixed results (Table 0.1). Many small tiger populations are completely isolated, critically endangered and facing a bleak future. Entire subspecies from Bali, Java and areas adjacent to the Caspian Sea have not survived; the Caspian and Javan tigers became extinct after the tiger's endangered status was widely recognised (Fig. 0.1). The South China tiger, if not already extirpated in the wild, is down to a few individuals and is slipping away (Tilson *et al.* 1997). There has been much anxiety for the tiger and its future in the general and conservation press. Concerned, knowledgeable people predict the tiger's demise, with some isolated small populations expected to blink out in the near future.

With these dire predictions have come proposals for actions to assure the tiger's future in the wild. But there are no hard statistics concerning the tiger's status compiled in any central database – no reliable index of the trend in tiger numbers, available habitat, or available prey, or assessments of the risks it faces in the various sectors of its range, or even a catalogue of conservation activities under way. Even a cursory review of the good science available on tiger natural history and the changing context of tiger conservation reveals how little hard information there is upon which to base planning and conservation actions to secure the tiger's future on its home ground. The Tigers 2000 Symposium at the Zoological Society of London (ZSL) was conceived and held as part of a remedy for this glaring shortcoming.

As convenors preparing for Tigers 2000, we discussed the significant questions facing efforts to save the tiger with our colleagues through the autumn of 1996, and again after the symposium while preparing these contributions for publication. We – the editors and contributing authors – hold strong feelings about tigers and believe that tigers should and can remain a part of the Asian landscape. We also understand that tigers are in trouble because of conflicts over values. We believe that live tigers are worth more than dead tigers and that landscapes with tigers are worth more than landscapes without them. But we also realistically wonder how widely this vision is shared.

Conflicts of visions may dominate history, but conflicts of interest dominate in the short-term. How do we prevail over short-term interests that leave landscapes devoid of tigers? We asked our colleagues: what road maps can we draw to secure a future for the tiger in the rapidly changing social and natural landscapes of Asia? There was a consensus that saving the tiger can only be accomplished through a series of partnerships, and that it depends on the people who live near and with tigers every day. It is these people who pay the highest price and who must be convinced that saving the tiger is worth their while. Otherwise wild tigers will not survive.

We came away from our discussions thankful for e-mail and faxes and with a healthy respect for the power of networking. We all think we know about tigers and what tigers need to survive. But do we really? We come from a school that teaches that before trying to fix a problem, define the problem. What do we know about the tiger's needs? Can we accurately map tiger distribution? Do we know what limits tiger distribution? Tiger numbers? Can we chart the market forces leading to the haemorrhage in tiger numbers during the late 1980s and early 1990s? Once we understand these forces, where, how and when can we effectively intervene to save and sustain the tiger? One clear lesson has emerged

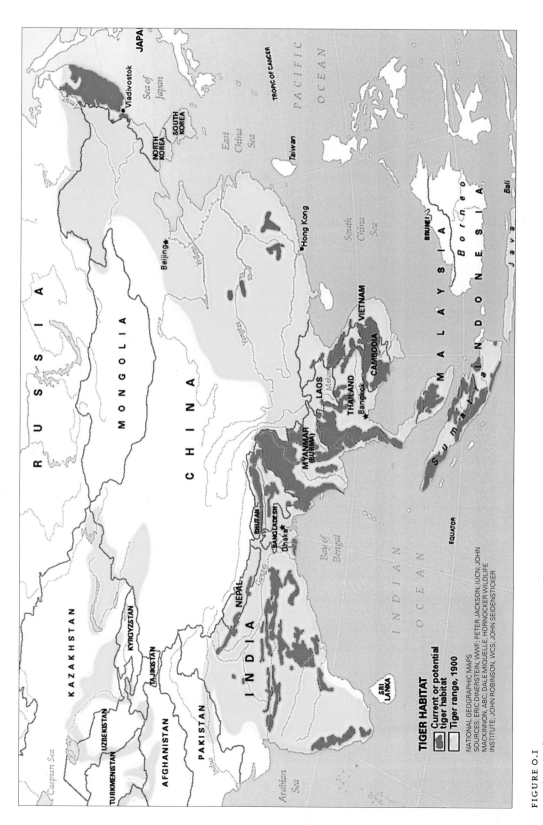

FIGURE O.I

Distribution of potential tiger habitat based on Level I and Level II Tiger Conservation Units as determined by E. D. Wikramanayake *et al.* (this volume). Priority Level III areas are not shown here but are shown in Figs. 18.2–4. Tigers do not currently inhabit all of these areas, as the detailed surveys of the structure of tiger populations in Nepal and Thailand by J. L. D. Smith and his colleagues (this volume) have shown. The estimated potential distribution of tiger habitat in south China was based on Koehler (1991), Marks (1998), Nowell & Jackson (1996), MacKinnon *et al.* (1996), Johnsingh (1996) and The World Bank (1995). The distribution of tiger habitat depicted in the Russian Far East is based on Matyushkin *et al.* (1996). Surveys during the winter of 1997/8 revealed a few Amur tigers in China along the Russian border (Dale Miquelle pers. comm.). *Map courtesy of the National Geographic Society.*

Table o.i. *The status of the tiger* Panthera tigris *(Linnaeus 1758) in May 1998*

Tiger subspecies	Minimum	Maximum	Source
Indian (Bengal) tiger *P. t. tigris* (Linnaeus 1758)	**3176**	**4556**	
Bangladesh	362	362	1
*Bhutan	67 (adults)	81 (adults)	2
China	30	35	3
India	2500	3750	4a, b
Myanmar, Western	124	231	5
*Nepal	93 (adults)	97 (adults)	6
Caspian (Turan/Hyrcanian) tiger *P. t. virgata* (Illiger 1815)			
Formerly Afghanistan, Iran, Chinese and Russian Turkestan, Turkey	Extinct 1970s		
Amur (Siberian/Ussuri/Manchurian/Northeast China) tiger	**360**	**406**	
***P.t. altaica* (Temminck 1844)**			
China	30	35	3
Korea (North)	<10	<10	7
*Russia	330 (adults)	371 (adults)	8
Javan tiger *P. t. sondaica* (Temminck 1844)	Extinct 1980s		
South China (Amoy) tiger *P.t. amoyensis* (Hilzheimer 1905)	**20**	**30**	3
Bali tiger *P.t. balica* (Schwarz 1912)	Extinct 1940s		
Sumatran tiger *P. t. sumatrae* (Pocock 1929)	**400**	**500**	9
Indo-Chinese tiger *P. t. corbetti* (Mazák 1968)	**1227**	**1785**	
Cambodia	150	300	10
China	30	40	3
Laos	Present		
Malaysia	491	510	11
Myanmar, Eastern	106	234	5
Thailand	250	501	12a, b
Vietnam	200	200	13
Totals	**5183**	**7277**	
Rounded totals (nearest 500)	**5000**	**7000**	

Table compiled by Peter Jackson, Chairman, Cat Specialist Group, The World Conservation Union (IUCN), from reports from range countries. Most estimates are educated guesses, but censuses in Bhutan, Nepal and Russia provided more reliable numbers.

*Figures for Bhutan, Nepal and Russia are for adult tigers counted. Tiger specialists consider such figures more realistic because many cubs are unlikely to survive to maturity.

Sources
1.   Jalil, S. M. (1998) Bengal tiger in Bangladesh. Unpubl. report to Year of the Tiger Conference, Dallas, USA.
2.   McDougal & Tshering (1998).
3.   Wang Wei (1998) Status of tigers in China and conservation strategies. Unpubl. report to Year of the Tiger Conference, Dallas, USA.
4a.   Thapar, V. (1994) pers. comm.
4b.   Project Tiger (1993).
5.   Uga, U. & Thang, A. (1998) Review and revision of the National Tiger Action Plan (1996) of Myanmar. Unpubl. report to Year of the Tiger Conference, Dallas, USA.
6.   Anon. (1997) Tiger Action Plan for the Kingdom of Nepal. Unpubl. report to Year of the Tiger Conference, Dallas, USA.
7.   Pak U-ll (1994) in litt.
8.   Matyushkin *et al.* (1996).
9.   Ministry of Forestry (1994).
10.   Samith, C., Sophana, V., Vuthy, L. & Rotha, K. S. (1995) Tiger and prey species. Unpubl. report presented at training course, International Centre for Conservation of Biology, Lanchang, Malaysia.
11.   Jasmi bin Abdul (1998) The distribution and management of the Malayan tiger in Peninsular Malaysia. Abstract presented at Year of the Tiger Conference, Dallas, USA.
12a. Rabinowitz (1993).
12b. Anon. (1998) Thailand's tiger action plan. Unpubl. report presented to Year of the Tiger Conference, Dallas, USA.
13.   Bao, T. C., Nhat, P., Dang, N. X., Phu, T. K., Hoi, N. V., Thanh, V. N., Luong, L. V., Thank, N. H. & Nguyen, D. D. (1995) Tiger Conservation in Vietnam: Status and problems. Unpubl. report presented at training course, International Centre for Conservation of Biology, Lanchang, Malaysia.

after 30 years of attempting to save the tiger: there is no magic bullet. Saving the tiger is an exercise in breaking down the larger problem into smaller focused ones with technically practical and politically feasible solutions. No one can do this alone. The future of the tiger lies in reaching out and forging and sustaining key partnerships.

In these waning years of the twentieth century, the relationship between conservation and science has been strengthened. Or, as Howard Quigley expressed it during our discussions, 'Good conservation is based on good science'. Of course good science is only one suture in the activities and programmes needed to save and sustain tiger populations. But it is a place to begin. Any discussion concerning the science of tigers that did not place the tiger in the context of conservation actions would yield little that was new or useful in securing the tiger's future in Asia. Accordingly, we sought and included contributions that reached beyond the science of this splendid great predator, and placed it in the swirling, shifting social and landscape mosaic that is Asia today.

The science of tigers was raised to a plateau above museum cabinets and hunters' accounts by George Schaller (1967) in his now classic *The Deer and the Tiger*, published just as there was growing realisation that the great cat was becoming endangered. Additional resolution in our understanding of tiger ecology and behaviour was achieved with a blending of ancient and modern technologies. Trained elephants allowed scientists to approach tigers in riverine forest/tall grass habitats in Nepal, capture them with chemical restraints, and attach radio-transmitters. Through long-term radio-tracking, from the ground, elephant-back, four-wheel-drive vehicles and aircraft, scientists gained unprecedented insights into tiger foraging, mating, rearing, dispersal and refuging systems (Sunquist 1981; Smith 1993). The challenge of adding to our understanding of tigers has been taken up by many individuals and research teams, and Tigers 2000 was an opportunity to review and synthesise these findings. We asked contributors in Part I to summarise what we know about the resilience of the tiger in genetic, morphological, behavioural and

ecological terms. We also link the tigers we study and the tigers in our minds. Most scientists who study tigers also understand that they have a high metaphysical content, as do those places where tigers lived and still live. In our search for tigers in nature, we face the challenge of recognising manifestations of the conceptions and images of the tiger that we carry in our minds (Cronon 1996). This strongly influences our valuation of tigers, our view of their place in nature, and the approaches we take to secure the tiger's future.

With the exception of efforts to identify and understand factors that led to some tiger extinctions, the study of tigers has largely ignored the effect of the presence of humans. The science of tigers has also largely been viewed within the 'equilibrium' or 'balance of nature' paradigm. Conservation action under this paradigm argues that one simply selects pieces of nature for protection, leaves them undisturbed, and they will retain their species composition and function indefinitely and in balance (Meffe & Carroll 1997). Anyone who has worked with tigers on their home ground knows, however, that it is impossible to isolate and protect tigers from human influences. In Part II, we asked contributors to focus on a

By blending ancient and modern technologies, scientists have gained new insights into the ecology and behaviour of wild tigers.

new generation of research that is developing an understanding of tigers living in large, dynamic landscapes. In moving towards an understanding of how tigers live in the context of these larger spatial and temporal scales, it is no longer possible to view the tiger as unaffected by the actions of humans. In this paradigm and context for establishing the ecological criteria for tiger conservation, we can expect to craft more realistic conservation solutions to sustain tiger populations in the future.

Contributors to Part IIIA–C recognise that human values ultimately put boundaries and constraints on our ability to manage landscapes where tigers live. In other words, to secure a future for the tiger across Asia, tiger conservation must be adaptable, contextually relevant and socially acceptable. Dale Miquelle and his co-workers used the term 'establishing political criteria' for tiger conservation activities to characterise this approach. On-the-ground tiger conservation activities are determined by issues of cultural history, valuation, management systems and policy processes. We asked contributors to Part IIIA–C to present their insights into the linkages and the layers involved in seeking to relieve first the social and economic forces that drive tiger poaching and the loss of tiger habitat and prey, and second the conflicts between people and tigers sharing habitat. It may seem trite to simplify this very complex environmental problem and say that contributors sought to frame tiger conservation activities in win-win terms for people and tigers. But we believe that at the end of the day that is what it comes down to, or the tiger will lose.

Tigers 2000 was a gathering of people, some of whom have spent all their adult lives working to understand and save this great cat. A good many had never met before. For two formal meeting days and a third informal day, we spoke, compared notes and impressions, did reality checks, examined what was happening, and confirmed that we knew and agreed on the problems and potential solutions for saving the tiger. For ourselves and for many of the participants, Tigers 2000 was a paradigm-defining event and a bright spot for the tiger's future.

Tigers are deeply embedded in the cultural history of Asia and appear in the West in many manifestations. Shiva as destroyer is pictured wearing a tiger skin and riding a tiger. His consort Parvati the Beautiful rides a tiger in her aspect as Destroyer of Evil. A disciple of Buddha rides a tiger to demonstrate his powers and ability to overcome evil (Lumpkin 1991). 'Dictators ride to and fro upon tigers which they dare not dismount. And the tigers are getting hungry' said Winston Churchill, drawing on a Chinese proverb to warn of impending dangers in Europe in 1936. We selected Riding the Tiger as the title of this work, not as a complex allegory from cultural history, but as a metaphor of the relationships and options we now face in securing a future for wild tigers.

We are at a precarious fork in the road. 'Riding the tiger' means moving forward. If we stop 'riding the tiger' by letting up on our concern, we will indeed lose the tiger and wild Asia, as we did in Java. 'Formerly a symbol of nature's power, the Javan tiger is now a symbol of paradise lost,' (McNeely & Wachtel 1988). Alternatively, we can make the difficult choices that will sustain tigers in the future in other reaches of their range. The future for wild tigers lies in recognising that tigers and people are inseparable and both are bound by the nature of their relationship with each other and their environment.

Those of us dedicated to saving the tiger may be a minority in an indifferent world, but we agree with anthropologist Margaret Mead's declaration: 'Never doubt that a small group of thoughtful, committed citizens can change the world. Indeed it is the only thing that ever has.'

Most remaining Amur
tigers live in the vast
Sikhote-Alin Mountains,
Russian Far East. In these
northern temperate forests
low prey density is the
norm and the exclusive
territories of adult tigers
often exceed 1000 km² for
males or 500 km² for
females.

PART I

Introducing the tiger

Overview *John Seidensticker, Sarah Christie and Peter Jackson*

> Tigers are so beautiful, so powerful, so secretive, so shrouded in myth that the daily reality of their lives can seem prosaic. Tigers do not 'roam' the forest as romantic writers like to have them do; instead they doggedly work carefully delineated territories, on the lookout for their next meal – and on the alert for any other predator that threatens access to it. They need meat, massive amounts of it, just to stay alive.
>
> (*Ward 1997*)

Understanding the factors that explain the life history adaptation of tigers and the consequences of these adaptations is the passport to understanding how we can expect tigers to respond to the extensive environmental changes occurring throughout tigerland. In Part I we ask: Is the tiger a hapless big carnivore that is too brittle to survive in Asia's rapidly changing landscape? Or, if it is resilient, what is the nature and extent of this resilience? Only with this understanding can we begin to set realistic goals for meeting the tiger's ecological needs in our tiger conservation programmes.

The tiger is the largest obligate terrestrial carnivore in all of the mammalian assemblages in which it occurs in Asia. It is a specialised predator of large ungulates. It is never found far from water but displays great adaptability in living in different climatic regimes, ranging from temperate oak-pine forests to tropical rain forests and mangrove swamps. The greatest ungulate species diversity and biomass in Asia are reached in areas where grasslands and forests form a mosaic and there is an interdigitation of many vegetation types. In these areas, the tiger also reaches its highest densities.

In these relatively closed habitats, the tiger lives and hunts these large ungulates alone. The efficiency of hunting and living alone in closed habitats has propelled the tiger's social organisation and is manifest in all its behavioural systems: food finding and feeding, mating, rearing, refuging and dispersal (Eisenberg 1981; Sunquist 1981). The other great *Panthera* cat, the group-living lion, is usually associated with more open habitats. Both were products of the environmental turmoil of the Pleistocene, evolving as predators to follow the radiations of large ungulates, particularly the Bovidae and Cervidae. The tiger evolved in Asia and has remained there, tied to its closed habitats. Aside from lions having relatively longer legs, it is virtually impossible to tell these two cats apart without their skins. They are the giants in their respective guilds. Lions and tigers are about four times the size of leopards, the next largest cat in their guilds today. There is strong sexual dimorphism and adult male lions and tigers can be nearly 1.7 times heavier than females.

In this section, Mel and Fiona Sunquist and Ullas Karanth review the life history adaptations of the tiger in the context of the mammalian assemblages in which it evolved and document what we know about the tiger's ecological and behavioural resilience. To explore the tiger's proximate limitations, they prepared a historical retrospective of the conditions that led to the extinction of three of the eight named subspecies. They conclude that this great cat evolved as the predator of the largest deer, wild cattle and wild pig, and where these essential prey have been extirpated, the tiger does not survive. But where large prey are abundant, the tiger survives and has a robust reproductive output. This is a theme that we will return to throughout this book. Large blocks of habitat remain in some regions that could harbour tigers today, but the absence of large ungulate prey in these areas limits tiger numbers and distribution. Stemming the loss of habitat *per se* will not save tigers. Tiger conservation is dependent on the conservation of tiger prey populations.

In historic times, the tiger lived across 70 degrees of latitude and 100 degrees of longitude, from the Russian Far East south through Indochina, the Indian Subcontinent, south of the Himalayas, and into the Indus Valley (Fig. 0.1). Apart from this primary distribution, tigers lived adjacent to the Caspian Sea and on the Greater Sunda Islands of Sumatra, Java and Bali, echoes of Pleistocene climatic changes. Andrew Kitchener describes the flimsy data on which the eight subspecies of the tiger were first designated, and the phenotypic variation found in the tiger over its range. The Sunda Island tigers, for example, have about half the mass of tigers of the same sex from north-eastern India and Nepal (Seidensticker & McDougal 1993). Sunquist *et al* and Kitchener both conclude that, given the tiger's dispersal powers, the Sunda Island and Caspian Sea tiger populations retained some dispersal connectivity during some Pleistocene intervals with the main continental tigers, which were continuous in their population structure in historic times. Human disturbance over vast areas of Asia when swidden cultivation was the norm did not hinder tiger distribution. Rather, this form of agriculture tends to augment habitat for tiger prey and thus benefits tigers. In some regions forest clearing to the limits of cultivable land truncated and restricted tiger distribution to inaccessible, often mountainous, areas, but did not create barriers for tiger movements (Seidensticker 1986; Marks 1998). Kitchener proposes a new alignment of tiger subspecies for us to ponder, based on this recognition of historic population structure. And perhaps we should adopt his recommendations in our conservation planning, if we are wise. Kitchener recommends that we shift our conservation planning for tigers from a taxonomic to an ecological basis and the way to do so is outlined by contributors in Part IIIA–C.

Joelle Wentzel and a team from the laboratory of Steven O'Brien, using samples from tigers from known geographical locations submitted by many of the participants in this symposium, have begun to characterise the extent and character of genetic variation found in the tiger. In a survey of short tandem repeat (STR) polymorphism loci (see Box 3.1), they did find significant differences between the named subspecies. They also found a remarkable lack of genetic variation across the tiger's vast range; tigers appear to have as low a genomic variation as any cat species so far sampled. O'Brien and his associates attribute this condition to a bottleneck in tiger numbers late in the Pleistocene.

What does this tell us about the future of wild tigers living in small populations, or recommend to those charting tiger recovery plans? Low variation *per se* does not appear to have any impact on the viability of natural populations of tigers. It does not appear to impact fitness, threaten population growth, or affect short-term persistence. But we are looking at the end of the time when there was great connectivity in tiger population structure across their wide range. As contributors in the following sections demonstrate, those habitat conditions no longer exist and will never exist again. We know that breeding between closely related individual tigers in zoos results in significant inbreeding depression (Ballou & Seidensticker 1987; Mace & Christie in prep.) and this should caution against interpreting these results as indicating that tigers may have evolved a genome that is relatively unaffected by inbreeding (Lacey 1997).

There are deterministic processes (Gilpin & Soulé 1986) at work in the form of an ongoing ecological shift occurring across most of the tiger's geographical range. Forest fragmentation is occurring in landscapes increasingly dominated by agriculture, and there is a general environmental deterioration in remaining forest tracts from excessive extraction of fodder and other products, including the tiger's essential ungulate prey. This is vastly diminishing the quality of areas where the tiger can live. Through most of the tiger's former range, fragments of habitat are separated by hostile terrain. Today's tiger population is a highly fragmented metapopulation. In a metapopulation, a population of populations, the movement of individuals between patches of habitat is highly restricted because of the hostile conditions that prevail between patches. There is a high probability of extinction for tigers living in small habitat patches that lack any dispersal connectivity to other areas where tigers still live. And in these small isolated patches we will witness the extinction of tigers. This is in contrast to cases where metapopulation structure enables the rate of recolonisation to exceed the rate of extinction in individual habitat patches (McCullough 1996). Tigers can become extinct in small fragments of habitat – already diminished in quality through deterministic factors – because of extrinsic factors, such as variation in the influence of other species (e.g. pathogens, predators or poachers) and catastrophe (floods, fires or droughts), because of intrinsic factors such as demographic stochasticity and genetic deterioration, or through all of these factors acting together. These patches are not subsequently recolonised because all corridors to habitats where tigers remain have been lost. What is of profound concern is that the structure of the tiger population today is so highly fragmented – into more than 160 areas between which there is no dispersal connectivity (see Part IIIA–C) – and, as contributors to the symposium demonstrate in Part II, tiger population structure is further fragmented within those areas. The extinction of tigers in more than half of these areas may be inevitable.

There are tigers – living, roaring tigers that leave tracks, and kill deer, wild pigs and cattle, and sometimes people – and then there are the tigers that live in our minds. Peter Jackson explores this vexing problem in his essay on the cultural history of tigers from his own cultural background and his long-term affinity for tigers and their conservation. A stern lesson in the history of large carnivore conservation efforts has been learned through the tendency of biologists, acting as conservationists, to seek answers only in narrow ecological terms; this is failing because in reality large carnivore declines are attributable to how these animals are valued by societies (Clark *et al.* 1996). Large carnivores evoke complex images in our minds that greatly affect our valuation process and the prescriptions we propose for their conservation. Tigers and people have lived in proximity in Asia since long before any recorded history. Humans competed with tigers for food; sometimes tigers killed people and, with improved technologies, people were able to kill tigers. There always has been tension between people and tigers over who has control of nature in the 'wilds' of tigerland. This is manifest in the varied interpretation of the tiger in Asian cultures and traditions. We return to this theme in the Epilogue, recognising that the tigers in our minds affect valuation and consequently largely define the context for tiger conservation and recovery efforts.

With abundant prey, more tigers pack into a given space. Five or more adult tigers live within 100 km² in some tropical moist forest and tall grass protected areas in India and Nepal.

Ecology, behaviour and resilience of the tiger and its conservation needs

Mel Sunquist, K. Ullas Karanth and Fiona Sunquist

Introduction

Over the last century the tiger's landscape has changed dramatically. An expanding human population has put increased pressure on the tiger's habitat, its prey, and on the tiger itself. Forests and grasslands have been lost, degraded and fragmented, and ungulate populations have declined precipitously, both in abundance and distribution. Tiger numbers have also declined, and almost all remaining populations are now small and isolated. In the last 25 years these changes have accelerated, increasing concern for the continued existence of the tiger.

Conservation of the tiger in this constantly changing, human-altered landscape requires that we have a clear understanding of the animal's limits and capabilities. How flexible is the tiger in its resource requirements? If it is flexible, what are the limits? Can tigers respond reproductively to increased rates of mortality? How effective are tigers at dispersing across a fragmented landscape? In other words, is the tiger a resilient species? In this chapter we will first review the tiger's evolutionary history, since it is against this background that conservation efforts must also be viewed. We will then consider the tiger's reproductive, dispersal and predatory capabilities, anticipating that information on these three aspects of its natural history will help sketch the tiger's resiliency profile (Weaver et al. 1996). Finally, we will examine the events associated with the recent extinctions of the Caspian and Javan tigers, looking for additional insights into how best to conserve the tigers currently remaining in the wild.

We want to thank our many colleagues who have enhanced our understanding of tigers and our own lives as well. We thank John Seidensticker, Sarah Christie and Peter Jackson for inviting our participation in the Tigers 2000 symposium and for review of the manuscript. We also thank Esso UK, the symposium sponsor, and our host, The Zoological Society of London. The figures were prepared by Dale Johnson.

Evolutionary history

The geographical distribution of the tiger once spanned Asia, from eastern Turkey to the Sea of Okhotsk (Fig. 0.1). Over the last 50 years its range has been greatly reduced, but tigers are still found in a broad variety of forest types, including dry deciduous, moist deciduous, semi-evergreen, wet evergreen, riverine, swamp and mangrove. They are also found in the coniferous-deciduous forests of eastern Russia, in the tall-grass habitats south of the Himalaya, and the tropical rainforests of Sumatra and Malaysia. They show a corresponding tolerance to variations in altitude, temperature, and rainfall regimes.

That tigers are found in such a diversity of forest types and climates indicates that habitat per se was not a critical element in the tiger's evolutionary history. Rather, the divergence of the tigris line from the Panthera stock likely followed the Pleistocene radiation of the cervids and bovids in Southeast Asia (Flerov 1960; Geist 1971), as the evolution of large-bodied forest ungulates (e.g. Axis, Rusa, Cervus, Bos) created a niche for a large-bodied, forest-edge predator.

The Pleistocene was an epoch of glaciations and fluctuating extremes of climate; at least four separate glacial periods alternated with warmer interglacial phases. The cold temperatures associated with the ice ages were most pronounced in northern latitudes; in the tropics the most obvious effect was changing ocean levels. During the glaciations water was locked up as ice, lowering sea levels and exposing vast new areas of dry land. As the climate warmed the ice sheets melted and withdrew, and rising oceans returned to cover the land bridges. In Southeast Asia, the islands on the Sunda Shelf – Sumatra, Java, and Borneo – were alternately joined then separated as the ice caps formed and melted. For large mammals, the Pleistocene marked a major period of turmoil. Speciation and extinction rates quadrupled over those of the Tertiary, and several groups of mammals underwent explosive radiations (Kurtén 1971; Geist 1983).

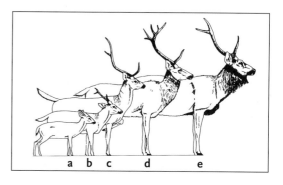

FIGURE I.I
Old World radiation of cervids, with changes in body size and antlers, for muntjac (**a**), hog deer (**b**), chital (**c**), barasingha (**d**) and sambar (**e**). Redrawn from Geist (1983).

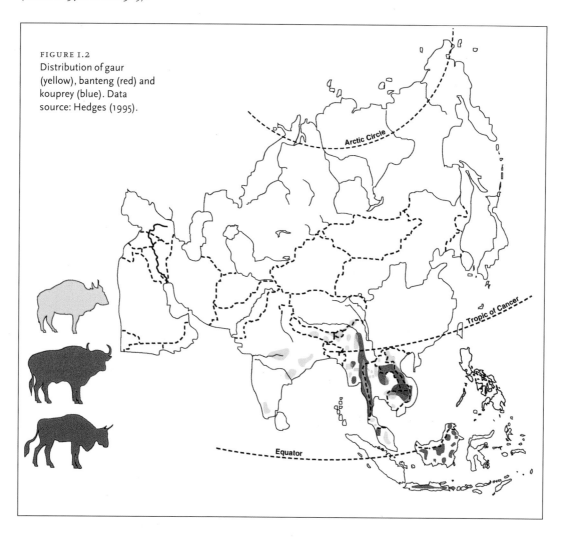

FIGURE I.2
Distribution of gaur (yellow), banteng (red) and kouprey (blue). Data source: Hedges (1995).

Deer flourished during the Pleistocene. From their evolutionary centre in Asia, small, ancestral, forest-living cervids, similar to present-day muntjac, radiated to occupy a variety of niches. Body size and antler complexity increased as the cervids diversified into forest-edge and grassland habitats (Fig. 1.1). The long sub-hypsodont molars of chital, hog deer and barasingha developed as these species became browser/grazers at the forest edge, or more specialised grazers in the savannahs and marshlands (Geist 1983). However, the majority of today's 14 or so cervid species still retain their forest-living affinities and inhabit dense forest, open woodland, or forest-edge habitats.

The bovid family also diversified into a huge variety of species – some 50 genera in Asia alone. The beginning of the Pleistocene marked the appearance of the true bovine, the cattle, bison and buffalo (Kurtén 1971). Colonising the more open habitats, these bovids developed high-crowned teeth, which were more resistant to the wear associated with feeding on silica-laden grasses. The wild cattle group is represented today in India and Southeast Asia by three species (Fig. 1.2). The banteng is an animal of open dry woodlands and forest glades; the kouprey, now almost extinct, is found in open dry forest; whereas the massive gaur prefers denser forest habitats, emerging at night to graze in clearings and glades (Wharton 1957, 1968; Schaller 1967; Hoogerwerf 1970).

An examination of the distribution of cervids in south Asia (Figs. 1.3 & 1.4) reveals that where annual precipitation is less than 500 mm conditions are generally too dry for most cervids (Fig. 1.5). Conversely, the diversity and abundance of cervids decline where conditions are too wet. Within the tropical forests of south Asia the terrestrial biomass of ungulates increases with annual rainfall up to about 1900 mm. Above 1900 mm/year the positive correlation between rainfall and ungulate biomass breaks down (Fig. 1.6). True tropical rainforest offers little primary productivity at ground level, and mammalian biomass is dominated by arboreal herbivores such as primates (Eisenberg 1980). For example, Gunung Leuser National Park in Sumatra, which typically receives over 2000 mm rainfall a year, has six species of primates but the few ungulate species (e.g. sambar, wild boar, muntjac) present occur at extremely low densities (Whitten et al. 1987; Griffiths 1996).

The greatest ungulate biomass in south Asia is reached in areas where grasslands and forests form a mosaic and the interdigitation of many different vegetation types supports a rich ungulate community (Eisenberg & Seidensticker 1976; Karanth & Sunquist 1992). Changing river courses, fire and other anthropogenic disturbances greatly increase the amount of edge habitat, which is preferred by many ungulate species. In addition. tiger density is positively correlated with prey biomass density (Fig. 1.7).

Reproductive capabilities

Where conditions are favourable, tiger populations can grow rapidly. Gestation is short, only 103 days; females breed relatively early, and they come into oestrus rapidly following loss or dispersal of young. Demographic parameters are well documented for tigers in Nepal's Royal Chitwan National Park (Sunquist 1981; Smith et al. 1987a; Smith & McDougal 1991). Their data show that females first breed at about three years of age (mean=3.4 years, $n=5$), after establishing residency. Litter size is commonly three (mean=2.98, range 2–5, $n=49$), but may be up to five, and the interbirth interval may be as short as 20 months (mean=21.6 months, range 20–24, $n=7$). A relatively short interbirth interval enhances the reproductive output of tigresses, especially if litters are large, and survival of young is high. In three cases when entire litters were lost shortly after birth, the interval between litters was 7–8 months.

Tiger populations can recover relatively rapidly from substantial losses, as long as the habitat and prey populations remain intact. Old shikar (hunting) records document in detail how literally hundreds of tigers were shot during royal hunts in the Nepalese terai. When the same areas were hunted a few years later an equally impressive number of tigers were killed (Smythies 1942).

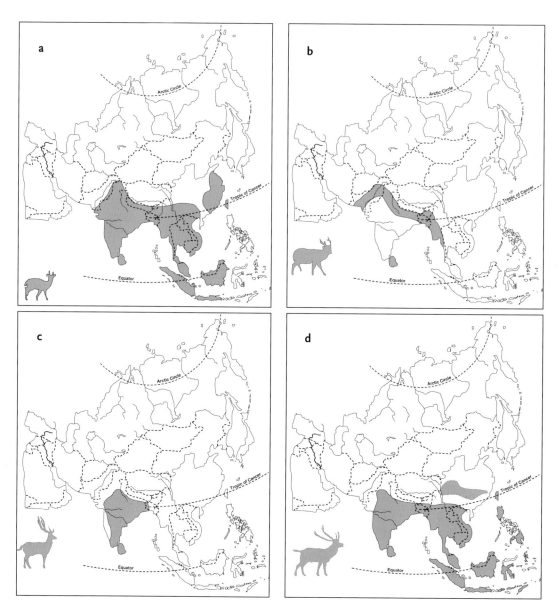

FIGURE I.3
Old World distribution of muntjac (**a**), hog deer (**b**), chital (**c**) and sambar (**d**). Data source: Whitehead (1972).

Hunts were carried out from December to February, involved hundreds of beater elephants, and typically covered an area of several hundred square kilometres. A hunt in Chitwan in 1935–36 produced 77 tigers; three years later 120 tigers were killed when the same area was hunted. Similarly, 47 tigers were killed in 1933–34 in Naya Muluk, about 270 km west of Chitwan; three years later 59 tigers were shot in same area.

The only available long-term data documenting tiger reproductive potential come from Russia, where estimates of size and structure of the Amur tiger population in Sikhote-Alin State Biosphere Reserve have been maintained since 1963

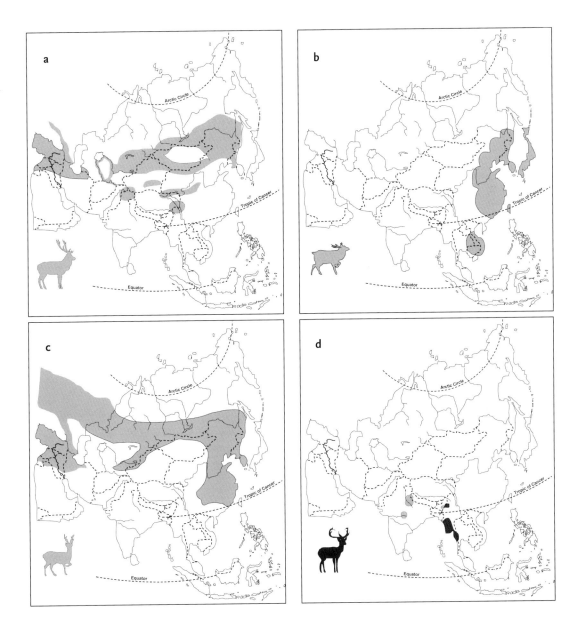

FIGURE 1.4
Old World distribution of red deer (**a**), sika (**b**), roe deer (**c**), and barasingha (light grey) and Eld's deer (dark grey) (**d**).
Data sources: Whitehead (1972), Bannikov (1978), and Danilkin (1996).

(E. N. Smirnov & D. G. Miquelle this volume). Between 1963 and 1965 there were no tigers in the area even though the reserve contained quality habitat and human impact was minimal. The first tiger appeared in the reserve in 1966 and during the colonisation phase that followed (1966–93) records show that the tiger population increased from one to 24–31 individuals. The growth rate averaged 6% per year over the 28-year period but varied temporally for different segments of the population. The adult segment of the population increased until 1985 and then stabilised, presumably because available

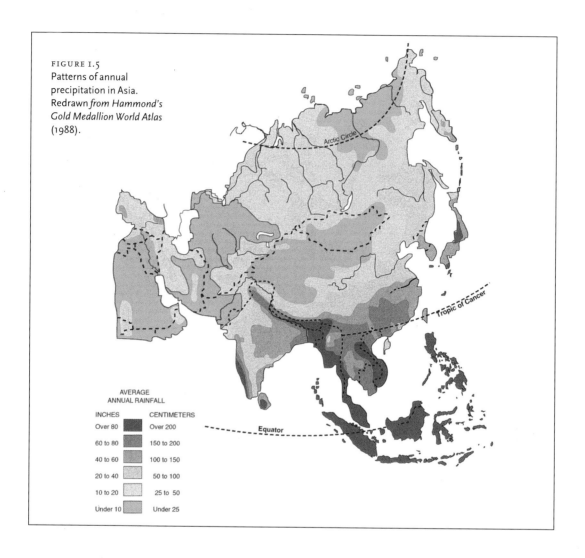

FIGURE 1.5
Patterns of annual precipitation in Asia. Redrawn *from Hammond's Gold Medallion World Atlas* (1988).

AVERAGE
ANNUAL RAINFALL

INCHES	CENTIMETERS
Over 80	Over 200
60 to 80	150 to 200
40 to 60	100 to 150
20 to 40	50 to 100
10 to 20	25 to 50
Under 10	Under 25

breeding territories were filled. Thereafter, continued growth was attributed to the appearance of more young and sub-adults in the population.

While the population growth data from Russia may be the only long-term information available, anecdotal accounts and the results of other studies show that the number of females with young, litter size and age of dispersal (related to interbirth interval) can be quite variable. For example, estimates of the percentage of females with young in Kanha Tiger Reserve, India, varied from 25% to 43% (Schaller 1967; Panwar 1979b). In Chitwan National Park, Nepal, estimates of the percentage of females

with young varied from 50% to 87% (McDougal 1977; Smith 1978). Similarly, the mean number of young per female varies from 1.2 (n=12) in Russia to 2.98 (n=49) in Nepal (Smith & McDougal 1991; E. N. Smirnov & D. G. Miquelle this volume).

The variation in these population parameters suggests that tigers can and do respond reproductively to changes in environmental conditions. In Kanha Tiger Reserve, for example, the suspension of stock grazing increased protection from poaching, and removal of several villages created more and better ungulate habitat and consequently better conditions for the tiger. Prior to these

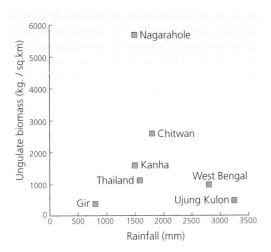

FIGURE 1.6
Annual precipitation and ungulate biomass. Data sources: Berwick 1974 (Gir); Schaller 1967 (Kanha); Srikosamatara 1993 (Thailand); Eisenberg & Seidensticker 1976 (Chitwan); Karanth & Sunquist 1992 (Nagarahole); Spillet 1967 (West Bengal); Hoogerwerf 1970 (Ujung Kulon).

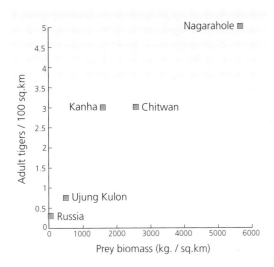

FIGURE 1.7
Prey biomass and adult tiger density. Data sources: Miquelle et al. 1996b (Russia), Hoogerwerf 1970 (Ujung Kulon); Schaller 1967 (Kanha); Eisenberg & Seidensticker 1976; Sunquist 1981 (Chitwan); Karanth & Sunquist 1992; Karanth 1993 (Nagarahole).

improvements there was only a single area within the park where tigresses were seen with young. At that time, Schaller (1967) estimated that 25% of females had young. Afterwards, reproduction was documented in three areas of the park, with as many as 43% of females being with young (Panwar 1979b). In Chitwan, during a period of population stability, the percentage of females with young in the prime floodplain habitat was an incredible 87% (Smith 1978). In Russia, the percentage of females with young did not change significantly during the colonisation of Sikhote-Alin, varying between 31 and 48%, but reproductive rates (cubs per adult female per year) did (E. N. Smirnov & D. G. Miquelle this volume). Litter size increased significantly from 1.2 in 1966–1975 to 1.9 in 1985–1993. This increase was attributed to differences in age of breeding females, there being smaller litters among the preponderance of young adult females during the initial stages, but increasing in the later phases as females matured. While mean litter sizes of tigers in Russia are smaller than those reported for Nepal, habitat quality and prey density are considerably higher in Chitwan. Nevertheless, the Russian data show how quickly Amur tiger populations can grow in newly colonised areas with healthy prey populations.

Dispersal capabilities

Another important aspect of resilience is a measure of dispersal capabilities. However, little is known about how dispersing tigers move, especially through fragmented landscapes. The only dataset on dispersal is from the work of Smith (1993) in Chitwan. He found that males dispersed about three times farther than females; most females were philopatric, settling next to their mothers. Dispersal distances in Chitwan were rather short. The average dispersal distance for males was 33 km, the longest was 65 km. The average dispersal distance for females was slightly less than 10 km, the longest was 33 km. This does not mean that tigers are incapable of dispersing greater distances. Prior to Smith's study, there was evidence to suggest that a subadult

In tropical Asia, waterholes undisturbed by people and livestock are in short supply and are critical resources for tigers and their prey.

male from Chitwan travelled 150 km to the Trijuga-Koshi-Tappu area in eastern Nepal (Sunquist 1981). Age of dispersal varied even within the same area, with some young leaving their natal area at 19 months, while others stayed to 28 months.

Anecdotal accounts in Heptner & Sludskii (1992) report extensive movements for the Caspian tiger. They recount that tigers were occasionally seen in steppe and desert areas, places altogether foreign to them. They also noted that tigers 'crossed time and again between the eastern bank of the Aral Sea to the lower Syr-Darya, which is about 500 km as the crow flies'. Expanses of desert were reportedly traversed by tigers travelling from northern Iran to the bay on the eastern Caspian Sea. Heptner & Sludskii also reported that, 'In 1945 it was conclusively established that tigers coming from the Amu-Darya to the Syr-Darya following wild boar, travelled about 1000 km in two to three months'.

While great distances were travelled by tigers in the Caspian region, it would be interesting to know whether they actually crossed open areas or, more likely, that their travels followed the narrow riparian courses that penetrated this otherwise desert environment. Smith (1993) found that tigers did not disperse across open cultivated areas (10–20 km wide), but they did travel through degraded forest habitat.

Ultimately, the question of dispersal capabilities rests on the animal surviving to reproduce. In the only estimate to date, Smith (1993) found that four out of four young females successfully established breeding territories, whereas only four out of 10 males survived to breeding age. While females are the key to reproductive success, the sex that is more vulnerable to extinction determines the probability of extinction in a given area or subpopulation. Dispersal is male-biased in mammals and usually the dispersing sex has a higher mortality rate than the philopatric sex, which was the situation in Chitwan.

Predatory capabilities

Tigers employ a hunting technique that relies on individual hunting tactics, concealment, a stalk, and sudden rush and dispatch of the prey. However, not all prey are attacked in the same way, nor are all prey killed in the same way. For example, some prey are first seized by the hindquarters, others by the neck, and killing bites may be delivered to the throat or back of the neck. Predatory behaviour may differ by prey size or species and by habitat, and some behaviours may change with experience. The plasticity in their prey capture and killing behaviour

affords tigers access to a wide range of prey types and sizes (Seidensticker & McDougal 1993), from the 20-kg muntjac to the 1000-kg gaur.

A review of tiger food habits from just a few selected Asian parks shows that the mean weight of prey killed by tigers can be quite variable (Table 1.1), differing by a factor of four. We used scat data, corrected for prey size (Ackerman et al. 1984), to derive the relative number of animals of different species killed. There was surprisingly little difference in mean mass of prey killed by tigers in Chitwan (61.8 kg), Kanha (66 kg), and Nagarahole (65.5 kg). In Huai Kha Khaeng the mean weight of prey killed was about 15 kg. This low value likely reflects the paucity of large prey at the site when the study was done.

A tigress requires 5–6 kg of meat a day for a maintenance diet (Sunquist 1981). This translates to 1825–2190 kg/year of meat but as 30% of each carcass is inedible a tigress needs to kill some 2373–2847 kg/year of meat on the hoof. Theoretically, this amount of meat could come from any size mammal. A tigress could just as well kill one 20-kg muntjac every 2–3 days or one 200-kg sambar every few weeks.

We can gain some insights into tiger-prey dynamics from a hypothetical scenario, where the only prey available is the 20-kg muntjac. At their highest recorded density – seven per square kilometre (Karanth & Sunquist 1992) – a 100 km² area of forest would contain 700 adult muntjac. Assuming that half were female, and that each female produced one young per year, 350 young would be added to the population each year, for a total of 1050. A tigress killing muntjac at a rate of one every 2–3 days would remove 122–183 individuals per year from this population. The amount of meat on the hoof removed is 2440–3660 kg/year, which, minus the 30% inedible portion, yields 1708–2562 kg/year. This figure is essentially a maintenance diet for a tigress.

However, reproduction increases energy requirements and a tigress feeding two large young needs approximately 50% more food. To meet this demand our hypothetical tigress would need to kill one muntjac every 1–2 days, or 183–365

Table 1.1. *Food habits of tigers in four Asian parks based on frequency of occurrence of mammalian species in scats*

Site	Species (mean body mass, kg)	Relative no. killed
Chitwan National Park, Nepal (*n*=123 scats)	*Axis axis* (55 kg)	27.8
	Cervus unicolor (212 kg)	15.3
	Axis porcinus (40 kg)	13.5
	Muntiacus muntjak (20 kg)	6.4
	Sus scrofa (38 kg)	10.8
	Hystrix indica (8 kg)	2.7
	Lepus nigricollis (3 kg)	6.6
	Semnopithecus entellus (9 kg)	16.9
	Mean mass (kg) of prey killed = 61.8 kg	
Kanha National Park, India (*n*=300 scats)	*Axis axis* (55 kg)	50.3
	Cervus unicolor (212 kg)	6.3
	Cervus duvauceli (150 kg)	5.7
	Bos frontalis (287 kg)	4.8
	Semnopithecus entellus (8 kg)	21.6
	Hystrix indica (8 kg)	10.3
	Sus scrofa (38 kg)	1.0
	Mean mass (kg) of prey killed = 66 kg	
Huai Kha Khaeng Wildlife Sanctuary, Thailand (*n*=38 scats)	*Muntiacus muntjak* (20 kg)	34.8
	Cervus unicolor (150 kg)	2.7
	Sus scrofa (38 kg)	6.7
	Macaca & Semnopithecus (8 kg)	10.8
	Hystrix brachyura (8 kg)	27.1
	Arctonyx collaris (10 kg)	17.9
	Mean mass (kg) of prey killed = 14.7 kg	
Nagarahole National Park, India (*n*=455 scats)	*Axis axis* (55 kg)	22.8
	Cervus unicolor (212 kg)	11.4
	Bos frontalis (287 kg)	7.5
	Sus scrofa (38 kg)	8.4
	Muntiacus muntjak (20 kg)	8.4
	Semnopithecus entellus (9 kg)	11.3
	Moschiola meminna (8 kg)	13.6
	Lepus nigricollis (3 kg)	1.5
	Hystrix indica (8 kg)	0.6
	Cuon alpinus (15 kg)	1.0
	Unidentified items	13.5
	Mean mass (kg) of prey killed = 65.5 kg	

Sources: Chitwan – McDougal 1977; Kanha – Schaller 1967; Huai Kha Khaeng – Rabinowitz 1989; Nagarahole – Karanth & Sunquist 1995.

muntjac/year. At this level of cropping, some adult muntjac are taken, there is no recruitment, and the muntjac population declines.

Thus, our hypothetical 'muntjac-only scenario' would provide a maintenance diet for a relatively low-density population of tigers (one tiger/100 km²). The inclusion of a few sambar and wild boar would presumably allow for occasional reproduction. This scenario may approximate the situation in true tropical rainforest, such as in the

Large deer make up nearly three-quarters of the tiger's diet in most parts of its range. An adult tigress needs nearly 3000 kg of meat-on-the-hoof each year; feeding cubs requires an additional 50%. Tigers can only meet this demand by killing large hoofed mammals – deer, wild pigs and wild cattle.

mountainous terrain of Gunung Leuser National Park, Sumatra, where tiger and prey biomass densities are naturally low. Above 600 m in the park tiger prey is largely muntjac and the occasional serow; in this terrain Griffiths (1996) estimated tigress home range sizes to be in the region of 137–190 km².

In many parts of the tigers' range where large cervids and bovids are heavily poached, tigers may be approaching an artificially induced 'muntjac-only scenario.' As prey populations continue to decline across the tigers' range an increasing number of tiger populations will come to exist at low densities and an impoverished prey base will support only occasional reproduction (K. U. Karanth & B. M. Stith this volume).

Large cervids make up nearly three-quarters of the biomass contribution to tiger diets in most parts of its geographic range (Fig. 1.8), although there are not many areas remaining where ungulate assemblages remain intact. Thailand, for example, has lost four of the six species of deer that occurred there in recent times (Rabinowitz 1989). In neighbouring Cambodia, the once incredible assemblage of gaur, banteng, water buffalo, kouprey, sambar, and elephant has been lost (Wharton 1968). Similarly, elephants, water buffalo, and barasingha no longer occur in Chitwan (Mishra 1982). These recent extinctions of large ungulates, particularly the cervids, have most certainly modified the large carnivore communities in south Asia, especially that of the tiger.

The tiger nevertheless persists in many areas and clearly it is an adaptable species. It has the ability to live in a diversity of habitat types and it tolerates a wide range of temperature and rainfall regimes. Tigers are also capable of producing relatively large litters; they are quick to recycle after the loss of a litter and, with a relatively short interbirth interval, their numbers can increase fairly rapidly. Tigers can

Nepal: Chitwan

India: Nagarahole

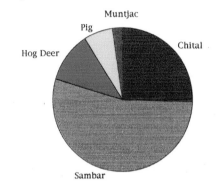

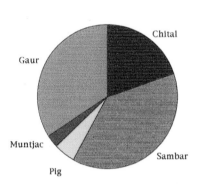

Thailand: Huai Kha Khaeng

Russia: Sikhote-Alin

FIGURE 1.8
Biomass contribution of
large cervids to tiger diets.
Data sources: McDougal
1977 (Chitwan);
Rabinowitz 1989
(Thailand); Karanth &
Sunquist 1995
(Nagarahole); Miquelle *et
al.* 1996b (Russia).

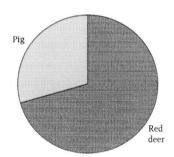

also capture prey differing enormously in size and they will change their hunting patterns and tactics to kill different prey. These characteristics are all hallmarks of a resilient species.

So what happened with the Caspian and Javan tigers? Why did we lose them? The historical records of the distribution of tigers in and around the Caspian show that the tiger was not found throughout the region, as is often depicted, but rather its distribution was dendritic, associated with watercourses, river basins, and lake edges (Heptner & Sludskii 1992). This was not surprising, as the region around the Caspian and Aral seas is essentially a desert, and typically receives less than 200 mm of rainfall annually (Fig. 1.5).

In addition to the historical records of its presence, the Caspian tiger's former distribution can be approximated by examining the distribution of cervids in the region (Fig. 1.4a, c). Red and roe deer, for example, were found in forests extending from areas around the Black Sea to the western side of the Caspian and around the southern end of the Caspian in a narrow belt of forest cover. Roe deer were also found in forested areas south of Lake Balkash, which approximates the easternmost distribution of the Caspian tiger (Heptner & Sludskii 1992). The Bactrian deer, another sub-species of red deer, was found in the narrow belt of forest habitat on the southern border of the Aral Sea, extending southward along both the Syr-Darya and Amu-Darya rivers (Bannikov 1978). The numerically dominant ungulate in the region was wild boar (Fig. 1.9). Pigs were found in forested habitats, along watercourses, and in reed beds and thickets of the Caspian and Aral seas. Where water-courses penetrated deep into desert areas, suitable

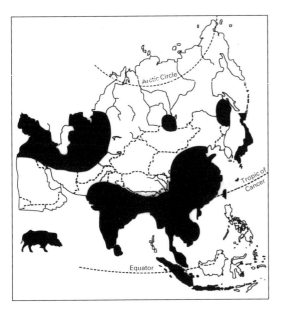

FIGURE 1.9
Distribution of wild boar. Data source: d'Huart (1991).

Georgia, Azerbaijan, and northern Iran, but they were basically tied to riverine habitat along watercourses. In Azerbaijan the last tiger was shot in 1948; in northern Iran the last tiger was shot in 1959 (Nowell & Jackson 1996).

In addition to persecution and habitat conversion, the tiger's disappearance from the Caspian region was also related to its reliance on wild boar (Heptner et al. 1989). Besides being fairly large prey, wild boar are also prolific; with an early age of first reproduction and large litters, wild boars can sustain heavy cropping rates, making them an ideal prey species (Mauget 1991). However, wild boar population dynamics often show sharp numerical changes over relatively short time periods. In the Caspian region, swine plague, foot-and-mouth disease, and other calamities such as floods and fires regularly caused large and rapid die-offs. Twice within a decade large numbers of young wild boar perished in floods along the Syr-Darya and Amu-Darya rivers. In some areas, wild boar populations dwindled to one-tenth or doubled or trebled in just a few years (Heptner et al. 1989).

One of the most important factors in the decline and extinction of the Caspian tiger was that it was already vulnerable due to the restricted nature of its distribution. Much of the cat's range appears to have been confined to watercourses embedded in a large expanse of desert environment, and as rivers began to be used as highways for colonisation the riverine habitats on which the tigers relied were gradually lost. Furthermore, with colonisation came intense persecution of both the tiger and its prey, which was greatly facilitated by their confinement to habitats adjoining the rivers. Red deer and wild boar formed the tiger's prey base, with wild boar being the principal item in the diet, but as deer numbers declined tigers had to rely increasing on wild boar. Though wild boar can normally sustain intense cropping they are subject to dramatic fluctuations caused by disease and catastrophic events. In the end, intense persecution and a reduced and fluctuating prey base was simply too much and the small numbers of tigers remaining were unable to persist against the odds. The last Caspian tiger was seen in the early 1970s (Heptner & Sludskii 1992).

wild boar and tiger habitat was often linear, only a few kilometres wide at most (Heptner & Sludskii 1992).

The demise of the Caspian tiger began with Russian colonisation of the area (Johnson 1991). Tigers disappeared from the region largely because of cultivation of the reedbeds combined with ruthless persecution. Large parties of sportsmen and military personnel hunted wild boar and tigers with reckless abandon. Just as the Mississippi and Ohio rivers in North America were used as avenues of transportation in the early colonisation of the American Midwest, so the vast network of rivers in this area made travel and colonisation of central Asian forest and steppe possible. Cotton and other crops grew well in the rich silt along the rivers, and agriculture replaced the extensive reedbeds and the wild boar that supported the tigers. Bactrian deer numbers also declined greatly as forests along the Amu- and Syr-Darya rivers were converted to cropland. The last record of a tiger in this area was an unconfirmed observation in 1968 along the lower reaches of the Amu-Darya near the Aral Sea (Nowell & Jackson 1996). On the western side of the Caspian, tigers occurred in the forested areas of

Adult tigers generally only come together to mate. They mate repeatedly and these intense bouts stimulate the tigress to ovulate. Even then, male and female treat each other with the utmost caution. After each mating the female twists around and takes a swipe at the male who must make a rapid getaway.

The circumstances associated with the extinction of the Javan tiger are somewhat different. Java, like the Caspian region, is outside prime cervid habitat. The region is subject to extremely high rainfall, with low terrestrial biomass, and most mammalian biomass consists of arboreal forms. Seidensticker (1986) has documented the decline and demise of the tiger in Java. In the 1800s tigers were widespread in Java, but during the 1900s much of the island's forests were converted to teak. By the 1940s the tiger had been eliminated from all but the most inaccessible habitat and by 1970 the species survived only in Meru-Betiri Reserve on the eastern south coast. This region had been protected from exploitation by its precipitous and dissected mountainous terrain, which was not particularly good tiger habitat. At one time deer had been numerous in east Java, but inadequate protection against hunting and a rapidly expanding human population caused deer numbers to decline drastically. By 1950 there were only a few places outside reserves or sanctuaries where deer survived in any number. Wild boar, once the principal prey of tigers in Java, were slaughtered en masse in a large-scale poisoning effort encouraged by the government (Hoogerwerf 1970).

In 1976, when Seidensticker surveyed Meru-Betiri for potential tiger and leopard prey he found the only abundant and common prey was primates. There were no cervids; bovids were extremely rare, and wild boar were widely distributed but at low density. A sample of three tiger scats contained the remains of two macaques, two porcupines, one bird and a palm civet. In a sample of 51 leopard scats, primates (65%) were the most frequently occurring prey, followed distantly by mouse deer (5.9%), pangolin (5.9%), civets (3.9%), porcupines (3.9%), bats (3.9%), colugos (3.9%), squirrels (3.9%) and muntjac (2%). Clearly the leopard was sampling what was available. This prey base was obviously not sufficient to sustain the three to five tigers estimated to be in the reserve and no tigers were seen in Meru-Betiri after 1976 (Seidensticker & Suyono 1980).

Tigers possess many attributes that can increase their chances of surviving in altered landscapes, but we must not lose sight of the evolutionary history of the species. The emergence of this large-bodied, forest-edge predator followed the radiation of the cervids and, today, deer make up three-quarters of the tiger's diet in most areas. Indeed, it is clear that cervids are vital to the tigers' survival in the wild. Persecution and the loss of large ungulates led to the extinction of the Caspian and Javan subspecies and additional tiger populations and subspecies will undoubtedly be lost as deer and other prey disappear. Tigers living in regions where high rainfall results in a naturally low cervid diversity are especially vulnerable. To ensure the tiger's survival we must not only find more effective ways to protect the cat and its habitat but also explore all avenues to protect, maintain, and increase cervid populations.

Tiger distribution, phenotypic variation and conservation issues

Andrew C. Kitchener

Introduction

The tiger once had one of the widest geographical distributions of any cat, stretching originally from almost 10° latitude south of the equator (Java and Bali) to more than 60° north (the Russian Far East) and through more than 100° longitude (Mazák 1996; Nowell & Jackson 1996). It is not surprising, therefore, that the tiger displayed a good deal of variation as a result of adaptations to the different climates, habitats and assemblages of prey found throughout its geographical range (Pocock 1929; Brongersma 1935; Mazák 1981, 1996). In recent years this has led to a consensus in the conservation community that there are eight distinctive subspecies of the tiger, which apparently vary in body size, characters of the skull, and coloration and markings of the pelage (Mazák 1981, 1996; Herrington 1987; Nowell & Jackson 1996).

The tiger is critically endangered; three of the eight putative subspecies are extinct and a fourth is close to extinction in the wild (Nowell & Jackson 1996). All populations are under severe pressure from loss of habitat and prey, and hunting for their fur and for their use in traditional Chinese medicine. Viable captive populations exist for two of the subspecies, while programmes are being developed for the remainder. The conservation strategy for the tiger is inextricably bound to knowledge of its geographical variation and, in particular, its subspecific taxonomy. However, in recent years it has become apparent that the subspecies taxonomy of many mammals – including some of the most well known and threatened felids – is very poorly known, which may have consequences for their conservation status and legal protection (Corbet 1966, 1970, 1978, 1997; O'Brien et al. 1990; O'Brien & Mayr 1991; Corbet & Hill 1992; Kitchener 1993, 1997; Miththapala et al. 1996; Pennock & Dimmick 1997).

In many cases subspecific names are based on single, often aberrant, individuals that were collected from poorly known localities or that were discovered in zoos during the nineteenth and early twentieth centuries. The definition of subspecies has changed from merely being used as a label for a local geographical variant to representing a morphologically and genetically distinct subpopulation, which has evolved in isolation, but which may subsequently hybridise with neighbouring populations to a limited extent (Corbet 1970, 1997; Mayr & Ashlock 1991; O'Brien & Mayr 1991). The advent of molecular techniques has shown that there is often a real discrepancy between traditionally recognised subspecies and genetically distinct populations, which are consequently often called evolutionary significant units (ESUs), because in many cases there have been no revisions of morphological subspecies since the nineteenth or early twentieth centuries (O'Brien & Mayr 1991; Corbet 1997). The time is ripe to reconsider the scientific basis of the putative subspecies of the tiger and the implications this may have for its conservation.

From a once almost continuous distribution throughout southern and eastern Asia, there are now about 160 distinct and fragmented populations of tigers, which have been designated Tiger Conservation Units (TCUs) by Dinerstein et al. (1997). Although Dinerstein et al. (1997) proposed an ecological approach to tiger conservation (i.e. TCUs should represent the entire range of ecological conditions in which the tiger currently survives), this does not obviate the need for an in-depth knowledge of the tiger's geographical variation, especially when considering the more intensive genetic management required by some populations. For example, even if adequate protection and habitat were provided for the South China tiger it would be unlikely to be viable in the long term due to its small population size (Nowell & Jackson 1996). Therefore, we need to establish how genetically and morphologically distinct the South China tiger is, so that we can develop appropriate management and conservation strategies, including genetic reinforcement from other tiger populations, as has occurred accidentally and intentionally for the Florida panther (O'Brien *et al.* 1990). Most TCU populations (61%) are not viable in the long term (Dinerstein *et al.* 1997) and may require similar metapopulation management. A poor knowledge of subspecies taxonomy and geographical variation means that we are ill-equipped to assist the conservation of the tiger, especially where a high level of genetic management is required.

In this chapter the evolution of the tiger and its subsequent colonisation of Asia are reviewed. The scientific basis of the eight putative tiger subspecies is examined and the possible causes of variation in today's tigers are reviewed. In a preliminary morphological study, patterns of variation in tigers are analysed against a background of the possible isolating mechanisms that may have influenced subspecies evolution. Finally, a revised approach to the evaluation of tiger subspecies is suggested and the consequences for current conservation strategies are discussed.

The evolution of the tiger

Evidence for the evolution of the tiger comes from fossil remains and molecular phylogenies (Brongersma 1935; Hooijer 1947; Hemmer 1967, 1987; Wayne *et al.* 1989; J. Wentzel *et al.* this volume). Cats of the genus *Panthera* probably evolved within the last five million years or so (Hemmer 1976; Collier & O'Brien 1985; Wayne *et al.* 1989). On the basis of a cladistical analysis of various skeletal and anatomical characters, fossil remains and biogeography, Hemmer (1981)

suggested that the original pantherine radiation occurred in eastern Asia, although there is fossil evidence to support an African origin for the lion and leopard (Hemmer 1976). However, the tiger is thought to have an eastern Asian origin (Hemmer 1981, 1987; Herrington 1987; Mazák 1981, 1996).

Reconstructing the fossil history of tigers is difficult owing to the existence of mainly fragmentary remains, risk of confusion with closely related species, and uncertain dating of finds. The oldest fossils are from northern China and Java (Hemmer 1971, 1976, 1987). Originally described as *Felis palaeosinensis* (Zdansky 1924), a small fossil tiger from Henan (formerly Honan), northern China is thought to date from the end of the Pliocene and the beginning of the Pleistocene and so may be up to two million years old (Hemmer 1967, 1987). However, it should be noted that these tigers were intermediate in size between modern Indian leopards and Sunda Island tigers, and may actually represent a large form of leopard or an ancestor of two or more of today's *Panthera* cats. Clearly the taxonomic status of *Felis palaeosinensis* would benefit from further investigation.

Early tiger fossils have also been recorded from the Jetis Beds of Java, which have recently been dated to between 1.66 and 1.81 million years old (Dubois 1908; Brongersma 1935; Hemmer 1971, 1987; Swisher *et al.* 1994). Therefore, by the beginning of the Pleistocene and possibly as long as two million years ago, the tiger already had a wide distribution in eastern Asia. It is commonly stated that the centre of evolution for tigers was northern China (e.g. Mazák 1981; Hemmer 1987; Herrington 1987), but the fossil evidence is equivocal given the wide distribution of the species at the beginning of the Pleistocene. It is also unnecessary to require or even envisage such a restricted locality for the tiger's evolution, since all that is required is sufficient temporal separation of a population to allow its genetic and morphological divergence from a sister or ancestral species, which could have occurred over a wide area of eastern Asia.

Abundant tiger fossils dated approximately from the early middle to late Pleistocene are known from

China, Sumatra and Java (Appendix 2), but tiger fossils only appeared in India, the Altai, northern Russia and elsewhere in the late Pleistocene (Brandt 1871; Lydekker 1886; Tscherski 1892; Dubois 1908; Zdansky 1928; Brongersma 1935; Loukashkin 1937; Hooijer 1947; Hemmer 1971, 1976, 1987). Recently, Herrington (1987) identified some fossil big cats from eastern Beringia as tigers, but so far none has been recorded from North America. Small tigers are also known from the late Pleistocene of Japan (Hemmer 1967).

Hooijer (1947) documented a decline in size of tigers during the Pleistocene until the present day, excluding those from the Russian Far East, which have stayed about the same size. This is not unusual in Pleistocene mammals (Guthrie 1984; Kurtén 1967) and probably reflects a decline in the seasonal productivity of the environment (Geist 1987a, b), and/or a decline in average prey size.

Holocene remains of the tiger have been recorded from Java and Borneo (Appendix 2; Hooijer 1947, 1963). The tiger is apparently now extinct on Borneo and the significance of this will be discussed later.

Molecular phylogenies based on a variety of methods confirm the close relationship between members of the genus *Panthera* and show that the tiger diverged more than two million years ago and before the divergence of the lion, leopard and jaguar (Collier & O'Brien 1985; Wayne *et al.* 1989; J. Wentzel *et al.* this volume). Therefore, the fossil and molecular histories of the tiger are broadly in agreement.

Although it is unclear where the tiger first diverged from other *Panthera* cats in eastern Asia, it is presumed to have spread north into northeast Asia, and south into the Sunda Islands and the Indian subcontinent (Hemmer 1987). The late arrival of the tiger in India is apparently supported by its absence from Sri Lanka (unless it became extinct there subsequently), which was cut off by rising sea levels at the beginning of the Holocene. There is some uncertainty over the origin of tigers in southwest Asia. Heptner & Sludskii (1992) proposed that tigers colonised this area from north-west India, but Hemmer (1987) and Mazák (1981,

1996) suggested a route from northeast Asia via central Asia.

Although absent today, on zoogeographical grounds the tiger should be an inhabitant of Borneo, but the evidence is equivocal and includes a skull, skins, teeth, wall paintings and photographs, which are usually assumed to be of external origin or influence (Hose & McDougall 1912; Pocock 1929; Peranio 1960; Gersi 1975; Medway 1977; Yasuma 1994). However, members of the Bisaya tribe claimed that the teeth in their possession were from indigenous tigers, although they had been inherited through four to seven generations (Peranio 1960). This suggests that the tiger may have been an indigenous species until 200 or so years ago, but is probably now extinct. The best evidence for the former existence of the tiger on Borneo is the tip of an unerupted upper canine that was found in the Niah Cave, Sarawak in deposits thought to date to the Neolithic of Borneo (Hooijer 1963; Medway 1977).

Tiger subspecies: distinct entities or poor sampling?

In determining subspecies there are two main non-random patterns of variation to be considered (Corbet 1970). First, discontinuous variation or morphologically distinct populations would seem to suggest former or current isolation of these populations so that they could evolve morphological and genetic characters unique to those populations (i.e. subspecies). Secondly, variation may vary continuously and gradually over the geographical range due to natural selection, which maintains a morphological and/or genetic gradient (i.e. no sub-species). This form of variation is known as a cline. There may also be some combination of these two patterns.

Corbet (1966, 1970, 1997) suggested that the current proliferation of subspecies names for many mammal species has arisen in many cases from inappropriate sampling of clines. This may also apply to the tiger. Excluding Linnaeus' type description which is not supported by a specimen, the first accepted tiger subspecies dates to 1815 when the Caspian tiger was described (Illiger 1815), but no holotype exists although Illiger's description does suggest that he had one or more skins at hand. Then in 1844 came Temminck's descriptions of the Amur tiger and the Javan tiger. Therefore, by the middle of the nineteenth century, four of today's accepted tiger subspecies had been described, based probably on only three preserved specimens. It was completely unknown whether variation between these geographical extremes was clinal or discrete.

However, many more specific and subspecific names have been described since (Pocock 1929; Brongersma 1935; Corbet & Hill 1992), and even today new subspecies are still being described or new names proposed which are clearly not valid (Kirk 1994; Kock 1995). Four of the recognised subspecies were originally described in the nineteenth century, three others in the early twentieth century and the last in the 1960s. The first seven subspecies to be described are based on only 11 specimens, and at least three subspecies names (*altaica*, *sondaica* and *amoyensis*) are based on specimens from poorly known or unknown localities (Table 2.1; Temminck 1844; Hilzheimer 1905; Brongersma 1935). The last named subspecies, *corbetti*, was based on an examination of skins and skulls of 25 specimens, but no mention of comparative material from other subspecies was made (Mazák 1968).

The characters used by authors to distinguish between subspecies were also minimal; e.g. Temminck distinguished between the Amur and Javan tigers on the basis of fur length (Temminck 1844; Mazák 1967). The most recent revisions of subspecies have used body size, skull characters, and patterns of striping and coloration of the pelage to distinguish between subspecies (e.g. Pocock 1929; Brongersma 1935; Hemmer 1969; Mazák 1981, 1996), but authors often point out the wide variation in coat colour and markings within populations (e.g. Pocock 1929; Brongersma 1935; Weigel 1961; Mazák 1967; Schroeter 1981; Heptner & Sludskii 1992). Against this trend, Hooijer (1947) suggested that most variation in tigers was clinal.

Today eight subspecies of the tiger are recognised by the conservation community (Table 2.1;

Table 2.1. *The authors and numbers of specimens on which the type descriptions of the eight putative tiger subspecies were based*

Subspecies	Authority	No. of specimens
tigris	Linnaeus 1758	0
virgata	Illiger 1815	u/k
altaica	Temminck 1844	1
sondaica	Temminck 1844	1
amoyensis	Hilzheimer 1905	5
balica	Schwarz 1912	1
sumatrae	Pocock 1929	2
corbetti	Mazák 1968	25

u/k – Unknown; no holotype is designated, although the type description suggests that at least one specimen was available for comparison.

Herrington 1987; Mazák 1996; Nowell & Jackson 1996). Herrington (1987) carried out a multivariate statistical analysis of measurements from 37 skulls of five subspecies, which showed that there were distinct morphological differences between the skulls of putative subspecies and eight animals of unknown (i.e. captive) origin. In particular, Chinese tigers had distinctive skulls, which Herrington (1987) equated with their putative ancestral status. There was some overlap between the shape and size of skulls of Indian and Indochinese tigers and between those of Indochinese and Sumatran tigers, which is suggestive of clinal variation. However, there are some problems with Herrington's (1987) analysis: sample sizes were mostly very small (ideally there should be at least 20 specimens of each sex per subspecies) and we are not told whether samples were controlled for sex, age and size (in order to determine shape differences). Therefore, apparent distinctive differences between putative subspecies could be due to the vagaries of inappropriate samples. Her conclusion that Chinese tigers appeared to have a distinctive and primitive skull shape must be treated with some caution.

This would seem to suggest that there is a very poor scientific basis for the eight currently recognised subspecies of tiger. Therefore, the current conservation strategy for tigers could be greatly undermined by these mainly *ad hoc* descriptions of tiger subspecies.

Variation in tigers

There are three main sources of variation that can be observed in tigers: body size, striping patterns and coloration of the pelage, and skull characters. I reviewed the range of variation in tigers using 40 skins and 77 skulls from specimens in the collections of the Natural History Museum, London, and the National Museums of Scotland, supplemented by data from the following sources: Ognev 1962; Mazák 1967, 1996; Mazák *et al.* 1978. Sample sizes were too small for each subspecies for a comprehensive understanding of variation in tigers and consequently my analysis may have suffered similar sampling problems to Herrington's (1987), but the results presented here do form the basis for future research and appear to contradict the current consensus on subspecies.

Body size

It is generally known that the largest tigers occur in the Russian Far East and the smallest are found in the Sunda Islands (e.g. Hooijer 1947; Mazák 1996). However, this apparent cline in body size is more complex, because tigers from the northern Indian subcontinent [greatest length of skulls for males: 332–376 mm (n=18)] are as large as Amur tigers [331–383 mm (n=9)], but the largest specimens are from northern China (maximum: 406 mm) (see also Loukashkin 1938; Mazák 1967, 1996). There are also intersexual differences in body size. Mazák's (1981) body and skull measurement data show that whereas female mainland tigers do not vary significantly in body size, male tigers show increasing body size (and hence sexual dimorphism) with latitude. Therefore, sexual dimorphism is apparently greater in Amur tigers than in Indochinese tigers. Similar intraspecific variations in sexual dimorphism may occur in other felids, e.g. leopards (S. Miththapala pers. comm.).

My analysis shows that size and sexual dimor-

phism increase with latitude as indicated by skull and tooth measurements. For males greatest length of skull increases with latitude [greatest length of skull (mm) = 325.6 + 0.791×latitude (°) r=0.620], but carnassial length increases only slightly with latitude [carnassial length (mm) = 34.66 + 0.035×latitude (°) r=0.444], but peaks occur at approximately 28°N (Nepal, northern India) and at about 45°N in the Russian Far East (Figs. 2.1 & 2.2). Within each putative subspecies there is also considerable variation in size (Figs. 2.1 & 2.2). There is a significant difference in the median greatest lengths of skulls for males of putative subspecies (Kruskal-Wallis H=29.46, P<0.0005), but these are mostly not between geographically neighbouring subspecies except between Balinese (sample size, n=2) and Javan (n=5), Indochinese (n=8) and Indian (n=18), Chinese (n=2) and Amur (n=9) (medians do not overlap at 95% confidence intervals). However, these differences occur mostly where sample sizes are very small, so that these are unlikely to be representative of each putative subspecies. There is a significant difference between the mean carnassial lengths for males of putative subspecies [analysis of variance (ANOVA) $F_{7,44}$=3.16, P<0.01], but there are no significant differences (least significant differences all P>0.05) between the mean carnassial lengths for geographically neighbouring subspecies, indicating that the variation is clinal. However, it should be noted that in all cases, except for Indian tigers, sample sizes were too small to be certain that they were representative of each putative subspecies.

Female tigers show a less marked cline in greatest length of skull [greatest length of skull (mm) = 280.8 + 0.468×latitude(°) r=0.651] and the upper carnassial is more or less the same size in mainland animals [carnassial length (mm) = 31.72+0.034×latitude(°) r=0.516] (Figs. 2.3 & 2.4). The median greatest lengths of skulls of female tigers of putative subspecies were significantly different (Kruskal-Wallis H=23.58, P<0.001), but these are not generally between geographically neighbouring subspecies, except for between Javan (n=2) and Sumatran (n=3)/Balinese (n=4), where the very small sample sizes may not be representative

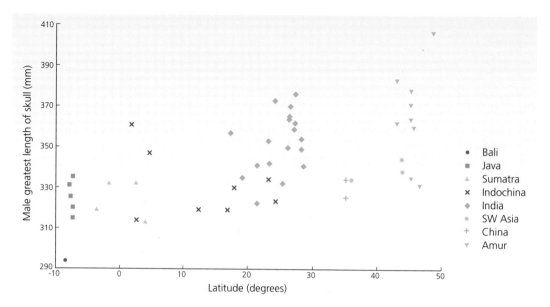

FIGURE 2.1
The relationship between the greatest length of male tiger skulls and latitude.

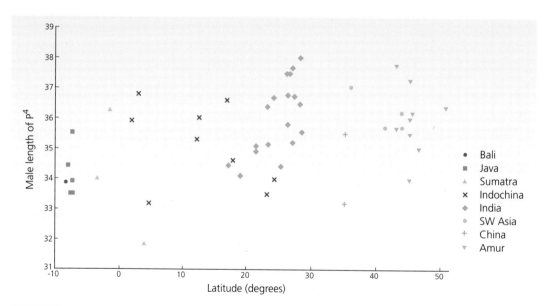

FIGURE 2.2
The relationship between the crown length of the fourth upper premolar (P^4) of male tigers and latitude.

of each subspecies. There was also a significant difference between the median carnassial lengths of females of putative subspecies (Kruskal-Wallis $H=20.04$, $P<0.005$), but there are no significant differences between geographically neighbouring subspecies, so that this variation is clearly clinal.

However, sample sizes were too small to be certain that they were representative of all putative subspecies. Only the Sunda Island tigers seem to have significantly smaller upper carnassials ($t=5.72$, $P<0.000005$).

Regression analysis of the effect of latitude on

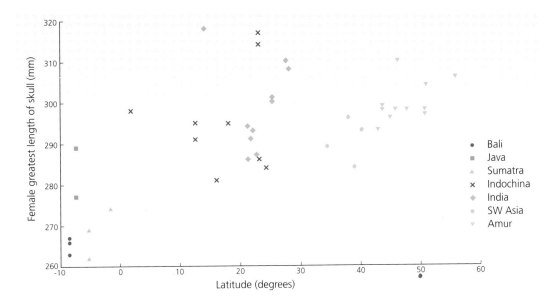

FIGURE 2.3
The relationship between the greatest length of female tiger skulls and latitude.

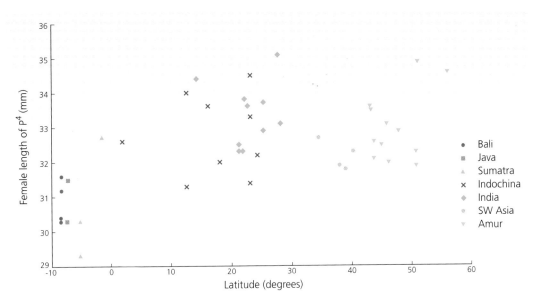

FIGURE 2.4
The relationship between the crown length of the fourth upper premolar (P4) of female tigers and latitude.

skull length and carnassial length shows differences between the sexes (Fig. 2.5). Although sexual dimorphism in skull length increases with latitude (male to female ratio of slopes = 1.69), sexual dimorphism in carnassial length does not (ratio=1.02).

Coloration and markings

The ground colour of tigers' pelages varies from a dark red to a pale yellow (Pocock 1929; Brongersma 1935; Mazák 1981, 1996). In part this seems to reflect habitat and/or humidity (i.e. Gloger's Rule; see Ortolani & Caro 1996); darker tigers are found

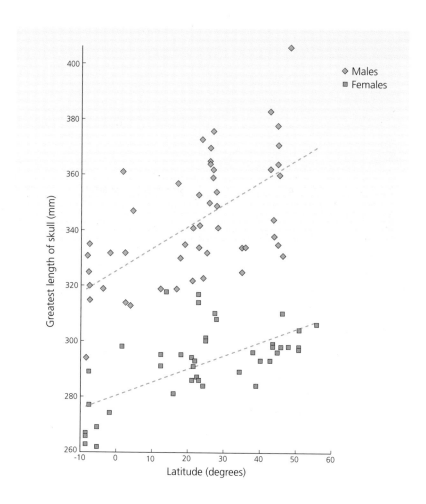

FIGURE 2.5
A comparison of the
relationship between the
greatest skull lengths of
male and female tigers and
latitude. The regression
lines for each sex show that
sexual dimorphism
increases with latitude (see
text for details).
■ – females; ◆ – males.

in the tropical rain forests of South East Asia and the Sunda Islands (Pocock 1929), whereas the Amur tiger is often very pale, especially in winter coat (Mazák 1967). However, variation within populations may be considerable. Mazák (1967) reported a female Amur tiger with a pelage ground colour similar to that of a Sunda Island tiger's, and the coloration of the pelages of Indian and Amur tigers is said to be very variable (Pocock 1929; Brongersma 1935; Schroeter 1981).

Stripe coloration may vary too. The stripes of Amur tigers have often been described as being dark brown rather than black (e.g. Mazák 1976, 1981, 1996). However, it should be noted that northern tigers experience far longer periods of daylight in the summer months, which may cause fading of the pigments found in their coats. Therefore, black

stripes may fade to brown and darker ground colours may become paler during the year. This is supported by recent observations by P. Jackson, who noted that moulting Amur tigers at the Hengdaohezi Tiger Breeding Centre, near Harbin, China, had black stripes in their new, darker, summer coat, which contrasted with the brown stripes of the pale winter coat. It should also be noted that variation within putative subspecies may be much greater than that recorded within captive animals, which may have been derived from relatively few founders from a restricted part of their distribution, which did not display the full range of variation in the wild population.

Tiger subspecies are often partly characterised on the basis of their striping patterns. For example, Mazák (1981) distinguished Sunda Island tigers on

Table 2.2. *The character states used to score the pelage coloration and striping patterns of tiger skins in the Natural History Museum, London. Intermediate scores of 2 were given to characters 3, 4 and 6 where it was not possible to assign a pelage to the character stated below*

Character no.	Score = 1	Score = 3
1	'U' stripe on tail	'U' stripe absent
2	Spots	No spots
3	Thick stripes	Thin stripes
4	Stripes on forequarters	Reduced striping on forequarters
5	Loops on flank	No loops
6	Ground colour dark	Ground colour light

the basis of the higher frequency of stripes which often end in a line of spots. Sumatran tigers are usually described as having thicker stripes than Javans, and Amur tigers are said to have thinner stripes than Bengal tigers (Pocock 1929, 1939; Brongersma 1935; Mazák 1996). However, Heptner & Sludskii (1992) showed that both Amur and Caspian tigers displayed a wide variety of striping patterns and ground colour variations.

The coloration and striping patterns of tiger skins was scored for seven characters. Each character was given a score of either one or three as shown in Table 2.2, with an intermediate character state of two, if necessary. The characters were scored subjectively relative to the extremes of variation that were evident in the sample, so that the results presented here should be treated with caution. However, scores of three were associated with light skins with thin and reduced striping. Scores of one were associated with dark skins with thick, abundant stripes. These correspond broadly with what is expected from the apparent clinal variation in coloration and striping described above. For each skin the character scores were totalled to give a specimen score. The means and ranges of specimen scores for the tiger skins are shown in Table 2.3. The ground colour was faded in Indian specimens which had been used as rugs and exposed to long periods of daylight, so that scores were corrected by subtracting one point from the specimen score to take account of fading.

Table 2.3. *Character scores for skins of tigers in the collections of the Natural History Museum, London and the National Museums of Scotland*

Putative subspecies	n	Mean	Range	SD
tigris	14	11.14	7–15	2.66
tigris[a]	14	10.14	6–14	2.66
corbetti	11	9.04	7–12.5	2.21
tigris × sumatrae	2	8.75	8–9.5	1.06
sumatrae	7	8.64	7–10.5	1.25
sondaica/balica	3	9.00	8.5–9.5	0.50
virgata	2	11.75	11–12.5	1.06
altaica	1	14.00	—	—

[a] Total scores reduced by 1 point for possible effects of fading of tiger skin rugs in daylight.
n – Sample size; SD – standard deviation.

It was evident that the pelages of putative tiger subspecies revealed much greater variation in coloration and striping pattern within subspecies and much more similarity between them than is usually described in the literature (Fig. 2.6; see below). For example, striping patterns and coloration attributed typically to Javan and Bali tigers (Hemmer 1969; Mazák *et al.* 1978) were found also in animals from Sumatra, Burma and South India (Fig. 2.7). The ranges of pelage scores for putative subspecies overlap completely with each other, although Sunda Island tigers generally have low scores (dark, well-striped skins) and Indian tigers have very variable scores (Table 2.3). There was no significant difference between the median

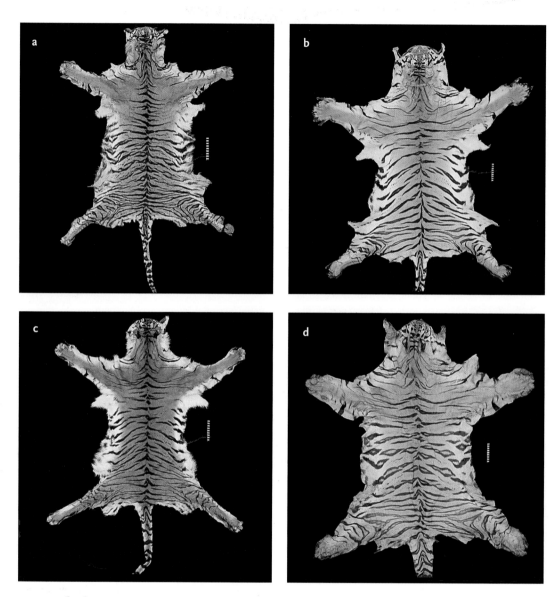

FIGURE 2.6a–d
The pelages of tigers showing the variation in striping pattern. **a.** Sumatra (BMNH 35.4.6.2); specimen score = 10. **b.** India (BMNH 1983.307); specimen score = 6–7. **c.** Malaya (BMNH 37.1.2.1); specimen score = 10.5. **d.** Bangladesh (BMNH 1882.12.10.1); specimen score = 7–8.

scores of each putative subspecies (Kruskal-Wallis $H=7.09$, $P>0.05$). However, sample sizes were too small for a more precise understanding of variation in the pelages of putative subspecies.

The number of stripes crossing the mid-flank was estimated by placing a ruler along the length of a photograph of the pelage perpendicular to the stripes and counting the number of stripes that touched the edge of the ruler. The mean number of stripes for each putative subspecies is shown in Table 2.4. There was a significant difference in the median number of stripes of putative subspecies (Kruskal-Wallis $H=19.13$, $P<0.05$), but this was generally not between geographically neighbouring

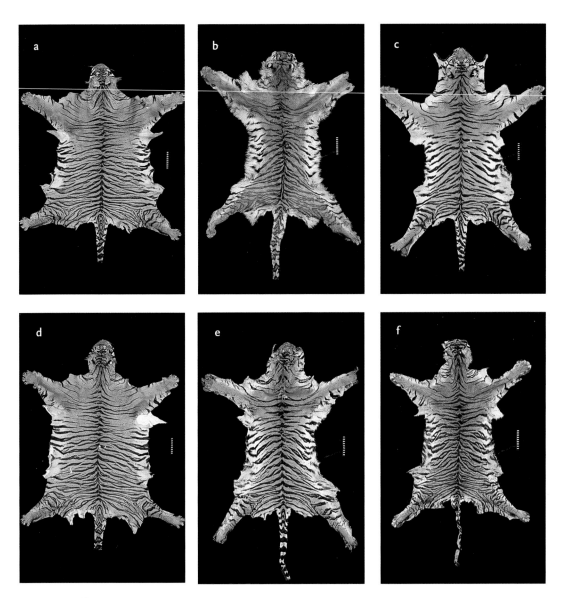

FIGURE 2.7a–f
The pelages of tigers from different putative subspecies, which show a striping pattern characteristic of the Javan tiger. **a.** Bali (BMNH 37.12.1.2); specimen score = 8.5. **b.** Afghanistan (BMNH 1886.10.15.1); specimen score = 11. **c.** India (BMNH 32.3.2.1); specimen score = 9–10. **d.** Java (BMNH 37.12.1.1) specimen score = 9. **e.** Annam (BMNH 33.4.1.204); specimen score = 8. **f.** Sumatra (BMNH 35.4.6.3); specimen score = 9.

subspecies, except between Balinese (*n*=4) and Javan (*n*=4), but sample sizes were too small to be representative of each subspecies. This indicates that there is clinal variation in the number of stripes, ranging from a maximum in Java and Bali to a minimum in Amur tigers (Table 2.4), but the ranges

overlap completely in all cases and again the samples are too small to know whether these are representative of populations.

Therefore, in summary, within the samples I was able to examine, variation within putative subspecies may be greater than variation between them.

Table 2.4. *Numbers of stripes on the mid-flank of tiger pelages in the collection of the Natural History Museum, London and from photographs in Mazák (1996)*

Putative subspecies	n	Mean	Range	SD
altaica	7	22.0	20–26	2.52
tigris	14	24.0	21–29	3.23
corbetti	11	26.4	21–31	2.54
sumatrae	7	27.0	20-34	5.94
virgata	4	26.8	24—28	1.89
sondaica/balica	8	30.5	23–37	4.34

n – Sample size; SD – standard deviation.

For example there is a striking similarity in striping pattern in the holotypes of the Amur and Javan tigers, which both show a reduction in striping in the forequarters and thin stripes (Temminck 1844; Mazák 1996). However, there were trends in the variation with darker ground colour and more stripes in the south east and paler ground colour and fewer stripes in the north. Indian tigers seemed to be particularly variable.

Skull characters

Two main skull characters have been used to separate subspecies (shape of the occiput and shape of the nasal bones) along with other lesser characters including development of the sagittal crest and degree of convexity of the dorsal surface of the skull (Pocock 1929, 1931a; Hemmer 1987; Mazák 1996). Hemmer (1987) stated that he could identify the skulls of tiger subspecies easily compared with the skulls of subspecies of lions and leopards, which reflected the long evolutionary history of tiger subspecies. However, even the skull characters that separate lions and tigers are not absolute, and leopard skulls are only certainly distinguishable on size (Pocock 1929, 1931a, b).

The most important skull character that has been used to separate tiger subspecies is the shape of the occiput, which is apparently characteristically narrow in the Javan and Bali tigers, and much broader in Caspian tigers (Pocock 1929; Brongersma 1935; Hemmer 1969; Mazák 1996). However, narrow occiputs can also be observed in

Indian and Indochinese tigers (Fig. 2.8). I measured the height of the occiput from the top of the foramen magnum, and the width of the occiput in tiger skulls of all putative subspecies (Fig. 2.9). The narrowness of the occiputs of male (Fig. 2.10) and female (Fig. 2.11) tiger skulls can be assessed by how far they fall to the right of the line indicating unity in Figs. 2.10 and 2.11. These show that the occipital region of Javan tigers is no narrower than that seen in many Indian tigers when skull size is taken into account. Some putative tiger subspecies do have very broad occiputs, including Caspian and Amur tigers, which may indicate a close relationship between these populations, but further research is required to see if this character really is discriminating (Fig. 2.9). It should be noted that female tigers do not appear to show much variation between putative subspecies (Fig. 2.11), so that occipital shape may only be useful for discriminating males of some subspecies.

Why do tigers vary in size through their range?

My analysis shows that tigers vary in body size, coloration and markings and in some skull characters throughout their range. Although the general increase in striping and darkness of ground colour may follow Gloger's Rule (Ortolani & Caro 1996), only body size seems to show any obvious pattern of variation with large animals in India and the Russian Far East and small animals in Southeast Asia and Indonesia. There are many possible influences on tiger body size, many of which are interrelated. These are described briefly below:

Bergmann's Rule

Tigers appear to follow Bergmann's Rule. At high latitudes lower temperatures mean that animals have to generate more heat to maintain a constant body temperature than at lower latitudes, all other factors being equal. However, small animals have a greater surface area to the volume of their bodies than large animals, which means that they lose more heat at the same ambient temperature. Therefore, larger tigers occur at higher latitudes because of the need to reduce relative body surface area in order to conserve energy in colder climates. Similar

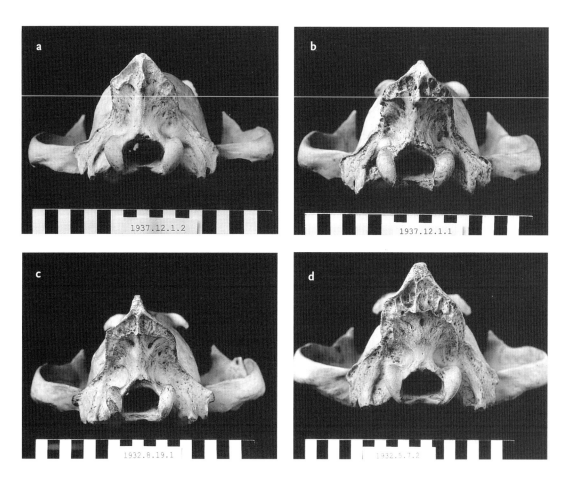

FIGURE 2.8a–d
The occiputs of the skulls of tigers showing the typical shape for Javan tigers in other populations. **a**. Bali (BMNH 37.12.1.2). **b**. Java (BMNH 37.12.1.1). **c**. Burma (BMNH 32.8.19.1). **d**. India (BMNH 32.5.7.2).

latitudinal variation in size has been observed for pumas (Kurtén 1976).

My analysis would seem to support this view (Figs. 2.1–2.5). We would expect Sumatran tigers to be smaller than Javan tigers because Sumatra straddles the equator, whereas Java is further south, but there are too few data to confirm this. However, large Indian tigers and very small Bali tigers would seem to confound this influence as the main cause of body size variation in some populations.

Guthrie's or Geist's Rule
Guthrie (1984) and Geist (1987a,b) noted that body size in large mammals was influenced by the seasonal amount of food available (the amplitude and duration of the seasonal productivity pulse) to support growth, so that in areas of high seasonal abundance of food, they would be able to achieve a greater proportion of their potential for annual growth, resulting in a greater adult body size. This should not be confused with overall total annual productivity or biomass, which is often higher in tropical ecosystems. Seasonal productivity is broadly correlated with latitude (i.e. spring and summer day length), so that body size increases with latitude (c.f. Bergmann's Rule). However, Geist (1987a) noted that in very high latitudes (53–60°N) where seasonal productivity is reduced due to a shorter growing season, body size diminishes again in deer and wolves.

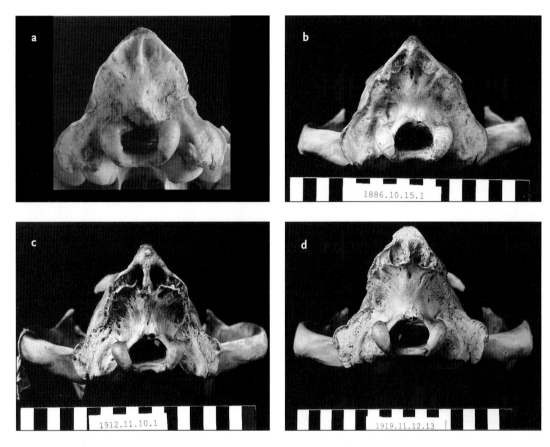

FIGURE 2.9a–d
The occiputs of the skulls of tigers. **a.** Amur (NMSZ 1989.70.12). **b.** Caspian (BMNH 1886.10.15.1). **c.** Sumatran (BMNH 1912.11.10.1). **d.** Sumatran (BMNH 1911.12.13).

Therefore, tiger body size may be influenced similarly except it does not show a strict latitudinal correlation. Presumably the seasonal productivity of habitats in northern India would be expected to match those in the Russian Far East to account for their similarity in body size, even though the biomass of utilised prey is much greater in many habitats in India and Nepal than in Russia, so that home ranges of female tigers may be only 20 km² in India compared with 500 km² in Russia (J. Seidensticker pers. comm.). However, other factors may be more important in influencing body size in tigers.

Island dwarfing

When species become isolated on islands they often evolve a smaller body size. This has been observed to occur in many species including woolly mammoths and giant deer. The cause of island dwarfing is not clear, but is probably a result of competition for limited food resources often in the absence of predators (Lister & Bahn 1994). Having only small prey to feed on, this could explain why the Sunda Island tigers (especially the Bali tiger) and the now extinct Pleistocene Japanese tiger are smaller than mainland tigers. However, Javan and Bali tigers had access to and killed full-sized banteng, which

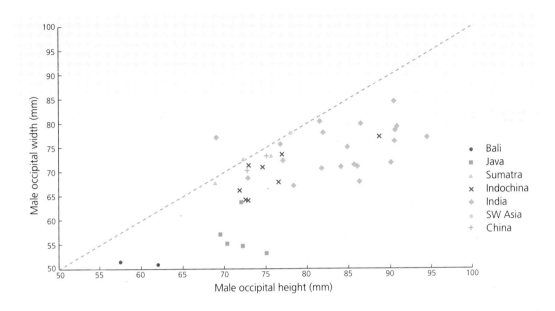

FIGURE 2.10
The relationship between height and width of the occiputs of male tiger skulls. The line indicates a 1:1 ratio between the two variables.

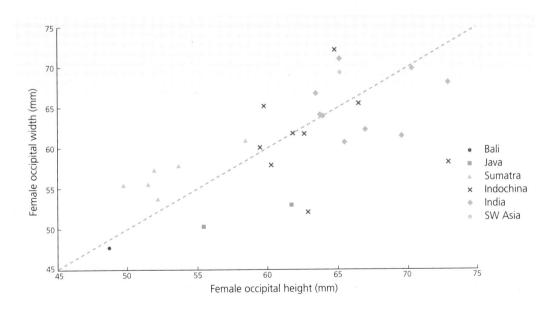

FIGURE 2.11
The relationship between height and width of the occiputs of female tiger skulls. The line indicates a 1:1 ratio between the two variables.

Table 2.5. *Mean weight of vertebrate prey indices (MWVP) for different populations of tigers based on a) scats and b) kills*

Country	Locality	Latitude	n	MWVP (kg)	Ref
a. scats:					
Sumatra	Way Kambas NP	5°02'S	82	26.22	1
Thailand	Huai Kha Khaeng	15°22'N	32	22.20	2
India	Eravilulam NP	10°14'N	18	186.58	3
India	Bandipur	11°39'N	27	93.00	4
India	Nagarahole NP	12°02'N	472	80.53	5
India	Kanha NP	22°16'N	326	74.72	6
Nepal	Royal Bardia NP	28°31'N	223	53.41	7
Nepal	Royal Chitwan NP	27°30'N	123	64.91	8
Nepal	Royal Chitwan NP	27°30'N	55	58.41	9
Russia	Sikhote-Alin	45°N	552	114.48	10
b. kills:					
India	Bandipur	11°39'N	16	90.75	4
India	Nagarahole NP	12°02'N	152	161.27	5
India	Kanha NP	22°02'N	228	93.07	6
Russia	Sikhote-Alin	45°N	59	93.61	11
Russia	Primorski	46.15°N	40	112.65	11

n – Sample size; MWVP – mean weight of vertebrate prey index. Prey body weight data: Schaller 1967; Medway 1978; Dinerstein 1980; Mishra & Kotwal 1990; Nowak 1991; Karanth & Sunquist 1995.
References: 1. Sriyanto, Bastoni, N. Franklin & R. Tilson (Sumatran Tiger Project), unpubl. data. 2. Rabinowitz 1989. 3. Rice 1986. 4. Johnsingh 1992. 5. Karanth & Sunquist 1995. 6. Schaller 1967. 7. Støen & Wegge 1996. 8. McDougal 1977. 9. Sunquist 1981. 10. Miquelle *et al.* 1996b. 11. Abramov 1962 (in Schaller 1967).

Sumatran tigers do not (Seidensticker & McDougal 1993), so that island dwarfing is unlikely to be the most important factor affecting tiger body size.

Mean weight of vertebrate prey index (MWVP)

The body size of tigers could be affected by the size of prey on which they feed. The mean weight of vertebrate prey index (MWVP) is a measure of the average prey size of predators. It has been found to be highly correlated with body size in jaguars and pumas irrespective of latitude (Iriarte *et al.* 1990; Hoogesteijn and Mondolfi 1996). MWVP is determined partly by the size of the most utilised prey (and so it is related to Guthrie's Rule above), but also by what species are available, including their vulnerability to predation, habitat effects and competition with other predators (Iriarte *et al.* 1990). It is calculated as a geometric mean of the numbers and body masses of a predator's prey:

$$\text{MWVP} = \text{antilog}_e[\Sigma(n_i.\log_e M_i)/N]$$

where:

n_i – number of individuals of species i.

M_i – mean body mass (kg) of species i.

N – total number of prey individuals of all species $(=\Sigma n_i)$.

Unfortunately, there are few good quantitative data on the diets of tigers outside India and the Russian Far East, where the biggest tigers occur. However, MWVP has been calculated for all available data sets (Table 2.5). In studies of scats of samples greater than 30 there appears to be higher MWVPs in higher latitudes [MWVP (kg) = 25.79+1.61×latitude (°), r=0.784], which are correlated with larger body sizes in tigers, but more research must be done to compare the diets of male and female tigers separately, because body size variation differs between them. Although felids are opportunistic predators, there are often intersexual differences in prey size (Kitchener 1991).

It is possible that female tigers may prey upon species that are roughly the same body mass throughout their range (e.g. small-to-medium-sized deer of various species and wild pig) and that their growth is restricted because they divert their energy into the production of cubs. However, males may exploit much larger prey too, including large bovids and moose (Schaller 1967; Miquelle *et al.* 1996b). In other words there may be resource partitioning between males and females that reaches an extreme in colder climates or at higher latitudes where prey availability may be less, or numbers of available prey may be fewer than in the tropics, or where size classes of prey may be restricted, or where the energetic demands of reproduction limit female growth.

Competitive character displacement

Dayan *et al.* (1990) have demonstrated that felids show competitive character displacement of the diameter of the upper canines. Similar observations were made in studies of mustelids (Dayan *et al.* 1989; Dayan & Simberloff 1994, 1996). The ratio of the diameter of the male upper canine to that of the female was constant within and between each felid species. In other words, each sex of each cat species appeared to be separate morphospecies. The size differences seemed to be constrained by the size of available prey, because canine size is probably related to the spacing of the cervical vertebrae in the preferred prey species in order to permit an efficient killing or nape bite (Leyhausen 1979). Thus, competition between sexes and species is resulting in resource partitioning of prey.

Similar competitive character displacement in the sizes of canines may explain the disjunct distribution of the leopard in southern Asia and provide further evidence for the former existence of the tiger on Borneo. The leopard occurs only on Java in the Sunda Islands, whereas the slightly smaller clouded leopard is found on Sumatra and Borneo, but is absent from Java, although there is fossil evidence for its survival until the Neolithic (Hemmer 1976).

Perhaps competition from clouded leopards and small tigers on Sumatra and Borneo squeezed out the leopard niche during the Holocene. This hypothesis could be tested by measuring the upper canine diameters of the Sunda Island cats, first to see if a 'gap' survives for the leopard on Sumatra, secondly to predict the size of a female Bornean tiger and, thirdly, to examine whether all three species could have survived on Java.

However, to some extent tigers may be unaffected by this phenomenon, because larger prey are typically killed by a suffocating throat bite rather than the nape bite (Sunquist 1981; Kitchener 1991), so that there is no constraint on canine size in relation to cervical vertebra size of the prey. Also, Seidensticker & McDougal (1993) have shown that tigers may kill buffalo by crushing the vertebrae with a nape bite. Therefore, tigers may not be affected by character displacement of canine size as are smaller felid species.

Geographical variation in tigers: a summary

From this preliminary analysis the pattern of geographical variation in tigers appears to be more complex than at first perceived. Body size is probably influenced by phenotypic responses to the environment, as well as being under genetic control and appears to vary clinally with increasing latitude, reaching peaks in the north of the Indian subcontinent and the Russian Far East. There may be several factors that produce the complex cline in the body size of tigers, including metabolic constraints, the size and availability of prey, competition with conspecifics and congeners, and variation in the annual productivity pulse of different habitats. Striping patterns, pelage coloration and skull characters are much more variable within putative subspecies than between them, suggesting that gene flow between populations may have continued well into the Holocene. However, there is a tendency towards dark, well-striped tigers in Southeast Asia and the Sunda Islands (c.f. Gloger's Rule), which suggests that environmental influences may be affecting tiger pelage characteristics locally. Perhaps some island populations show only a proportion of the total range of variation of mainland populations because the former were based on a few founders. Given current levels of knowledge it is, therefore, difficult to see what characters can be used to distinguish between the eight putative subspecies of tiger without more research. The preliminary study presented here suggests that there is in fact little evidence to support subspeciation in the tiger, based only on a morphological analysis. However, in common with other studies of tiger subspecific taxonomy, this study suffers from having too small sample sizes for most of the putative subspecies, so that we cannot be certain that they are representative of the variation observed within wild populations. Therefore, conclusions presented here are tentative.

Isolating mechanisms and subspecies

If eight subspecies of tiger were to be recognised, it should be possible to identify zoogeographically how the ancestral populations became isolated for them to evolve. Most of our knowledge of the evolutionary history of tigers is confined to the last two million years, when we have a good knowledge of environmental change. Until recently it was believed that the Pleistocene consisted of four stadials or ice ages interspersed with warmer periods or interstadials (Kurtén 1967; Sutcliffe 1985). However, oxygen isotope analysis from deep sea cores has shown that the climatic fluctuation was far more complex, with up to 28 stadial/interstadial cycles during the 1.7 million years of the Pleistocene (Shackleton & Opdyke 1973; Stringer & Gamble 1993). Therefore, this highly fluctuating climate would have been an important factor in affecting the evolution of putative tiger subspecies.

Looking at tiger distribution today, there seem to be certain environmental constraints on its distribution (Nowell & Jackson 1996; Dinerstein et al. 1997). Tigers appear to avoid high mountains, deserts and other open habitats. As seen earlier, the evolutionary history of the tiger has been dominated by the ice ages of the Pleistocene. So what factors of the stadials (ice ages) and interstadials (warmer periods) promote or reduce the possibility of subspecies evolution?

Stadials reduce sea levels, increase deserts and reduce forests, especially at higher elevations (Sutcliffe 1985). Reduced sea levels increase the possibility of gene flow by making islands accessible from the mainland (e.g. Morley & Flenley 1987; the Sunda Islands). Deserts may create barriers to tigers, but even a reduction in forests is unlikely to create much of a barrier to gene flow. Thus, during stadials gene flow is likely to be facilitated rather than impaired. Only the tigers in southwest Asia were likely to have been isolated by an expansion of the deserts in central Asia during the stadials.

Interstadials increase sea levels, reduce deserts and increase forests (Sutcliffe 1985). Higher sea levels would isolate populations on islands, but other effects of interstadials are likely to promote gene flow between tiger populations by expanding forests.

Tigers are large animals with a high degree of mobility (they may disperse for more than 1000 km and swim for up to 29 km across rivers or 15 km across the sea (Gargas 1948; Heptner & Sludskii 1992; Mazák 1996) and live in a wide variety of habitats (Dinerstein et al. 1997), as long as there is sufficient cover and prey. The only other significant barrier to the spread of the tiger would be competition with the lion, which tends to occur in drier, more open habitats. Therefore, it is hard to imagine any significant barriers to tiger gene flow, except a significant increase in deserts during stadials, which may have isolated Caspian tigers, and an increase in sea levels during interstadials, which may have isolated the three putative Sunda Island tigers and the Japanese tiger. However, the distances between the Sunda Islands are so small (12.5–20 km maximum swimming distances, taking into account small islands between) that tigers could have continued to swim between them even during the last 8000 or more years since they became isolated after the last Ice Age. Hemmer (1971) examined fossil tigers from the Pleistocene of Java and found that their local evolution seemed to have been interrupted by colonisations of larger tigers from the mainland during stadials.

Can you tell a tiger without its coat? Skull characters that separate lions from tigers are not absolute and leopard skulls are only certainly distinguishable on size. There is little evidence for discrete tiger subspecies and much of the variation seen in tigers is clinal.

Three models of tiger subspecies

My preliminary analysis of morphological variation in tigers suggests that there is little evidence for discrete subspecies and shows that much of the variation is clinal. This is supported by a crude zoogeographical assessment, which shows that there were few barriers to gene flow during much of the Pleistocene history of the tiger. On the basis of the limited study so far, I would like to propose three different models for the geographical variation in tigers, which would be testable by a variety of techniques. These models are not definitive, but provide a framework for future research that is not reliant on the current subspecies classification, which I believe is seriously flawed.

Model 1: Gene flow has been so extensive (until into the Holocene at least) that no subspecies can be recognised, although there is clinal variation due to environmental, climatic and ecological factors.

Model 2: Isolation of tigers on the Sunda Islands since the last Ice Age (c. 8000 years) has resulted in

the evolution of a distinctive subspecies. Mainland animals have a complex cline influenced by several environmental and ecological variables, but retain full gene flow. However, animals from peninsular Malaya are of uncertain status, since they could easily cross to Sumatra, but may have been isolated from other Southeast Asian tigers by the Isthmus of Kraa. The proposed subspecies are:

Panthera tigris tigris – Mainland Asia. Larger body size (greatest skull length mostly >335 mm (males), mostly >280 mm (females); variable ground colour and numbers of stripes, although tendency towards lighter ground colours and fewer numbers (mean <27) of stripes.

P. t. sondaica – Sunda Islands, possibly peninsular Malaya. Smaller body size (greatest skull length mostly <335 mm (males), mostly <280 mm (females); darker ground colour and greater number of flank stripes (mean >27).

These characters are provisional and exact diagnoses are not possible given the small sample sizes and the high degree of overlap in some of these characters.

Model 3: A combination of desertification during stadials and fluctuating sea levels during the Pleistocene has resulted in the isolation of populations resulting in three subspecies:

> *P. t. tigris* – Mainland Asia, possibly Sumatra. As above; model 2.
>
> *P. t. virgata* – Southwest Asia. Greater numbers of stripes compared with *tigris* (mean > c. 26), longer fur and broader occiput.
>
> *P. t. sondaica* – Java and Bali, possibly Sumatra. Small body size (see above); greater number of stripes (mean >30) and darker ground colour; narrow occiput (except Sumatra).

It is possible that Sumatran tigers represent mainland animals that colonised Sumatra before the end of the last Ice Age, or even hybrids (i.e. *P. t. tigris* x *P. t. sondaica*).

Although a good deal of speculation is involved here, it is to be hoped that it is at least informed speculation based on a global view. These models represent hypotheses that can be tested by a combination of morphological, molecular and zoogeographical techniques.

Conservation relevance

In this chapter I have shown that assumptions about the validity of the eight subspecies of tiger recognised by the conservation community today are incorrect, because of poor samples and a lack of knowledge of the degree of variation within putative subspecies. My analysis has shown that much of the morphological variation in tigers is clinal; i.e. tigers vary continuously throughout their geographical range and they do not appear to form discrete subspecies, but this conclusion is tentative given the small sample sizes for each putative subspecies. Today, the tiger's distribution has been fragmented into about 160 population segments (Dinerstein *et al.* 1997), so what implications does my analysis have for the conservation of the tiger?

Tigers are currently represented by populations, presumably with co-adapted gene complexes for surviving in a wide range of different habitats from boreal forest to tropical rainforest. Dinerstein *et al.* (1997) have proposed that an ecological approach should be taken to the conservation of the tiger rather than a taxonomic one; by maintaining viable populations in the widest variety of habitats and ecological conditions, any co-adapted gene complexes that promote the survival of local tiger populations will also be preserved. However, there are problems with this approach and it does not obviate the need to understand the geographical variation in, and possible subspecies of, the tiger.

Most TCU populations (61%) are considered not to be viable in the long term (Dinerstein *et al.* 1997). Many will probably be lost despite our best efforts and others that are considered to be secure today may become threatened or fragmented in the future. Whether or not these are considered to be either ecologically and/or taxonomically important, they may need to be managed as metapopulations with other TCUs and captive populations through the construction of habitat corridors, translocation, use of artificial reproductive techniques and re-introduction. Problems arise as to which populations should be linked in this way, and whether the variation within the species is best represented as a cline or a distinct subspecies. For example, does a mangrove forest tiger have the same co-adapted gene complexes and morphological adaptations and/or characters as a rainforest tiger? Do all members of genetically homogenised captive populations show the same degree of variation as their wild-living putative subspecies?

The answer from this preliminary study is that most variation is likely to be clinal and consequently this opens up considerable opportunities for the management of the smaller TCUs, given that recorded tiger dispersal may result in effective gene flow between populations more than 1000 km apart (Mazák 1996). Therefore, critically endangered South China tigers could readily be genetically reinforced by animals from northern Southeast Asia and possibly the Indian subcontinent. The most important conservation outcome is that tigers continue to survive in China, where they continue to perform their vital role as top predator.

In the meantime and following criticisms of the evolutionary significant unit (= subspecies in my terminology) as the basis for conservation management and legal protection (Pennock & Dimmick 1997), and the problems of hybridisation between putative subspecies (O'Brien & Mayr 1991), it is important that we develop and implement conservation and management strategies that maintain the tiger as a top predator throughout as much of its remaining distribution as possible, rather than being distracted by traditional subspecies names that have little or no basis in fact or science.

Therefore, a thorough understanding of the morphological and genetic variation in tiger populations is required to provide a sound basis for answering the difficult conservation questions of the next millennium.

Research in four areas is urgently required:

1   A morphological database of tiger skins and skeletons to record variation from captive and museum specimens throughout the world.

2   The collection of substantially more skins, skeletons and molecular samples from natural mortalities of wild and captive tigers, especially wild-caught founders of captive populations. These should be held within a limited number of institutions to facilitate research. Rosenthal (undated) surveyed 161 major museums in North America, Europe and Asia and found only 34 skulls, 24 skeletons and 59 skins of the Amur tiger, most of which were of captive origin or had no data. This sample is barely adequate for taxonomic research and the analysis presented here is also affected by small sample sizes.

3   A combined morphological and molecular study of the pattern of variation in tigers.

4   A Geographic Information System study to investigate the environmental limits on tiger distribution today, in order to reconstruct the colonisation by, and possible isolation of, subpopulations during the Pleistocene. A preliminary study is currently being carried out by the National Museums of Scotland in collaboration with the Department of Geography, Edinburgh University.

The National Museums of Scotland would welcome the opportunity to develop these proposals with other institutions and individuals.

3

Subspecies of tigers: molecular assessment using 'voucher specimens' of geographically traceable individuals

Joelle Wentzel, J. Claiborne Stephens, Warren Johnson, Marilyn Menotti-Raymond, Jill Pecon-Slattery, Naoya Yuhki, Mary Carrington, Howard B. Quigley, Dale G. Miquelle, Ronald Tilson, Jansen Manansang, Gerald Brady, Lu Zhi, Pan Wenshi, Huang Shi-Qiang, Leslie Johnston, Mel Sunquist, K. Ullas Karanth and Stephen J. O'Brien

Introduction

Increasing awareness of biodiversity depletion over the past two decades has highlighted concerns about conservation of endangered species and their included genetic diversity (Avise 1994). Conservation genetic analyses offer insight into recent natural history, present status and future survival implications of endangered populations (O'Brien 1994b). A basic supposition of conservation genetics is that the preservation of genetic variability within rare or endangered species will enhance the chances of its survival over time (Avise 1994). Because of the highly endangered status of the tiger, it is important to assess the genetic variation of surviving populations to assist in developing management plans for the species (Smith & McDougal 1991).

Previous studies, including biochemical and morphological analyses, have provided information about the evolution, genetic heterogeneity and subspecies identification within *Panthera tigris*. Based on allozyme studies, slight differences in allelic frequencies were established between Sumatran, Indian and Amur tigers and moderate levels of heterozygosity were estimated (Newman *et al.* 1985; Goeble & Whitmore 1987; O'Brien *et al.* 1987). The tiger belongs to the Pantherine lineage that includes the five species of roaring cats (O'Brien *et al.* 1996; Johnson & O'Brien 1997). The divergence of this lineage of Felidae has been assessed to have taken

place two to three million years ago. The combination of fossil evidence and cranial measurements indicates that the species originated in southern China (Herrington 1987).

The distribution of tiger subspecies at one time spanned most of Asia (Fig. 0.1), occupying diverse habitats including tundra, tropical forest, mangrove swamps and deciduous forests. Tigers roamed as far north as Siberia, as far west as Turkey, to the east on the Island of Sakhalin, and down to the Southern islands of Indonesia. However, numbers of free-ranging tigers have declined by 95% during this century (Jackson & Kemf 1996), and remaining populations are in general small and fragmented (Figure 0.1).

It is important to assess genetic variability in the small populations of the putative subspecies, notably South China and Amur tigers (Smith & McDougal 1991), because low variability has been found to be detrimental to endangered species by compromising their adaptive potential (Lande & Barrowclough 1987; O'Brien 1994a, b; Frankham 1995). A small population is also susceptible to the effects of inbreeding depression (Soulé 1980). The consequences of a population bottleneck have been clearly shown in the African cheetah. Its demographic crash resulted in the genetic uniformity of the species along with the undesirable traits of low fecundity and disease vulnerability (O'Brien *et al.* 1985). In the case of the cheetah, the bottleneck was estimated to have occurred during the late

Pleistocene, approximately 10 000 years ago during the mass extinction of vertebrates (Menotti-Raymond & O'Brien 1993). Genetic studies of other endangered megavertebrates such as Asian lion (Gilbert *et al.* 1991), Florida panther (Roelke *et al.* 1993), European bison (Hartl & Pucek 1992), South African wildebeest (Corbet *et al.* 1994) and northern elephant seals (Hoelzel *et al.* 1993) have all indicated depletion of genetic variation because of demographic reduction during historic times. The demographic contractions suffered by the tiger are caused by human encroachment, habitat destruction and illegal poaching (Kenney *et al.* 1994).

For centuries, many Asian cultures have revered tigers as gods with healing power and used the bones and body parts of the animals for their potions. The high prices paid in the black market have led to increased poaching of tigers. In Russia, the construction of the Chinese Eastern Railroad in the late 19th century caused an influx of Russian settlers and guns into the east. By the 1930s, the population of Amur tigers had been reduced to 50 individuals (State Committee of the Russian Federation for the Protection of the Environment 1996). This bottlenecked population now numbers between 415 and 476 animals (Matyushkin *et al.* 1996) The growing human population of Sumatra has fragmented the remaining Sumatran tiger habitat to isolated pockets on the island. Because of habitat fragmentation, gene flow between populations has been interrupted since migration corridors between the small isolates have been blocked.

The concept of subspecies has been recognised as a taxonomic category based upon phenotypic differences and geographic distribution (Mayr 1963; Avise & Ball 1990; O'Brien & Mayr 1991). Subspecies names, or Latin trinomials, are usually meant to reflect patterns of geographic isolation unless challenged by other measures. Subspecies may be sympatric (overlapping), parapatric (abutting) or allopatric (geographically separate) (Mayr 1982). Because of their wide range, isolated populations of tigers are traditionally assigned subspecies designations according to geographic and morphological descriptions (Mazák 1981).

While subspecies recognition of isolated populations has been both imprecise and controversial, their importance for conservation is critical since subspecies are frequently the units for legislative protection of endangered taxa.

To establish objective molecular genetic verification of subspecies classification, we initiated a study of rapidly evolving gene families among the five living tiger subspecies. Our survey was based on a select group of 28 voucher specimens, tigers with verifiable origins from a specific geographic location or direct descendants of parents whose origin was established. Our protocol provides viably frozen immortal cell lines as a living repository for their genetic materials, but also avoids being misled by captive specimen misidentification. The results of our analysis, to be published in detail elsewhere (Wentzel *et al.* in prep.), led to two primary conclusions:

▶ Tigers of the five subspecies are genetically very similar, suggesting an historic genetic reduction and gene flow among all tiger subspecies until the late Pleistocene.

▶ Geographic isolation and genetic drift of rapidly evolving nuclear short tandem repeat (STR) loci provides a basis for discerning population genetic differentiation sufficient to identify the subspecies origin of tigers.

In this chapter we summarise the conclusions of that study and ascertain the subspecies affiliation for 17 captive tigers of unconfirmed subspecies.

Low genomic variation among tigers

The mitochondrion of vertebrates contains a 16 000 base pair (bp) cytoplasmic chromosome that encodes 13 protein coding genes, and 24 other loci, mostly specifying transfer ribonucleic acid (tRNA) and control region sequences (Anderson *et al.* 1981; Lopez *et al.* 1996). In mammals, mitochondrial deoxyribonucleic acid (mtDNA) evolves more rapidly than nuclear genes due to differences in DNA repair systems between nuclear and cytoplasmic compartments (Rand 1994; Lopez *et al.* 1997).

Box 3.1   Molecular biology in conservation biology
Sriyanie Miththapala

In the last two decades, new methods in molecular biology have developed rapidly, providing conservation geneticists with numerous, precise tools for clarifying obscure phylogenies, describing the effects of stochastic changes, discerning relationships and identifying subspecies and population substructures (Mace *et al.* 1996; Wayne & Koepfli 1996). However, such work must be interpreted not only by molecular geneticists but also by field biologists and managers with scant knowledge of the jargon of molecular biology. Methods mentioned in J. Wentzel *et al.* (this volume) are explained here with a view to clarifying technical jargon and describing, in layperson's terms, the methodologies used.

Examination of variation in *mitochondrial deoxyribonucleic acid* (mtDNA) is a popular method in molecular biology. Unlike nuclear DNA, mtDNA is a closed, circular molecule (some 16 000 kilo bases (kb) long). The specific genes contained within the mtDNA molecule are generally conserved across taxa – whether the mtDNA is from a frog or a human, most of the same genes are present in this molecule – and include 22 transfer ribonucleic acid (RNA) genes, two 12S and 16S ribosomal RNA genes, 13 protein coding genes and a control region (Avise & Lansman 1983; Mace *et al.* 1996). The rate of change within mtDNA is some five- to tenfold faster than that in coding nuclear DNA, with the result that analysis of mtDNA allows for assessment of recent demographic events – i.e. description of phylogenetic history (Mace *et al.* 1996). Within the molecule, the control region, which initiates replication and transcription, evolves four to five times faster than the rest of the molecule and hence is likely to show the greatest degree of variation (Taberlet 1996).

Within this mtDNA molecule are sites where certain enzymes – called restriction enzymes – can cleave the circle, with the result that linear strands are produced. The number of linear fragments produced by digestion by a restriction enzyme is equal to the number of sites in the molecule. The number and position of these restriction sites may vary from individual to individual, population to population, subspecies to subspecies and species to species. When these strands, called *restriction fragment length polymorphisms* (RFLP) are size-separated by electrophoresis, different band patterns can be visualised. The total pattern obtained from a given number of restriction digests is called a haplotype. Visualisation and differentiation of these patterns allow for discrimination between taxa – in this case, subspecies.

The ultimate source of genetic variation is the sequence of bases in a DNA molecule. Analysis of sequences, therefore, allows for direct comparison between taxa at several levels, depending on the gene examined. The greatest differences in sequences would be expected in the control region.

The *major histocompatibility complex (MHC)* is a complex of several hundred nuclear genes (in mammals) about 3500 kb long, which are tightly linked. Within this complex are several classes, notably MHC class I and class II genes, which initiate and direct immune responses in vertebrates against foreign pathogens (Edwards & Potts 1996). The MHC genes activate white blood cells to combat infection, and the DRB locus in the complex distinguishes 'self' from 'non-self' proteins. In order to be able to combat a variety of pathogens, MHC genes are exceptionally polymorphic. Thus, it has been argued that assessment of variation at the MHC complex is essential for conservation genetics (Hughes 1991).

Single strand conformation polymorphism (SSCP) is a very sensitive, consistent and reliable technique that allows for identification of single nucleotide differences between alleles of individuals (Girman 1996). In J. Wentzel *et al.* (this volume) SSCP techniques were used to examine the DRB locus of the MHC.

Scattered throughout the nuclear genome of eukaryotes are tandemly repeated short units – usually 1–10 bp long (Bruford *et al.* 1996) which are highly polymorphic in the number of repeats and hence their length. These *short tandem repeat (STRs)*, or microsatellites, can be amplified through the *polymerase chain reaction (PCR)* to produce enough copies for detection of the different allele sizes through electrophoresis. This methodology has proven to be invaluable for conservation genetics, because differences ranging from across taxa down to a single individual can be identified (Bruford *et al.* 1996).

The use of different molecular biology techniques, assessing variability at highly polymorphic loci and at loci which evolve at different rates, safeguards against premature conclusions and exemplifies standards which should be followed in future research.

Over the past decade, measures of mtDNA sequence variation have provided valuable insight into the patterns of population differentiation, particularly among closely related populations at the subspecies and species levels (Avise 1994; Avise & Hamrick 1996). We have used mtDNA to quantify genetic variation among cheetahs, lions, Florida panthers and leopards (Menotti-Raymond & O'Brien 1993; Roelke *et al.* 1993; Miththapala *et al.* 1996). In addition, DNA sequence and restriction fragment length variation have proved exceedingly useful in phylogenetic reconstruction of the Felidae radiation (Janczewski *et al.* 1995; Johnson *et al.* 1996; Masuda *et al.* 1996; Johnson & O'Brien 1997).

When tiger specimens were examined using three measures of mtDNA variation [16S ribosomal RNA (rRNA) gene sequence, 330 bp; control region DNA sequence 329 bp; and restriction fragment length polymorphism (RFLP) using 28 restriction enzymes] the tigers showed little genetic diversity (Fig. 3.1). No variation was observed among sampled tigers for the 16S RNA gene or for mtDNA-RFLP variation. The greatest amount of variation for these mtDNA markers in comparable studies were

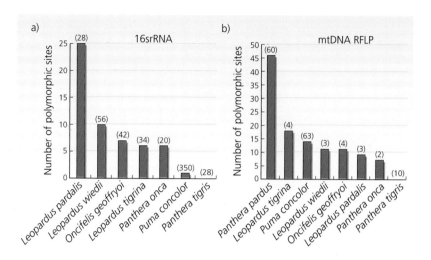

FIGURE 3.1a, b
Two estimates of mitochondrial deoxyribonucleic acid (mtDNA) variation among tiger voucher specimens compared to other cat species (Wentzel *et al.* in prep.). **a.** Number of polymorphic nucleotide sites observed for 16S rRNA sequence. **b.** Restriction fragment length polymorphism (RFLP) of ten tigers from mixed subspecies (from Table 1 in Johnson *et al.* 1996). Numbers in parentheses are number of individuals typed.

observed in ocelot, which revealed 25 polymorphic residues and 31 haplotypes for 16S RNA, and in leopard, which showed 46 polymorphic sites on 18 haplotypes for mtDNA-RFLP (Miththapala *et al.* 1996; Wentzel *et al.* in prep.). For the control region, the most rapidly evolving segments of mtDNA loci, tigers had only three polymorphic sites and three haplotypes (Fig. 3.2). Pumas and ocelots display 10 times this level of control region variation (M. Culver & W. Johnson pers. comm.).

A second measure of overall genomic variation was supplied by a screen of the tiger major histocompatibility complex (MHC). The MHC is an assemblage of tightly linked loci that encode cell surface antigens which present foreign peptides to the T-cell receptors of immune lymphocytes (Klein 1986). As a consequence of consistent outbreaks of debilitating infectious agents in free-living populations, abundant genetic diversity of the MHC is thought to provide a broad repertoire for recognition of divergent pathogens. A wide breadth of immune recognition is accomplished by an accumulation of gene duplications, periodic gene conversion as a

source of variation, and selective maintenance of abundant MHC allele heterozygosity within the populations (Yuhki & O'Brien 1990a).

The tiger sample also showed reduced variation for the MHC-class-II DRB locus compared to other species. Our study, which employed a single-strand conformation polymorphisms (SSCP) ascertainment of tiger variation (Fig. 3.2), revealed six alleles (*DRB-A–F*) among 22 voucher specimens that were successfully genotyped. The homologous *DRB* gene segments in domestic cat and human revealed 61 and 126 allele transcripts, respectively, in sampling of 37 cats and 251 humans (Yuhki & O'Brien 1997; Bodmer *et al.* 1994). This level of *DRB* variation, less than 10% of the amount seen in domestic cat, is consistent with a history of demographic contraction for all tigers, effectively reducing the amount of variation by an order of magnitude compared to other cat species.

We also examined the pattern and extent of nuclear gene variation in tigers based upon a survey of 20 feline-specific dinucleotide STR (or microsatellite) loci (Menotti-Raymond & O'Brien 1995;

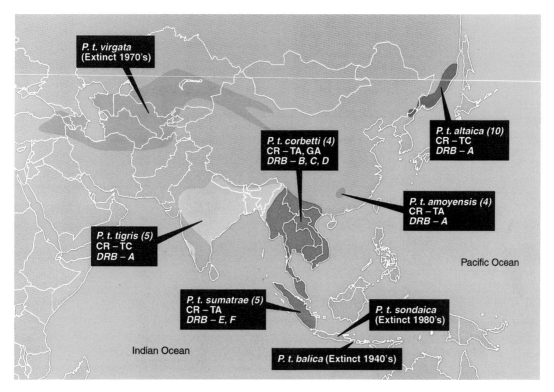

FIGURE 3.2
Approximate gegoraphical range of the eight traditional tiger subspecies, showing genetic markers isolated from known-origin specimens. In parentheses is the number of voucher specimens tested (see text).
CR = mitochondrial DNA control region haplotypes observed: TA, TC, GA or TG. These indicate nucleotide residues of the only two variable sites detected in tigers (Wentzel *et al*. in prep.). *DRB* indicates *MHC-DRB* alleles observed in this population. Note that Amur and South China tigers were genetically fixed for *DRB-A* and that the only Indian tiger typed successfully was also *DRB-A*.

Menotti-Raymond *et al*. 1997) (Fig. 3.3). A moderate amount of variation was observed among the voucher specimens, revealing 112 STR alleles and an average overall heterozygosity of 0.41 for all tigers and 0.37–0.44 for the samples of separate subspecies. A phylogenetic analysis of composite STR genotypes of vouchers revealed significant clustering consistent with geographic locale (Wentzel *et al*. in prep.). The significant substructure was also observed when FST values (i.e. the fixation index, or the value which represents the amount of population subdivsion found between subpopulations) were computed between voucher subspecies as well as relative genetic distances estimated by the proportion of shared alleles between populations (Bowcock *et al*. 1994) (Table 3.1). We have developed a parametric maximum

likelihood strategy for evaluating 'subspecies' or population affiliation based upon composite STR genotype (Stephens *et al*. in press). Using estimated STR allele frequency for subspecies derived from the 28 voucher samples, we were able to derive a robust statistical measure of subspecies affiliation using a 'log of odds' or 'lod score' quantitative estimate. Each voucher specimen genotype was tested for subspecies affiliation using allele and genotype frequencies of the remaining 27 voucher tigers to evaluate the procedure. The results affirmed the efficacy of the affiliation test since, for each voucher tiger's composite STR genotype, the deduced subspecies affiliation was unequivocal, statistically significant and concordant with the actual subspecies origin.

A group of 17 captive animals, whose subspecies

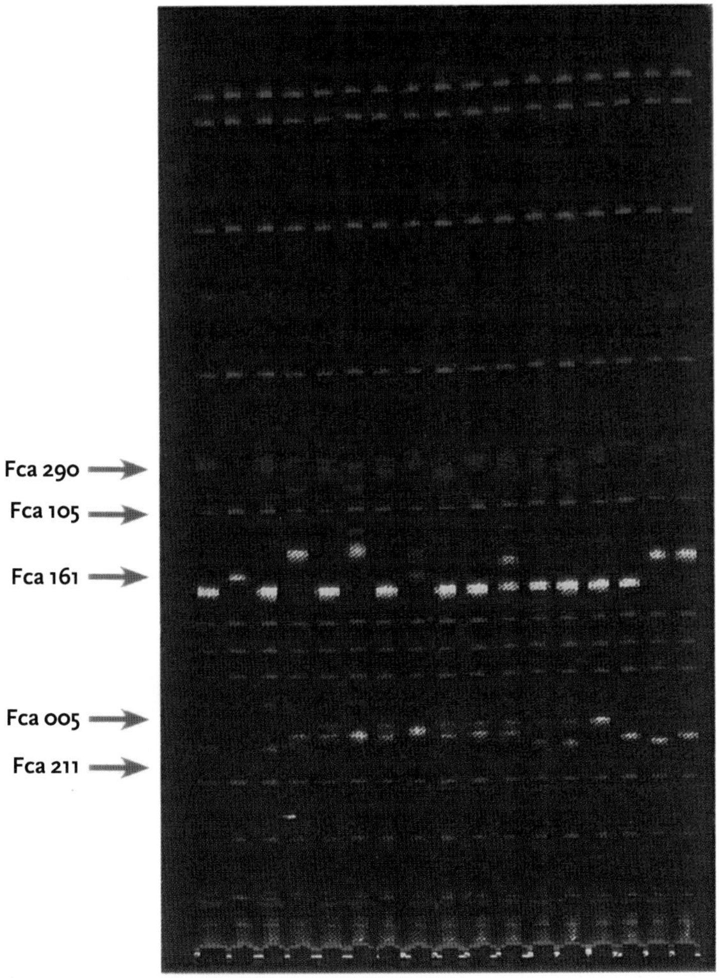

FIGURE 3.3a
Short tandem repeat (STR) alleles amplified from five separate loci. Fluorescent-dye-labelled polymerase chain reaction (PCR) products are multiplexed and electrophoresed on a polyacrylamide gel. Alleles are represented by blue, green and yellow bands in each lane. An internal lane standard (red fragments) is used to size the PCR fragments. Sizes of alleles are given as weights in terms of number of base pairs.

affiliation was in question, were genotyped for a subset (7–15) of the diagnostic STR loci as well as for *DRB* (Fig. 3.3). For 15 of 17 tigers, the composite STR genotype produced a definitive affiliation (Table 3.2). Those 15 tigers, all registered in the International Tiger Studbook and suspected Sumatran tigers, had appropriate STR composite genotypes for that subspecies (Table 3.2). One tiger (Pti-160), from Kiev Zoo, was thought to be an Amur tiger, but its STR genotype was more closely aligned with vouchers from Sumatran tigers. Pti-224, a tiger of unknown origin from Tiger World Missing Link in Texas, had an STR genotype that indicated Indochinese origin. The results from the Kiev and Texas tigers produced lower lod scores than the

other captive individuals and had signature alleles from more than one subspecies. These results indicate that at least one hybridisation event occurred in each tiger's pedigree. Precise determination of their subspecies origin is equivocal for these two individuals and their ancestry may include individuals from different subspecies. Of the 15 tigers identified as Sumatran, 13 maintained the *DRB-E* allele found exclusively among Sumatran vouchers (Fig. 3.2). In addition two new *DRB* alleles, *DRB-G* and -*H*, were discovered among the captive Sumatran tigers, indicating that the captive population contains additional genetic variation (at least for *DRB*) not represented in the voucher specimens. The Texas captive tiger had two *DRB* alleles, *DRB-I*

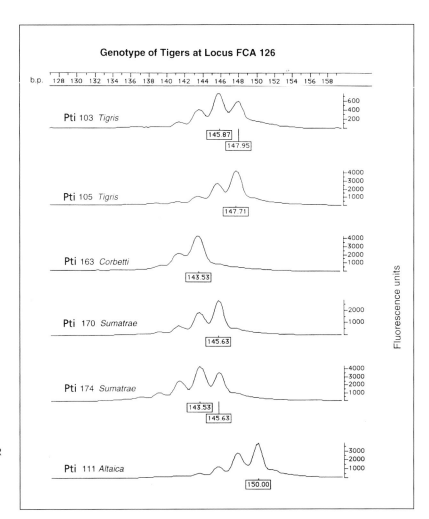

Genotype of Tigers at Locus FCA 126

FIGURE 3.3b
Densitometric scan of STR
profile of tiger DNA, using
Genotyper software from
an ABI-373 automated
DNA sequencer.

and *DRB-J*, not seen in any vouchers including those from Indochinese tigers. The additional *DRB* alleles found in the captive population support the notion that captive individuals may include a sampling of genetic variation that extends beyond the endemic diversity remaining in the wild populations.

Conclusions

The results reveal a dramatic reduction in tigers' mtDNA and MHC-DRB variation compared to other felids (Wentzel *et al.* in prep.), but moderate genetic variation in nuclear allozyme (Goebel & Whitmore 1987; O'Brien *et al.* 1987) and STR loci. The discordance between mtDNA and nuclear loci is likely the consequence of an historic demographic contraction that reduced tiger populations to small numbers of individuals (tens or hundreds) from a single geographic locale. Since mtDNA is maternally inherited, a moderate founder effect could result in a predominant monomorphic mtDNA lineage without constricting nuclear genetic variation appreciably (Avise 1994). The proposed contraction likely preceded the late Pleistocene (*circa* 10 000 years ago) since the Indonesian island subspecies has the identical mtDNA type to mainland subspecies (Fig. 3.2) and the island of Sumatra was last separated from the

Table 3.1. *Evidence for phylogeographic substructure among tiger subspecies voucher specimens based on 20 single tandem repeat (STR) loci (see text). FST[a] (above diagonal) and genetic distance[b] (below diagonal) between voucher tiger subspecies populations*

Genetic distance	FST					
	Indian tiger	Indochinese tiger	Sumatran tiger	Amur tiger	South China tiger	Ple–Ser
Indian tiger	—	0.21	0.37	0.30	0.35	—
Indochinese tiger	0.77	—	0.35	0.33	0.26	—
Sumatran tiger	1.26	1.19	—	0.41	0.36	—
Amur tiger	0.86	0.92	1.15	—	0.37	—
South China tiger	1.06	0.83	1.10	0.93	—	—
Ple–Ser[c]	1.81	1.60	1.53	1.92	1.45	—
Ple–Gir[d]	2.10	1.95	1.62	2.22	1.81	0.89

[a] Average FST for all tiger subspecies = 0.39.
[b] Proportion of shared alleles (Bowcock *et al.* 1994).
[c] African lion from Serengeti.
[d] Asian lion from Gir Forest in western India (Wildt *et al.* 1987; Gilbert *et al.* 1991).

continent about 10 000 years ago (Umbgrove 1949). Consistent with such a date is that tiger mtDNA variation is somewhat less than that observed in cheetahs (Menotti-Raymond & O'Brien 1993), which are also estimated to have experienced severe demographic contraction around the same time. The phylogeographic structure of the five subspecies revealed by composite STR genotype is also consistent with a late Pleistocene genetic reduction.

The population genetic substructure was used successfully to identify the subspecies affiliation of 17 captive tigers of uncertain ancestry (Stephens *et al.* in prep). Although the test is efficacious for pure subspecies members, resolution of hybrids between subspecies appears more difficult (Table 3.2). The procedure should be useful for first generation F1 hybrid identification; however, with more complex subspecies intermixing the power of ascertainment would be less certain.

The historic demographic reductions discerned here, plus the knowledge of major 20th century impacts on tiger numbers and habitat, have important implications for conservation of the species. An encouraging note is that the STR estimate of species-wide nuclear variation shows moderate amounts of residual diversity. However,

there is evidence for moderate inbreeding and loss of heterozygosity within two subspecies. Both Indian and Amur tigers were genetically homozygous for MHC-*DRB-A* allele (Fig. 3.2, Wentzel *et al.* in prep). This level of genetic uniformity is remarkable, reminiscent of the severe effects of population bottlenecks observed in cheetahs, Asian lions and Florida panther (Yuhki & O'Brien 1990b; Roelke *et al.* 1993; O'Brien 1994a). Genetic consequences of these events have not been reported explicitly for tigers; however, they may be contributing to fitness reduction in these populations.

The genetic and morphological basis of subspecies partition of tigers has been studied previously. Herrington (1987) reported that multivariate analysis of 45 cranial characters studied in 45 tigers discriminated subspecies unequivocally. But allozyme studies (Newman *et al.* 1985; Goebel & Whitmore 1987; O'Brien *et al.* 1987) and mtDNA sequences (Fig. 3.2) did not reveal subspecies-specific characters or evidence of substructure among subspecies. A comprehensive evaluation of previous subspecies analyses (A. C. Kitchener this volume) found little morphological evidence in support of subspecies genetic differentiation. A suite of 20 STR polymorphic loci, when analysed as

Table 3.2. *Subspecies affiliation of captive tigers*

Pti-#	Location DRB genotype	Named subspecies	International studbook no.	Composite STR ç subsp. affiliation	Lodscore range* (no. loci)	DRB genotype
	Captive tiger information			Results from genetic analyses		
Pti-079	San Diego Zoo	*sumatrae*	312	*sumatrae*	4.7–16.0 (13)	EE
Pti-096	Phoenix Zoo	*sumatrae*	719	*sumatrae*	5.1–16.5 (14)	EE
Pti-100	Omaha Zoo	*sumatrae*	527	*sumatrae*	6.6–18.7 (12)	EG
Pti-101	Omaha Zoo	*sumatrae*	380	*sumatrae*	4.3–12.2 (8)	—
Pti-107	Colorado Springs	*sumatrae*	234	*sumatrae*	5.1–12.9 (10)	—
Pti-164	National Zoo Park	*sumatrae*	495	*sumatrae*	5.4–17.8 (14)	EE
Pti-169	TSI, Indonesia	*sumatrae*	T9568	*sumatrae*	8.3–20.0 (14)	EE
Pti-173	TSI, Indonesia	*sumatrae*	T9506	*sumatrae*	5.8–13.8 (13)	EE
Pti-176	TSI, Indonesia	*sumatrae*	T9257	*sumatrae*	4.0–11.8 (14)	EE
Pti-177	TSI, Indonesia	*sumatrae*	532	*sumatrae*	5.6–15.9 (13)	EE
Pti-178	TSI, Indonesia	*sumatrae*	884	*sumatrae*	2.6–8.1 (12)	EF
Pti-179	TSI, Indonesia	*sumatrae*	T9605	*sumatrae*	4.5–14.1 (12)	EE
Pti-180	TSI, Indonesia	*sumatrae*	908	*sumatrae*	4.8–10.9 (7)	EE
Pti-183	TSI, Indonesia	*sumatrae*	917	*sumatrae*	3.1–6.6 (11)	EF
Pti-186	TSI, Indonesia	*sumatrae*	868	*sumatrae*	3.6–14.1 (13)	EF
Pti-160	Kiev Zoo	*altaica*	3261	*sumatrae***	3.9–4.2 (15)	HF
Pti-224	Tiger Missing Link Texas	unknown	—	*corbetti***	2.5–3.6 (11)	IJ

*Lod Score – Log of the likelihood ratio for the most probable subspecies versus the second most likely subspecies, derived from the captive specimen's composite STR genotype and the database of allele frequencies estimated from sampled voucher specimens. Range of values make use of a parametric correction for alleles not observed in each subspecies, but presumed to be present at set frequencies of 0.15–0.01 in the population at large (Stephens *et al.* in prep). A lod score of 3.0 in comparing *P.t. sumatrae* vs. *P.t. tigris* means that the tiger is 10^3, =1000, times more likely from *P.t. sumatrae* than from *P.t. tigris*. A lod score of 4.0 means it is 10 000 times more likely.
**These two tigers gave low STR lod scores and contained signature alleles from two different subspecies, making subspecies affiliation uncertain.

a multivariate composite genotype, did reveal moderate structure. Taken together these results suggest that tiger subspecies are actually very closely related and have probably undergone gene exchange until recently, followed by genetic drift in geographically isolated populations. These observations provide a minimum of molecular genetic support for classifying the subspecies as evolutionary significant units, except for the geographic isolation component which has occurred relatively recently, probably less than 10 000 years ago.

4

The tiger in human consciousness and its significance in crafting solutions for tiger conservation

Peter Jackson

Introduction

The tiger has always been regarded by humans with awe. Admiration for its beauty has been combined with underlying fear of its massive power. It has been a living presence for millions of people in Asia, from Turkey to the Russian Far East, and from India through Southeast Asia to Bali, in the southern tropics of Indonesia. In the forests and grasslands it shared with people it was worshipped; and its crude image exists today in jungle shrines.

Lions and leopards appeared in European cave art as long as 30 000 years ago, but the first known representation of the tiger appears to have been on the seals of the Indus civilisation, 5000 years ago, in what is now Pakistan. Later, when the Aryan peoples brought the Hindu religion to the Indian sub-continent they settled in vast forests where the tiger reigned, seldom seen, but always present. The tiger was absorbed into Hinduism and became a potent image as the mount ridden by the great female deity, Durga, while one of most important of the gods, Siva, sits on a tiger skin. As Buddhism evolved from Hinduism and spread through Asia, it carried with it consciousness of the tiger as a spiritual and cultural image, which adorns splendid murals in temples in Bhutan, China, Thailand and Tibet. Throughout tropical Asia and north to the Sea of Okhotsk, folk tales were woven round the tiger. Men were credited with the ability to turn into tigers, like the were-wolves of Europe, and spirit tigers were believed to roam the forests.

In the Chinese calendar, every 12th year is dedicated to the tiger. Its forehead markings resemble the Chinese character 'Wang' meaning 'King'. At the advent of the Year of the Tiger these markings are painted on the foreheads of children to promote vigour and health. The children also wear tiger caps and tiger slippers for the occasion. In Korea, the tiger became the symbol of the Mountain Spirit, and the White Tiger the Guardian of the West. In modern times, South Korea chose the tiger as the symbol of the 1988 Olympic Games in Seoul. Malaysia has two tigers supporting its National Crest. Laos chose the tiger for a special collectors' stamp. India, home of half the world's surviving tigers, replaced the lion with the tiger as its national animal when conservation started in the early 1970s. In Pakistan, where no tigers are now found, the political party which fought for independence from British rule, the Muslim League, still has the tiger as its election symbol and displayed it during its victorious campaign in 1997. Bangladesh has the image of the tiger on banknotes. Because the tiger is a symbol of power, Hong Kong, Malaysia, Singapore, South Korea and Thailand have been dubbed 'Asian tigers' because of their rapid economic advance.

For the people who live in and around the forests of Asia, the tiger continues to exert a heavy presence. Ancient beliefs survive in tribal societies, including that of the passage of human souls to tigers. Javan shadow puppet plays, popular in villages, begin and end with the image of the universe as a mountain (*gunungan*) surmounted by the Tree of Life, with a tiger and a banteng below it. The tigers in the mangrove forests of the Sundarbans in India and Bangladesh have a fearsome reputation. They have been – and still are – responsible for many deaths.

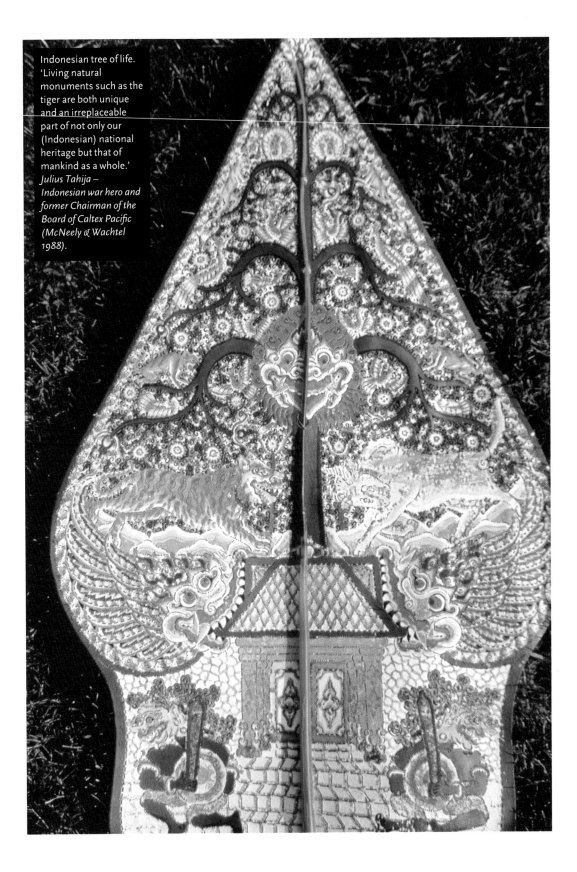

Indonesian tree of life. 'Living natural monuments such as the tiger are both unique and an irreplaceable part of not only our (Indonesian) national heritage but that of mankind as a whole.' *Julius Tahija – Indonesian war hero and former Chairman of the Board of Caltex Pacific (McNeely & Wachtel 1988).*

Woodcutters, honey collectors and fishermen seek protection with offerings to the jungle gods before entering the area. Nevertheless, the tiger is respected. Even when a tiger has killed a human being, and itself been brought to book, local people in India and Nepal will pay homage and cast flowers on the tiger's body.

Far from the tiger's world, Europeans have also been fascinated by the great cat, even though the lion, closer at hand in North Africa, had pride of place in early times. The Romans paraded – and killed – tigers in the arenas. The Latin poet, Martial, described tigers drawing chariots, which are depicted in mosaics carrying Bacchus, the Greek God of Wine. The Emperor Nero kept a large stable of tigers.

Shakespeare invoked the ferocity of the tiger. In one of his most famous plays, Henry V encouraged his men to attack the breach in the walls of Harfleur fortress in France with the words, 'Imitate the action of the tiger, stiffen the sinews, summon up the blood, disguise fair nature with hard-favour'd rage.'

The opening lines of the poem by the 19th century English poet, Thomas Blake, conjure up a vivid image:

> Tyger, Tyger, burning bright, in the forests of the night.
> What immortal hand or eye, hath framed thy fearful symmetry.

The tiger is often seen as a blood-thirsty killer, the result of centuries of horror stories and hunters' boasts. Rudyard Kipling made the tiger, Shere Khan, the villain in the Jungle Book. The French primitive painter, Henri Rousseau, made a snarling tiger the centrepiece of his depiction of a tropical storm, an image constantly propagated by pictures in the media today. The tiger's bad reputation was reinforced in Britain by an extraordinary coincidence which occurred in India in the 18th century. Tippu Sultan, a Muslim ruler in southern India, who was known as 'The Tiger of Mysore', worshipped the tiger. His banner carried the words 'The Tiger is God', and his throne was decorated to resemble a tiger. His soldiers had tiger-striped uniforms, and tiger images and stripes on their weapons.

Tippu fought many battles to defend Mysore from the British, and he and his father were defeated on one occasion by General Sir Hector Munro. Some years later, far away in the Sundarbans mangroves of Bengal, a young hunter was snatched in broad daylight by a tiger and killed – it was Sir Hector's son. When Tippu died in a losing battle in 1799, the British forces found in his palace a life-size model of a tiger mauling an Englishman. Turning handles produces tiger growls and groaning noises as the victim's arm moves vainly up and down. Ceramic copies became popular. The model can be seen today in the Victoria and Albert Museum in London.

In modern advertising, the tiger has become an international symbol, its power widely invoked, notably by Exxon/Esso urging motorists to, 'Put a Tiger in your Tank'. Petrol stations around the world portray tigers, real and cartoon, to help raise funds for conservation. Recently, the city of Berlin filled an advertisement with 16 tiger cub heads to proclaim itself, 'The best place to capture young tiger markets (of Central and Eastern Europe)'. The Detroit Tigers baseball team is one of many sports teams using the image to express their power. Cincinnati's footballers, the Bengals, invoke the famed tigers of India. The tiger has been linked too with sex. Early in this century, people were scandalised by a novel in which a woman quivered sensuously on a tiger skin her lover had given her. The author was satirised in the rhyme:

> Would you like to sin,
> with Elinor Glynn,
> on a tiger skin?

Colman's, the mustard people, recently revived memories with billboards depicting a semi-nude young woman reclining on a tiger skin and breathing, 'C'mon Colman's, light my fire'. Everywhere, the tiger's strikingly beautiful pelt has been coveted to adorn palaces and dwelling houses, and made into garments to clothe fashionable women. Its teeth and claws are used as amulets and pendants.

Human consciousness in the approach to tiger conservation

Today, there may be barely 5000 tigers scattered through the diminished forests of Asia. Their survival is threatened by poaching for their bones and other parts for Chinese medicines – a deadly tribute to the enduring belief in the power of the tiger. Many of the medicines have been found to contain no tiger products at all, but the psychological impact of the tiger remains.

Striking images indicate the powerful effect the tiger has in the countries where it lives, and in those far away. But there is a difference in the way people view the calls for its conservation. Away from the places where the tiger lives, people tend to see the tiger as a magnificent animal, a symbol of wild nature which must be conserved for its own sake; it appears to be only a question of raising funds to save the tiger from poachers, and its forests from destruction. But people who live in tiger range countries, whose numbers are growing exponentially, need more cultivable land and better access to the products of the forest; for them the tiger appears to have little value. Traditional reverence for tigers does not always still exist. The Udegai people in the Russian Far East are often cited as revering the Amur tiger. But their communities and culture were largely destroyed by the Russian communists. Today, at Krasni Yar, a village by the Bikin river where many Udegai live, there is a large sign portraying the head of a tiger, which has been almost obliterated by gunshot.

If tigers are to be conserved, local people's feelings and needs must be a paramount consideration. Unless they support conservation, the tiger is doomed. They are not necessarily hostile to the tiger; they have greater problems with deer and wild boar, which ravage their crops. A local tiger can even be seen as a protector against these pests. But people resent being excluded from forests and grasslands, which have been set aside for tigers and other wildlife, and which could provide them with basic necessities, such as firewood, building materials, fruits and medicinal plants, and grazing land for livestock. These necessities have been so over-exploited elsewhere that they are in short supply. When people invade reserves, it becomes the unfortunate task of forest and wildlife guards to act as unsympathetic policemen to protect the tiger and its ecosystem. If people's hostility is to be eliminated so that they can co-exist with tigers and other wild animals, they must be ensured the resources they need from land outside reserves. At the same time, the need for conservation of natural areas and their wildlife as a support for the well-being of humans must be explained to them.

The tiger is still alive in the consciousness of Asian peoples, many of whom retain respect for its place in culture and religion. This should be a powerful factor in enlisting public support, and should be used to convince political leaders that it should not be allowed to become extinct in their countries. The charismatic image of the tiger can play a leading role in education. It can be used to demonstrate the dependence of all creatures, including humans, on interaction within the web of life. The tiger depends on deer, wild boar and other animals for sustenance. These prey animals, in turn, depend on vegetation to survive. All need water and space to live. Humans, like tigers, depend on the interplay of species and a similar food base. Their animal protein may come from domestic stock, but that stock also depends on vegetation, water and living space. Water is a key element. Forests both depend on rainfall and control supplies by absorbing the rain and releasing the flow of water needed for drinking, agriculture and fisheries. A tiger cooling itself in a pond could represent the vital importance of water to all life, drawing attention to the tiger's role in maintaining healthy forests and thereby ensuring the provision of water for surrounding areas.

Local people should also be taken into reserves to see their beauty and the wild animals that live there; hopefully to see a tiger, for few have done so. Such visits should be part of programmes to educate people in the working of ecosystems and their importance for human welfare. Scientific research to learn the secrets of the tiger's life is of crucial importance to conservation, because more must be learned about the tiger and its needs, so as to give guidance to governments of tiger range countries on how best to go about their conservation programmes. But, in the ultimate analysis, all will be lost if the people who live with tigers cannot be persuaded to see conservation of nature as in their own interest. This is a task that needs immense and widespread effort, and it will be expensive in terms of finance and dedicated effort. It is a task that non-governmental organisations in tiger countries are best fitted to carry out because they can communicate in the language and traditions of their people. They need support from non-governmental organisations (NGOs) in wealthier countries.

There is a significant story in the Hindu classic, the Panchatantra, a compilation of animal tales dating back millennia, and known throughout Southeast Asia. One day, a poor Brahman found a tiger drowning in a well and calling for help. At first, fear led him to refuse to help the tiger out. But when the tiger promised not to eat him, the Brahman decided that, even if disaster befell him, giving help would lead to his salvation. So, he took the risk and saved the tiger, which rewarded him with gold ornaments, and the man gained royal recognition, prosperity and happiness. The story could serve as a parable for today: if the tiger is saved, it will ensure the maintenance of healthy natural ecosystems, not just for the tiger, but for the benefit of everyone.

Tiger ecology: understanding and encouraging landscape patterns and conditions where tigers can persist

Overview *John Seidensticker, Sarah Christie and Peter Jackson*

> Top carnivores are predestined by their perch at the apex of the web to be big in size and sparse in numbers. They live on such a small portion of life's available energy as to always skirt the edge of extinction, and they are the first to suffer when the ecosystem around them starts to erode.
>
> (*Wilson 1993*).

Ecosystems around tigers are being eroded by human activities. Understanding and encouraging landscape patterns and conditions where tigers can persist is the challenge we face in saving the tiger.

Our approach to understanding tiger ecology has developed from one that is largely descriptive to one that uses experimentation, comparison and modelling, and defines, quantifies and tests hypotheses. However, many have argued that tigers, with their high metaphysical content, are or should be considered to be beyond the realm of science. Some also argue that to risk a tiger's life using chemical restraint or to subject a tiger to the indignities of individual marking or a radio-collar is to degrade and disrespect this splendid carnivore. To subject the tiger to the disciplines of science, this point of view argues, is to take away its mystery and power and the special place tigers hold in our minds.

These views are to be respected and many times they have prevailed in debate. Permits to pursue the science of tigers are usually obtained and maintained with great difficulty and sometimes at great personal cost by those who seek to know the tiger a little better scientifically. But at sites throughout the tiger's range, scientists have moved forward in our efforts to add accuracy, precision and predictability to our understanding of the tiger, in the face of anthropogenic environmental changes with conse-quences that will lead to the tiger's extinction if unchecked.

Ecology at its base is the study of the distribution and abundance of organisms. The contributions in this section reflect a major shift that has occurred in ecology. In the 1960s and 1970s 'balance of nature' or 'equilibrium paradigm' (ecological systems with definable stable points such as climax communities) questions formed the basis of scientific accounts of tiger ecology (e.g. Schaller 1967). But today we know that change at some scale is a universal feature of ecological communities and that ecological systems are not in dynamic equilibrium and have no stable point. Ecologists now think in terms of 'the flux of nature' where regulation of ecological structure and function is not internally generated. For example, natural disturbances such as fires, floods, droughts, storms and outbreaks of disease are frequently of overriding importance. Ecosystems also consist of patches and mosaics of habitat types and are not uniform communities (Meffe & Carroll 1997). There is a tension in these approaches because the ecological thinking underlying most tiger reserve management is 'balance of nature' in concept (Seidensticker 1997). But if the persistence of many local populations of tigers depends on dispersal from elsewhere, rather than on local self-maintaining systems of checks and balances, we must obviously consider broad regions in order to understand the tiger's persistence (Minta & Kareiva 1994).

In Part II, contributing authors focus on processes that maintain tigers in broad landscapes. They examine the factors that control the macro- and micro-distribution and abundance of tigers and explore the capacity that tigers have to respond to changes, especially anthropogenic changes, in landscapes. Our contributing authors have approached this task with a full and inventive toolbox. Finding tracks of tigers and prey in the snow in vast mountain landscapes and along dusty forest tracks is as important as it ever was and has been augmented by following a tiger's ways and means using radio-telemetry. Chance encounters with tigers and direct observations of tigers at kills have been strengthened by the non-invasive approach of counting tigers using 'camera traps', distinguishing one from another by their distinctive individual markings (Box 10.1), and using sophisticated 'capture-release-recapture' algorithms to determine tiger densities within statistical limits of confidence. This information is linked to remote-sensing data via geographic information systems (GIS) to provide new insights into the structure of the ecosystems of which the tiger is a part. Harnessing the power of computers to process mounds of data, using appropriate algorithms, and making sense of the results with statistical confidence have allowed these scientists to explore tiger ecology at previously inaccessible spatial and temporal scales.

We do not believe that the tiger comes away from this inquiry any less magnificent a creature. We come away with a profound awe for the tremendous landscape scales and the sheer power, energy and endurance the tiger requires for life. We also come away with the knowledge of just how dependent the tiger is on people to keep a place for it in this world if it is to survive and be a part of our future.

Evgeny Smirnov and Dale Miquelle provide an account of the Amur tiger's recovery after being extirpated from the mountains adjacent to the Sea of Japan in Sikhote-Alin Zapovednik (strict nature reserve), 450 km northeast of Vladivostok in the Russian Far East. This is a story of collaboration and mutual respect by a determined Russian scientist and his equally determined American colleague as they worked together to place these long-term efforts within the realm of analysis of population dynamics and statistical prudence. This paper was not presented at the symposium. We have included it here because it sets the stage for understanding what happened within this tiger population over time and how long-term systematic observations can, with the proper statistical treatment, lead to a detailed understanding of tiger population dynamics. It also demonstrates the synergistic power of international collaboration.

The Sikhote-Alin reserve tigers were extirpated in the mid-1960s and the population was re-established through re-colonisation by tigers from other reaches of the 1000-km-long mountain system bearing the same name. This provided an opportunity to measure, for the first time, the tiger's capacity to increase – the most basic ecological parameter. Smirnov and Miquelle found that rather than acting in a density-dependent manner, as we would expect in the 'balance of nature' paradigm, with reproductive and survival rates decreasing as the region became saturated, both tiger litter size and reproductive rate increased as the population of adults stabilised. This is the first documentation of how tiger metapopulation dynamics actually work.

Dale Miquelle and his associates next examine the key components in landscapes that are critical for the survival of tigers. They ask if the micro-distribution of tigers is keyed to the distribution of specific components of habitat structure or the distribution of a specific complex of prey species. This is a daunting task given the vast extent of the Sikhote-Alin Mountains. They combined the results of snow-tracking tigers and prey, radio-tracking locations of tigers, and habitat structure gleaned from satellite imagery, tied these together with the power of a geographical information system, and teased out a new understanding of the Amur tiger's relationships to habitat and prey. In seeking to quantify the tiger's ecological needs in this changing world, an understanding of tiger habitat requirements is essential.

These scientists conclude that prey density/prey distribution is the key factor driving first and second order site selection by tigers (Fig. 6.4). It is not essential to define tiger habitat except in terms of distribution and abundance of key prey species. Protecting habitat at the local level for ungulates and giving local people incentives to support higher populations of key prey species are the cornerstones of tiger conservation. In Part IIIC, these authors go on to place these concepts in a policy framework – a proposed land-use plan – that includes the tiger in the future of the Russian Far East.

This central idea, that prey populations are a critical determinant of tiger population viability, is echoed by Ullas Karanth and Bradley Stith, who point out that the inventory of habitat available for tigers is quite extensive (see also Part IIIC). Rather than habitat loss, as is so often asserted, or even loss of tigers to poaching, they submit that the depletion of prey is the major factor driving the decline of wild tigers throughout much of their range. Using a stage-based, demographic model they simulate the effects of poaching and prey depletion on tiger population viability. Their results suggest that, by reducing cub survival, prey depletion may have a major impact on tiger population viability, and that tiger populations may be more sensitive to depressed prey populations than to low-intensity tiger poaching. Without arresting prey depletion, they argue, the decline of tiger populations may continue even if poaching is controlled. They conclude that tiger conservation must focus on specific strategies to enhance and monitor the ungulate prey base for tigers. The congruence of this work and that of Miquelle and his associates is remarkable. These workers – from India and the Russian Far East – had never met or exchanged ideas before the symposium.

Nagarahole National Park is 644 km² centrally located in a 24 000 km² tract of tropical moist and deciduous forest (TCU 55, Fig. 18.2). Today it harbours an abundance and biomass of large ungulates unparalleled in southern India, and the tiger thrives here. This is a recovering ecosystem. Before 1970 the area was degraded, ungulate populations were depressed, and the tiger was rare. Serious protection was initiated in about 1970, and intensified after 1974 when the reserve's status was

upgraded from Wildlife Sanctuary to National Park. By 1997 ungulate prey had increased fourfold in the dryer forest types and eightfold in the moist forest zones. Tigers increased in number from about 15 to 50 during this period. Ullas Karanth, Mel Sunquist and K. M. Chinnappa trace the trajectory of this recovery in this, the best-documented such case from the Indian reserve system. Karanth and his associates argue that sustainable tiger conservation efforts can be accommodated within the conceptual framework of sustainable landscapes that include areas free of resource extraction, rather than the prevailing paradigm of entire forest tracts blanketed in sustained use. They conclude that enduring tiger conservation strategies can only emerge through constructive, data-driven dialogue between those who champion the livelihood rights of the people living in tiger habitats and those who advocate the right of wild tigers to survive on this planet. We the editors want to note that the present beauty and the future of Nagarahole owe much to the vision and persistence of the remarkable Mr. Chinnappa and his student Dr. Karanth. Together they have taught that vision, persistence, thinking at the right social and spatial scales and constructive dialogue are keys to the tiger's future.

Raghu Chundawat, Neel Gogate and A. J. T. Johnsingh from the Wildlife Institute of India have begun to study the ecology of the tigers living in the 543 km² Panna Tiger Reserve nestled within 8600 km² of tropical dry forest in central India (TCU 24, Fig. 18.2). Chundawat and his associates have documented the vulnerability of the tiger in this wide swathe of remaining tiger habitat in India. There is considerable pressure and deterioration throughout this forest type from livestock grazing and the collection of fodder. For example there are 13 villages and 9500 cattle inside Panna and 79 villages outside it within 10 km of the boundary. Cattle have replaced wild ungulates as the predominate components of the potential prey base for tigers, but they are not taken by the tiger in proportion to their abundance. Sambar and nilgai are the tiger's preferred prey. Large home ranges, especially for adult males (243 km²), and extensive travel distances each day characterise tiger movements here compared with areas that have much higher prey densities in India and Nepal. The disturbance-free core of Panna is only 55 km². This is essential habitat for a tigress rearing cubs, and acted as a source producing young tigers that were dying in the sink of the hostile environment around Panna. Water is a critical resource in the Indian tropical dry forest throughout much of the year, and people and their livestock live close to most permanent water sites. These are essential habitat elements for tigers as well and this places the tiger in direct and detrimental competition with people. The tiger here is also at great risk from stochastic changes; extreme drought or a disease outbreak could decimate its population.

The tropical dry forest in India is fragmented into 23 isolated habitat patches or TCUs (Fig. 18.2), separated by hostile conditions where tigers cannot survive. We are faced with the question: how much degradation of habitat is too much for a tiger population to tolerate? Continued monitoring of tigers in Panna, especially movements and habitat-specific demographic rates, is essential for an understanding of the great risk confronting tigers here. Small reserves with little protection and the hostile disturbance regimes characterise this region. We came away with the distinct impression that the critical level of habitat degradation has been passed and the tiger may soon disappear throughout this forest type in India. With this study, however, we know enough about the tiger's needs in tropical dry forests to provide policy makers and managers with the information they require if they are to nudge human land-use toward conditions where tigers can persist.

At the other extreme in the tiger's range, in the tropical moist forests that once blanketed much of the Greater Sunda Islands, the tiger was nearly gone before anyone took note or moved to include it in Indonesia's future. The Bali tiger may have made it into the 1950s but was certainly gone by 1960. The Javan tiger was present in very low numbers until the late 1970s and then blinked out (Seidensticker 1987). Sumatran tigers were widespread throughout many areas on the island and were heavily persecuted in the 1970s. Nowhere were they as

numerous on the island as we had come to expect from our experience in the National Parks in India and Nepal (4–17/100 km²) (Seidensticker 1986; Karanth & Nichols 1998).

Neil Franklin and his associates are the first research team to establish tiger density estimates for Sumatran habitats, using an approach that is non-invasive and statistically reliable. This team started at the southern end of the Island in the isolated Way Kambas National Park, a thrice logged-over lowland rainforest interspersed with waterways and extensive *Imperata* grassland. This habitat complex is found over wide areas in the Sumatran lowlands and harbours relatively good populations of wild pigs, sambar, muntjac and large macaques, the tiger's principal prey in Way Kambas. Within this habitat and with this prey base and the protection provided by the park, tigers appear to be thriving in Way Kambas, and in the shadow of a large agriculturally based human population. There is considerable potential tiger habitat in Sumatra (>100 000 km²), nearly as much as there is in the Russian Far East. Surveys are now underway by this team and others to reveal the population structure of the Sumatran tiger and how to plan for its future.

The Way Kambas tiger monitoring project is important because it is a window into the dynamics of tigers living in relatively small habitat patches (~1000 km²), which is the norm for many tiger populations today. An initial assessment suggests that this tiger population is separated by at least 150 km of hostile habitat from other potential tiger populations. Also, during mid-1997, a significant portion of this isolated reserve burned during an extended dry period, courtesy of El Niño. Will the Way Kambas tigers persist in the short-term, and is this population viable in the long-term? Only further monitoring will tell what the future holds for tigers here.

The distribution and status of the tiger in Indochina have long been the subject of discussion in efforts to save the tiger because so little was known about conditions in this war-torn region. As hostilities have ended and closed areas slowly opened to outsiders, Alan Rabinowitz has been first on the scene to look at the situation. The science of tigers begins with an inventory of what we actually know. In an extensive account, Rabinowitz combines all the known reports on tiger occurrence through this region and categorises them according to their reliability. He concludes that there is simply no basis for the numbers of tigers that have been reported by governments to be living in this region. The good news is that there are extensive tracts of forest remaining over the region. But the tiger's prey has been extirpated from many forest areas. Without prey, there can be no tigers. The killing of tigers is also commonplace. With care and insight, prey populations can recover, and so too can the tiger. Rabinowitz concludes that Indochina has the potential to be a key to the tiger's future. What will it take to make this so?

David Smith and his associates begin to answer this question from one area within the broad outline offered by Rabinowitz. By engaging the extensive system of forest rangers living throughout the remaining forest areas in Thailand, they have compiled the first outlines of the metapopulation structure of Thailand's tigers. They hypothesise that Thailand's tigers are actually fragmented into 15 populations; 5 of these are linked to others in neighbouring countries and the remaining 10 are isolated. These workers have used this assessment as the basis for recommending the maintenance and creation of critical habitat linkages to maintain or re-establish connectivity between tiger populations as a major component in Thailand's tiger conservation plan. They point out that some of the largest tracts of remaining forest are along international boundaries, and emphasise the importance and opportunities of cross boundary co-operation to maintain tiger populations in Thailand. This is a recurring theme throughout tigerland.

The terai forests and tall grasslands that lie at the base of the Himalaya in India, Nepal and Bhutan once supported an ungulate biomass rivalled only by that in the Great Rift Valley in Africa. This is tiger country like no other. A female tiger in many areas here needs only 20 km² to live and raise her young, compared with the 500 km² a female needs in the temperate forests of the Russian Far East. Windows into what this natural spectacle once was are on display today in the Corbett National Park in India and the Royal Chitwan National Park in Nepal. Changes to the terai have been slow in coming (Seidensticker 1976), but in the last 35 years those changes have been relentless. The first radio-tracking studies of tigers were done in Chitwan. When those studies began, tigers probably still had a continuous distribution extending >1000 km from far-eastern Nepal through the terai, west to Corbett and Rajaji National Parks on the banks of the Ganges in India. The lessons of conservation biology have taught us to stop looking inward and look beyond the boundaries of refuges, to look for the connections and examine the structure of populations (Meffe & Carroll 1997). In surveys that have been ongoing for nearly a decade and were completed after the symposium, Smith and his associates found that the structure of the tiger population in Nepal and this portion of northern India has shifted from continuous to discontinuous. They found three tiger populations in Nepal ranging in size from 15 to 50 breeding adults, and a fourth in Dudhwa, just across the border in India. There remains a continuous habitat linkage from Chitwan in central Nepal to Sukhla Phanta in the west, but the integrity of this continuity is threatened in many places and the tiger's prey is much reduced. Smith and his co-workers conclude that these habitat connections, necessary to help ensure the long-term survival of the tiger in Nepal, can be achieved with a programme of forest restoration that benefits and is managed by local people.

The importance of prey as a defining component of 'habitat' as the primary issue in securing the tiger's future may seem obvious on reflection, a fundamental reality of bioenergetics. But you cannot determine the status of prey populations, any more than you can tiger numbers, from satellites blinking overhead. Actually establishing where suitable habitat with adequate prey occurs on the ground can only be done on the ground. Tiger conservation strategies that explicitly recognise the prey component of habitat are now emerging (see Part IIIC). Dale Miquelle summarised this best in his presentation when he told us that the key to tiger conservation is 'to love thy ungulate'.

We must learn to set aside the idyllic 'balance of nature' as a defining concept in tiger conservation strategies. In the modern world, human disturbance is pervasive and defines nearly all ecosystems in tigerland. Human management is an essential counterbalance to this in the future. The persistence of tigers in most areas is dependent upon habitat linkages and our challenge is to understand and encourage landscape patterns and conditions where tigers can persist.

5

Population dynamics of the Amur tiger in Sikhote-Alin Zapovednik, Russia

Evgeny N. Smirnov and Dale G. Miquelle

Introduction

Very little is known about population dynamics of tigers. Nearly all information comes from one population of Indian tigers in Royal Chitwan National Park, Nepal (McDougal 1977; Sunquist 1981; Smith & McDougal 1991), or from the study of captive individuals (Sadleir 1966; Kleiman 1974; Seal *et al.* 1987b). Information on population growth rates is largely lacking.

Despite the lack of knowledge, such information is important to conservation efforts for this endangered species. As the tiger is presently threatened with extinction due to poaching throughout much of its range (Jackson 1993b; S. R. Galster & K. V. Eliot this volume), estimates of its reproductive potential are critical to establishing the extent of illegal harvesting that can be sustained (e.g. Kenney *et al.* 1995). Little information exists to make such assessments.

Information on the distribution and status of Amur, or Siberian, tigers in Sikhote-Alin has been collected since creation of the protected area in 1935 (Matyushkin *et al.* 1981). Here we present estimates of the size and structure of that population based on a long-term monitoring programme that has been conducted in Sikhote-Alin State Biosphere Zapovednik since 1963. The purpose of this study was to use existing information to estimate observed population growth rate and reproductive parameters and to consider some of the implications for tiger conservation.

We thank H. B. Quigley and M. G. Hornocker of the Hornocker Wildlife Institute, and A. Astafiev and M. Gromyko of Sikhote-Alin State Biosphere Zapovednik for the logistical, administrative and financial support necessary to conduct this work.

Zapovednik scientists and forest guards contributed their many records of tiger observations, as well as their time, to the annual censuses; their efforts are greatly appreciated. Funding for the Siberian Tiger Project was provided by the National Geographic Society, the Exxon Corporation, the National Fish and Wildlife Foundation, the Save the Tiger Fund, the National Wildlife Federation and the Wildlife Conservation Society.

Study area and methods

Lands designated as 'Zapovedniks' receive the highest level of nature protection in Russia; human access is limited to forest guards and scientists. Sikhote-Alin State Biosphere Zapovednik is located close to the northern border of Primorski Krai (Province), with nearly all its territory situated within Terney Raion (District) (Fig. 5.1). Size of the Zapovednik has changed dramatically over time, reaching 1 157 000 ha when borders were delineated in 1938. In 1951 the Zapovednik was reduced to a low of 99 000 ha; presently it is approximately 400 000 ha. Portions of Sikhote-Alin Zapovednik border the Sea of Japan, but its central feature is the Sikhote-Alin Mountains, a low range (most peaks are below 1200 m) that parallels the Sea of Japan and bisects Primorski and Khabarovski Krais. Sikhote-Alin Zapovednik is situated close to the centre of remaining Amur tiger population (Matyushkin *et al.* 1996) and, with Lazovski Zapovednik, probably represents some of the best conditions remaining for Amur tigers due to its large size and the protection afforded to both tiger and prey populations.

Yearly surveys to estimate tiger numbers in Sikhote-Alin Zapovednik and adjacent lands

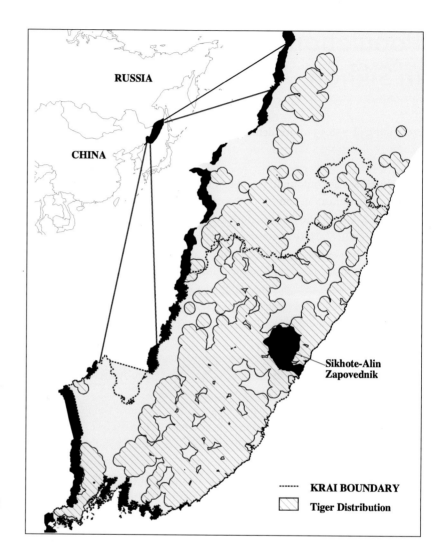

FIGURE 5.1
Location of Sikhote-Alin
Biosphere State
Zapovednik in Primorski
Krai, Russian Far East.
Scale: 1cm = 70km

RUSSIA

CHINA

Sikhote-Alin
Zapovednik

------- **KRAI BOUNDARY**

▨ **Tiger Distribution**

together covered a total area of approximately
500 000 ha. Information on tiger abundance has
been collected since 1963, but since evidence of
tigers was not found during the first three years,
estimates are based on data collected during the
28 years from 1966 to 1993 inclusively.

Methods for estimating tiger numbers, sex and
age structure are based on two types of information.
Since 1971, annual surveys have been conducted by
20–28 people along 300–400 km of permanently
established transects. These surveys are conducted
one to three times per year after recent snowfall.
Location, number and size of all tiger tracks are

recorded. Secondly, forest guards located at permanent stations throughout Sikhote-Alin Zapovednik
spend extensive time patrolling and, together with
all scientists working in Sikhote-Alin Zapovednik
(collectively averaging approximately 50 individuals
per year), they record all observations of tiger tracks.
All records from each winter count and reports are
tabulated chronologically and by area. Nearly all
information comes from tabulation of tracks in
snow, the season of snow cover usually lasting
approximately three to four months (December–
March).

Tracks of tigers are placed into four sex and age

categories based on the width of the pad on the front paw (Abramov 1961; Matyushkin & Yudakov 1974; Yudakov & Nikolaev 1987; Matyushkin *et al.* 1996): tigers with pad widths equal to or greater than 10.5 cm are considered adult males; pad widths of 8.5–10.5 cm are considered adult females; pad widths of 7–8.5 cm are defined as subadults (yearlings and some two-year-olds fall into this group); and pad widths less than 7 cm are defined as first-year cubs. The confounding problem with this method of identification comes with young males: the track size of subadult males overlaps with that of females, but by the time most subadult males disperse, their track size is already greater than that of adult females. Prior to dispersal, most subadult males are identifiable because they are associated with a family unit. The subadult classification is most poorly defined because it includes some

yearlings still in association with a family unit, as well as some yearlings and two-year-olds that have dispersed. Despite these problems, definition of these broad categories has been verified with measurements of freshly killed specimens (Yudakov & Nikolaev 1987), and with 19 individuals captured within Sikhote-Alin Zapovednik for the Siberian Tiger Ecology Project (Smirnov *et al.* 1997).

Records of all tracks are plotted, and population size and structure are derived by estimating the number of individuals in the study area, overlapping and adjacent individuals being distinguished by the size of their tracks and their temporal-spatial distribution. Given that home range estimates of resident females and males exist (Yudakov & Nikolaev 1987; Salkina 1993; D. G. Miquelle *et al.*, unpubl. data), it is possible to estimate the number of duplicate tracks of one individual based on

Amur tigers live at such low density that 'camera-trapping' is rarely successful. This rare photograph was taken with a home-made camera-trap triggered by a trip wire in Lazovsky Zapovednik in 1992.

size and temporal-spatial distribution. Repeated observations of similar-sized tracks in a limited region are considered one individual. Females with cubs are the most distinctive group due to track size, consistent group size and reduced home range size (Smith 1984; D. G. Miquelle *et al*. unpubl. data), and therefore provide a check on estimating the frequency of repeated track sightings of specific individuals.

We believe the procedure is conservative in estimating population size because: (1) there is a probability of missing tracks of individuals, especially in remote sections of Sikhote-Alin Zapovednik; and (2) repeated reports of tracks in an area are conservatively interpreted, i.e. tracks of similar size temporally separated in one drainage will usually be attributed to one individual, when in fact they could theoretically be made by two or more. Though statistical confidence limits cannot be applied to the count method, a range of values is given for each sex and age group for each year that reflects the uncertainty of interpreting existing records. Because the same methodology has been applied throughout the entire period of study, we believe the values accurately reflect trends in population size and structure, though we cannot assess how well they reflect true population size. Initial estimates based on radio-collared tigers in Sikhote-Alin Zapovednik support the interpretations of population structure and size reported here, and also suggest that the procedure provides a conservative estimate of population size (Smirnov & Miquelle 1995).

Population density is estimated for the entire study area (500 000 ha) for total population size, and for the adult population only. The adult population includes all individuals except young still associated with a family unit. The observed exponential rate of increase, *r*, is calculated by regression analysis (Caughley 1977) for the total population count, and for adults only. Slopes were tested to determine if growth rates varied significantly from each other, and from zero (Zar 1984). Because the count is not instantaneous, some mortality and natality may occur during the count period, introducing an unknown amount of error to the method.

However, most births occur outside the period when snow is on the ground (Seal *et al*. 1987b; E. N. Smirnov *et al*. unpubl. data), and the bias associated with mortality should have been generally consistent throughout the period of study, thereby not biasing estimates of growth rate.

Tracks of females with and without cubs provided estimates of the following reproductive parameters: litter size, percentage of adult females with cubs and reproduction rate. This methodology, similar to that used by Smith & McDougal (1991), does not estimate litter size at birth; it provides a conservative estimate of litter size because young are generally not detected before two to three months of age, when cubs become mobile and follow mothers to kills (Smith 1984; D. G. Miquelle *et al*. unpubl. data). Many litters are considerably older before they are detected in winter counts. Therefore, some mortality no doubt occurs before litters are registered. The reproduction rate was estimated as number of cubs per adult female per year, averaged over years. This estimate is slightly different than that reported elsewhere for large carnivores (Craighead *et al*. 1974; Wielgus & Bunnell 1994) where reproduction rate equals the mean litter size divided by mean birth interval. We use this calculation because mean birth interval is poorly known for Amur tigers.

The study period was divided into three approximately equal periods of 9–10 years: Period 1=1966–1975; Period 2=1976–1984; and Period 3=1985–1993. Changes in three reproductive parameters, i.e. reproduction rate, percentage of females with cubs/year and litter size, were assessed over these three time periods using analysis of variance with protected least significant difference (LSD) tests (SAS 1985). All means are reported ± standard deviations.

Results

Population estimates are based on 5203 reports of tracks or visual sightings from 1966 to 1993. The number of reports ranged from 19 in 1966 to 339 in 1985 ($\bar{x}$=185.8±88.3).

Table 5.1. *Estimates of size, sex and age structure of the Amur tiger population in Sikhote-Alin State Biosphere Zapovednik, 1966–93*

Year	Adult females	Adult males	Subadults	<1-year-old cubs	No. of litters	Total count
1966	2	0–1	0	1	1	3–4
1967	2–3	1	1	2	2	6–7
1968	3	1	2	1	1	7
1969	3–4	1	2	2	1	8–9
1970	3–4	1	2	2	2	8–9
1971	3–4	1	2–3	0	0	6–8
1972	3–4	1–2	1–2	1	1	6–9
1973	4–5	2	2–3	3	2	11–13
1974	4–5	2	3–4	1	1	10–12
1975	4–5	2–3	4–5	1	1	11–14
1976	4–5	2–3	3–5	1–2	1–2	10–15
1977	5–6	2	2–3	5	2	14–16
1978	5–7	2	6–7	2	2	15–18
1979	4–6	2–3	6–8	0	0	12–17
1980	5–7	2–3	2–4	6	4	15–20
1981	6–8	2–3	6–8	2	1	16–21
1982	5–6	2–3	4–6	5	2	16–20
1983	5–6	3	2–4	4	2	14–17
1984	7–9	3–4	5–6	5	3	20–24
1985	7–9	3–4	5–6	6	4	21–25
1986	6–8	3–4	5–6	7	4	21–25
1987	6–8	4–5	5	10	4	25–28
1988	7–9	3–4	8–10	4	2	22–27
1989	7–9	4–5	5–6	5	2	21–25
1990	8–10	3–4	4–5	4	4	19–23
1991	7–9	3–4	4–5	8	4	22–26
1992	6–8	3–5	10	8	4	27–31
1993	7–10	3–5	5–6	10	6	25–31

Between 1963 and 1965 there was no evidence of tigers consistently residing within the study area. Although it is likely that animals were establishing home ranges during this period, the first evidence of resident tigers during the 1960s came in 1966, when two females, one with a litter of one, were reported (Table 5.1). Total population size increased, with some fluctuations, during the entire 28 years of study (Fig. 5.2). However, the adult population appeared to stabilise beginning in 1983, when the adult female and male counts became fairly consistent for the next 10 years (Fig. 5.2 and Table 5.1).

The maximum estimated density of the Sikhote-Alin Zapovednik tiger population was reported in 1992 and 1993, when the total population density was 0.62/100 km². The maximum adult population estimate, reached in 1993, was 0.3/100 km². The sex ratio of the population averaged 2.4 females/male (SD=0.54, n=28).

For the 28-year period, the observed rate of increase for the entire population was 0.06 (Fig. 5.3). The rate of increase appeared to be slightly higher during the 1970s and early 1980s and then dropped slightly (Fig. 5.3). From 1966 to 1984 the observed rate of increase of the adult population (r=0.064) was not significantly different from that of the total population (t=0.956, P > 0.05). However, from 1985 to 1993, the total population continued

Table 5.2. *Reproduction rate (cubs/adult female per year), percentage of females with cubs, and litter size of Amur tigers in Sikhote-Alin State Biosphere Zapovednik, Russian Far East, during three time periods between 1966 and 1993*

Period	Reproductive rate			Percentage of females with cubs			Litter size		
	n	$\bar{x}$	SD	n	$\bar{x}$	SD	n	$\bar{x}$	SD
1966–1975	10	0.39[a]	0.27	10	37	22	12	1.2[a]	0.39
1976–1984	9	0.57[b]	0.35	9	31	18	17	1.8[b]	0.81
1985–1993	9	0.90[b]	0.33	9	48	15	34	1.9[b]	0.92
Overall value	28	0.61	0.37	28	39	20	63	1.7	0.85

Periods with different letters are significantly different ($P < 0.05$)

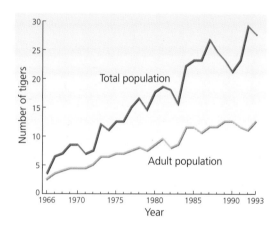

FIGURE 5.2
Size of Amur tiger population (total population and adult component) in Sikhote-Alin Biosphere State Zapovednik from 1966 to 1993.

FIGURE 5.3
Population size (natural logarithm of n) and regression equations defining exponential rate of increase for total population, and adult population in Sikhote-Alin State Biosphere Zapovednik, Russian Far East, 1966–1993.

to increase (Fig. 5.2), while the observed rate of growth of the adult population ($r=0.007$) was not significantly different from 0 ($F=1.45$, $P=0.263$).

The increasing divergence in total population and adult population size (Fig. 5.2) appears to be due not only to a stabilisation in the adult population, but to an increase in the reproductive rate. The proportion of females with cubs over the entire study period averaged 38.8% (SD=19.6). However, the reproduction rate increased over the three time periods ($F=6.13$, $P=0.007$) (Table 5.2), with a

significant difference between Period 1 (1966–75) and the two later periods (Table 5.2).

The increase in reproduction rate (cubs per adult female per year) was related to an increase in litter size (Tables 5.2 and 5.3). During Period 1, most reported litters (83%) consisted of a single cub, whereas during Period 3 more than 50% of the litters had two or more cubs and 30% of the litters had three or more (Table 5.3). Litter size during Period 1 was significantly lower than during the other two periods ($F=3.31$, $P<0.05$) (Table 5.2).

It is not clear if the increase in reproduction rate

Table 5.3. *Distribution of litter sizes reported for Amur tigers in Sikhote-Alin State Biosphere Zapovednik, Russian Far East, for three time periods between 1966 and 1993*

Period	Litter size				Total
	1	2	3	4	
1966–1975	10	2	0	0	12
1976–1984	7	6	4	0	17
1985–1993	16	8	9	1	34
Total	33	16	13	1	63

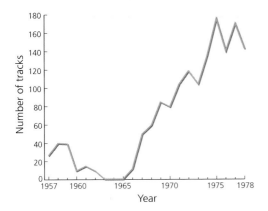

FIGURE 5.4
Frequency of tracks located within Sikhote-Alin Zapovednik, 1957–1978 (from Matyushkin *et al.* 1981).

was also related to an increase in the percentage of females with cubs. Although there were no significant changes throughout the period of study (F=1.91, P=0.16), the percentage of females with cubs in Period 3 (48%) was considerably higher than in either of the earlier periods (Table 5.2).

Discussion

In 1939–40 Kaplanov (1948) estimated that there were 10–12 tigers within Sikhote-Alin Zapovednik,

but that these animals were confined to some 30 000 ha of the Zapovednik. During this period the size of the Amur tiger population apparently reached its lowest level; Kaplanov (1948) suggested that the isolated population within the Zapovednik may have represented one-half of the total population remaining in the Russian Far East. With the outlawing of hunting (in 1947), and tighter controls on capture of cubs (1956), the overall population in the Russian Far East began an expansion phase that continued into the 1980s (Matyushkin *et al.* 1996). However, locally, the protection provided by Sikhote-Alin Zapovednik was severely dampened in 1951 when size of the Zapovednik was cut to 99 000 ha.

Matyushkin *et al.* (1981) plotted the frequency of tracks within Sikhote-Alin Zapovednik from 1957 through 1978 (Fig. 5.4). Even though the size of the Zapovednik was restored to 310 100 ha in 1960, the population of tigers continued to collapse. The localised extinction of this population was attributed to illegal hunting – new logging roads into former protected areas in the 1950s eliminated *de facto* protection – and continued capture of cubs. In the winters of 1962–63 and 1963–64, 14 cubs were captured within former Zapovednik territory. Attempts to capture cubs often resulted in the death of the female. Apparently, both the reproductive and recruitment components of this population were simultaneously eliminated by the mid-1960s.

The growth of the Sikhote-Alin Zapovednik population since 1966 therefore apparently represents a colonisation episode. Poaching of both prey species and tigers was rare throughout 1991 due to tight control of firearms, regular patrolling of the Zapovednik and no economic incentives (access to the international market for tiger skins and bones was virtually non-existent due to the closure of borders). Given these conditions, the colonisation episode occurred in what was probably high-quality habitat for Amur tigers with minimal human impact. Therefore, the observed rate of increase in the Sikhote-Alin Zapovednik population is probably close to the maximum rate of increase that might be expected during a colonisation phase. However, this estimate is not equal to the intrinsic rate of increase, i.e. the exponential rate at which a population grows when no resource is limited (Caughley 1977) for two reasons: (1) the age distribution was not stable during this period, a necessary assumption for estimating intrinsic rate of increase (Caughley 1977); and (2) dispersal probably played an important role in reducing the rate of growth (see later). Nonetheless, this estimate is valuable because it provides an index of how quickly Amur tiger populations could grow in newly colonised areas with healthy prey populations. Tigers have been extirpated from much of their former range and, if reintroduction becomes a possibility, this estimate provides an indication of potential population growth rate.

These data represent the first estimates of population growth rates that exist for the tiger, so it is impossible to make comparisons with other subspecies or other studies. Population growth was not as rapid as might be expected, given the nearly optimal conditions that existed during much of the study period. The population grew at a fairly consistent rate over 18 years, and did not mimic the 'classic' sigmoidal growth curve. Dispersal may have played an important role in reducing population growth rate. Matyushkin et al. (1981) reported that the region to the north of Sikhote-Alin Zapovednik was still uncolonised in the mid-1970s, and Pikunov (1988) reported the region to be sparsely populated with tigers in 1985. Only recently is there evidence to suggest that tigers have colonised the entire region of Terney Raion (Smirnov & Miquelle 1995). Therefore, areas outside the Zapovednik may have acted as dispersal sinks. Given the present level of poaching activity (S. R. Galster & K. V. Eliot this volume), regions adjacent to the Zapovednik may continue to act as sinks, especially since dispersing individuals may be especially vulnerable to poaching activity (Garshelis 1994). It remains to be seen whether population size within Sikhote-Alin Zapovednik stabilises in what appears to be a saturated population of adult animals. Presently, it appears that the colonisation phase lasted approximately 19 years, with the adult population approaching a threshold around 1984.

Estimates of population and reproductive parameters vary only slightly in comparison to information on other subspecies. For Indian tigers in Kanha Tiger Reserve in India, Schaller (1967) reported a sex ratio (females:males) in the adult population of 4:1 and 5:1. Smith and McDougal (1991) reported a sex ratio of 2.25:1 in Royal Chitwan National Park, Nepal. The sex ratio of the entire Amur tiger population can be estimated for at least four time periods from census data: in 1956–57 it was estimated at 1:1 (Abramov 1960); in 1959 at 1.3:1 (Abramov 1965); in 1969–70 at 1.7:1 (Yudakov & Nikolaev 1970), and in 1978–79 at 1.5:1 (Pikunov et al. 1983). All these estimates suggest a lower sex ratio than the 2.4:1 found in Sikhote-Alin Zapovednik, which is nearly identical to the intensive work done in Chitwan (Smith & McDougal 1991).

Schaller (1967) reported that 25% of the females in Kanha had cubs during his study, and reported approximately 33% females with cubs based on hunting reports of Indian tigers. A later study in Kanha reported 38% (25–43%) of females with young (Panwar 1979b). In Chitwan National Park in Nepal, three estimates of reproduction exist. McDougal (1977) reported an average of 50% females with cubs over four years in a region that contained two to three adult resident tigresses (calculated from data, p. 96), Sunquist (1981) reported 75% of resident females with young in 1976, and Smith (1978) reported 87% in 1978. Additionally, McDougal (1977) questions the accuracy of Schaller's figures, and suggests the true value to be much higher. All estimates from Chitwan are based on resident females, a social status that could not be exactly determined by our counting techniques (or those of Schaller). Abramov (1977) estimated the percentage of females with cubs as 20–25% for the Primorye population, a slightly lower estimate than the 38% we report for Sikhote-Alin Zapovednik. Differences in counting techniques make it difficult to determine whether reproduction rates of Amur tigers are actually lower than those of Indian tigers in the wild, or whether the apparent difference is a methodological artefact.

Sadleir (1966) reported that the average size of 79 litters raised in zoos was 2.8. The average size of

49 litters in Royal Chitwan National Park, Nepal was 2.8 (Smith & McDougal 1991). Using a technique similar to that of Smith and McDougal (1991), the average litter size in Sikhote-Alin, 1.7, is substantially less than that of the Indian tiger in the wild.

Our observations of reproductive parameters are counterintuitive to what would be expected if density-dependent effects became important as the tiger population approached carrying capacity. Reproductive rates and/or survival rates (especially of young) should have decreased as the region became saturated (e.g. Beier 1993; Garshelis 1994). However, both litter size and reproductive rate increased through the period of study, and young actually became a larger percentage of the population as the adult population stabilised. Several factors may explain what appears to be an inverse density-dependent phenomenon. First, young females dispersing into the Zapovednik during colonisation may have experienced lower reproductive output, a trait common to young mothers in many mammalian species (first litter size of two radio-collared tigresses in Sikhote-Alin was one). A preponderance of young females would explain the small litter size and low reproductive rate observed during Period 1. Secondly, as already noted, during the earlier stages of colonisation dispersal may have played an important role dampening the population growth rate, and reducing the percentage of subadults in the population. However, as adjacent regions became colonised, a greater percentage of subadults may have remained within the study area, since adjacent areas no longer provided the advantage of no competition for space. The fact that subadults made up a smaller percentage of the population in Period 1 (41%) compared to Periods 2 and 3 (58 and 53% respectively) supports this hypothesis. The data, as presented here (Table 5.1), do not allow assessment of whether the subadult male segment of the population (that most likely to disperse long distances) increased disproportionately, which would substantiate the hypothesis that fewer subadults were dispersing. However unlikely, available information does not rule out the possibility that survival of subadults did actually increase in an inverse density-dependent manner.

We believe that the existent population density may well reflect carrying capacity for this region, given the present density of ungulates. The same prediction has already been made for this population at lower densities. Between 1972 and 1977 Matyushkin et al. (1980) estimated the tiger density in Sikhote-Alin Zapovednik to be 0.13–0.32/100 km^2, and stated that the population probably represented a maximum for the conifer-broadleaf forests of central Sikhote-Alin mountains. However, densities have increased substantially since that time. Although not reported here, the density of tigers does vary among different regions within Sikhote-Alin Zapovednik. Highest densities occur along the coast of the Sea of Japan, and may reach 0.8/100 km^2. Lazovsky Zapovednik, some 300 km to the south, is considered better tiger habitat due to higher prey densities (Matyushkin et al. 1980). Zhivotchenko (1981) estimated tiger density at 1.4/100 km^2 between 1973 and 1979. Matyushkin et al. (1980) reported tiger density in the Lazo region as 0.6–0.9/100 km^2 for 1975, and more recently Salkina (1993) reported a density of 0.7/100 km^2 for 1991, with maximum density in one area reaching 1.03/100 km^2. Despite the claim that habitat quality is better there, the differences in tiger density appear relatively small. Both sets of estimates, based on populations largely protected within Zapovedniks, likely represent the highest densities reached by Amur tigers in the existing range.

Despite the fact that these values may represent maximums for the Amur tiger, they are dramatically lower than that reported for other subspecies. While in some degraded habitats in Asia population densities are as low as 1.0/100 km^2 (Rabinowitz 1993), tiger densities can reach as high as 11.6 individuals/100 km^2 (Karanth 1991). Low densities of Amur tigers, even in quality habitat, coupled with low reproductive potential in comparison to other subspecies, demonstrate the problems of conservation for this subspecies. Very large tracts of land must be preserved to ensure survival of a viable population, and population responses to poaching or other sources of mortality may not be as rapid as in regions to the south. Sikhote-Alin Zapovednik, the largest protected reserve in remaining Amur tiger habitat, contains only 10–15 adults at its present high density. It is unlikely that more space for resident adults exists (see D. G Miquelle *et al.* this volume, Chapter 19), or that the size of the Zapovednik can be increased sufficiently to ensure population viability. The key to ensuring survival of the Amur tiger will be the development of a connected network of protected and managed lands that will provide for a minimum population, and management of additional lands through a zoning process that will allow for human uses of the landscape that are compatible with tiger conservation (D. G. Miquelle *et al.* this volume, Chapter 19). Conservation of this subspecies will therefore depend on developing a strategy that not only increases the amount of protected lands, but also seeks ways to mitigate the impact of resource extraction on non-protected lands.

Hierarchical spatial analysis of Amur tiger relationships to habitat and prey

Dale G. Miquelle, Evgeny N. Smirnov, Troy W. Merrill,
Alexander E. Myslenkov, Howard B. Quigley, Maurice G. Hornocker
and Bart Schleyer

Introduction

The tiger is a habitat generalist. Having successfully radiated throughout Asia, eight subspecies inhabited a wide range of habitats, including the tropical rainforests of the Sunda Shelf Islands (Seidensticker 1986), the tall grasslands and riverine forests of North India and Nepal (Schaller 1967; Sunquist 1981), the mixed deciduous, dry evergreen, and dry dipterocarp forests of Thailand (Rabinowitz 1989, 1993), the mangrove swamps of the Sunderbans (Hendrichs 1975), the temperate and boreal forests of the Russian Far East (Matyushkin *et al.* 1996) and, until recently, the 'reed jungles', riparian thickets and montane forests extending into the Middle East (Heptner & Sludskii 1972). Despite an apparent lack of specificity in habitat requirements, some authors have suggested that there are key components of habitat structure that are critical to survival (Sunquist 1981; Sunquist & Sunquist 1989; Karanth 1991; Rabinowitz 1989, 1993). In contrast, others have argued that, aside from some gross structural features, habitat parameters are relatively unimportant for tigers (Miquelle *et al.* 1996b) and that a specific complex of prey species is the key component of habitat for all tiger subspecies (Seidensticker 1986; Miquelle *et al.* 1996b). In either case, because habitat loss is a critical factor threatening the tiger with extinction, it is not a moot point to consider habitat requirements. Some minimum set of habitat components is obviously essential for the survival of this species.

Habitat selection can occur at a number of spatial and temporal scales (Orians & Wittenberger 1991; Pedlar *et al.* 1997). Johnson (1980) recommended a four-tiered approach to considering habitat selection at different levels of resolution (Fig. 6.1). First order selection describes the geographic range of a particular species; second order selection describes the range of habitats that are incorporated into home ranges of individuals; and third order resolution describes the selection of habitats within an individual's home range. Finally, fourth order represents selection of individual food items. Collectively, analyses at these various spatial scales describe not simply changes in resolution, but different sets of constraints that act on populations and individuals at different scales.

We believe that this construct is a valuable means of considering tiger habitat requirements, but argue that prey should be considered as one component of the habitat, and should be included as a parameter in analyses at the other three spatial scales. Fourth order selection (i.e. prey selection) has been examined extensively for the Amur tiger (Abramov 1962; Abramov *et al.* 1978; Zhivotchenko 1981; Yudakov & Nikolaev 1987; Matyushkin 1992; Miquelle *et al.* 1996b). Here we provide an examination of habitat use at the other three levels of selection.

Specifically, for tigers, we argue that prey density and distribution, not habitat parameters, are the key factors driving first and second order site selection, but that other factors are important at the third order. In the absence of human-induced mortality and disturbance, the geographical distribution of all tiger subspecies as well as the distribution and relative density of tigers within a subspecies range are driven primarily by prey distribution and density; but within an individual's home range,

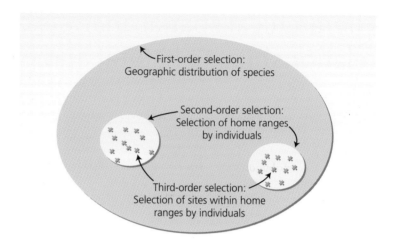

First-order selection:
Geographic distribution of species

Second-order selection:
Selection of home ranges
by individuals

Third-order selection:
Selection of sites within home
ranges by individuals

FIGURE 6.1
Schematic view of 3 spatial scales for
assessing habitat selection (after
Johnson 1980).

other proximate factors temper the strength of this relationship. We assess this general concept with a series of testable hypotheses and an analysis of habitat relationships of the Amur tiger at each spatial scale.

The northern limits of the geographical range of tigers in Russia have historically occurred below 50° latitude in the Amur Basin (Heptner & Sludskii 1972; Kucherenko 1985). Yet forests extend far to the north, without obvious physical barriers, expanses of water, or tall mountains. What limits Amur tiger distribution to the north? Consideration of first order selection by the Amur tiger is particularly illuminating because it provides, on a gross scale, an assessment of habitat requirements at the limits of the range of this species. We address three general questions that relate to tiger distribution (i.e. first order selection): (1) can tiger distribution be predicted accurately by assuming tigers select directly for specific habitat types? (2) can tiger distribution be accurately predicted by mapping actual distributions of prey species? and (3) can tiger distribution be predicted accurately by mapping distribution of preferred habitat of key prey species within their known range? These questions have relevance when considering country and range-wide assessments of tiger distribution (e.g. Dinerstein

et al. 1997), and in understanding what limits tiger distributions. We assess habitat selection of the Amur tiger at the first order level (geographical distribution) with four testable hypotheses and an analysis based on data from a recent range-wide survey (Matyushkin *et al.* 1996), and large-scale habitat mapping.

At the second order spatial scale, tigers select location and size of home ranges from a range of available habitats and habitat parameters available within the landscape. In the Sikhote-Alin Mountains of the Russian Far East, where the majority of Amur tigers remain (Matyushkin *et al.* 1996), home ranges tend to occur within a single or associated set of basins (Yudakov & Nikolaev 1987; Miquelle *et al.* in prep.). We analysed habitat selection with radio locations of tigers and composition of home range habitat parameters in comparison to all existing potential habitats within a series of major drainages along the eastern slopes (or eastern 'macroslope', as it is referred to in Russian literature, e.g. Matyushkin 1992) of the Sikhote-Alin Mountain Range. Five hypotheses and an analysis of site selection at this level of resolution provide an indication of primary parameters driving home range selection.

At the third order of selection each individual

selects an array of habitats within its home range. Analysis of third order selection represents a fine-grained assessment of what components of a home range are most preferred. Selection of home range site within existing habitats (second order selection) and selection of habitats within a home range (third order selection) should reflect critical needs of the individual. For any carnivore, it has been hypothesised that habitat selection should be related primarily to distribution and density of prey (Sunquist & Sunquist 1989), but does this selection occur at the second or third level of resolution? Prey distribution should vary by season, especially in a northern ecosystem. Therefore, several questions arise: (1) do Amur tigers select habitats by mimicking habitat preferences of prey? (2) do tiger home ranges shift seasonally in response to shifting prey distributions? (3) do parameters that define prey habitat preferences also define tiger habitat preferences? and (4) how does home range size relate to habitat/prey parameters? Although these questions may appear merely academic, they relate directly to space requirements – and therefore conservation strategies – for viable tiger populations. We constructed eight testable hypotheses for an analysis of third order selection based on data collected from radio-collared tigers in Sikhote-Alin Zapovednik (nature reserve) in the northeastern portion of Amur tiger range. We assess seasonal fidelity to home range, preferences for specific habitat parameters, and habitat use in relation to habitat availability and prey distribution.

We thank N. N. Reebin, A. V. Kostirya and numerous forest guards for assistance in data collection. The forest-cover map for Sikhote-Alin Zapovednik was developed by A. Shevlakov and M. N. Gromyko, the forest-cover map for the eastern macroslope was developed by S. M. Krasnepeev and V. A. Rosenberg, the landscape map based on satellite imagery was developed by A. Murzin, and the habitat type map for Khabarovski Krai was developed by A. N. Kulikov and V. Kulikov, all with support from the USAID Environmental Policy and Technology Project. The National Geographic Society, the Save the Tiger Fund, National Fish and Wildlife Foundation, the National Wildlife Federation, Exxon Corporation, the Wildlife Conservation Society, the Charles Engelhard Foundation, and Turner Foundation, Inc. provided funding for the Siberian Tiger Ecology Project. Support for development of this article was provided by the USAID Russian Far East Environmental Policy and Technology Project.

Study areas

Geographical range of the Amur tiger

The geographical range of Amur tigers in the Russian Far East stretches south to north for almost 1000 km throughout the length of Primorski Krai and into southern Khabarovski Krai east and south of the Amur River (Fig. 6.2). This region is represented mainly by the Sikhote-Alin Mountain Range, although tigers also occur within the Eastern Manchurian mountain system, which crosses into Russia from China at several places in southwest Primorye. In both regions, peaks are generally 500–800 m above sea level, with only a few reaching 1000 m or more. This region represents a merger zone of two bioregions: the East Asian coniferous-deciduous complex and the northern boreal (coniferous) complex, resulting in a mosaic of forest types that vary with elevation, topography and past history.

Over 72% of Primorye and southern Khabarovsk is forest covered. Typical tiger habitats are Korean pine/broadleaf forests with a complex composition and structure. This forest formation is considered by many to be key habitat for the Amur tiger (e.g. Kucherenko 1985). The large majority of these forests have been logged selectively at various times in the past, and human activities, in association with fire, have resulted in conversion of many low-elevation forests to secondary oak and birch forests. Above 700–800 m, spruce-fir forests prevail in central Sikhote-Alin. This elevational boundary for a predominantly coniferous forest type increases to the south, and decreases northward until, at 47°20' latitude, coniferous forests occur along the coastline (Fig. 6.2).

As with the plant communities, the faunal

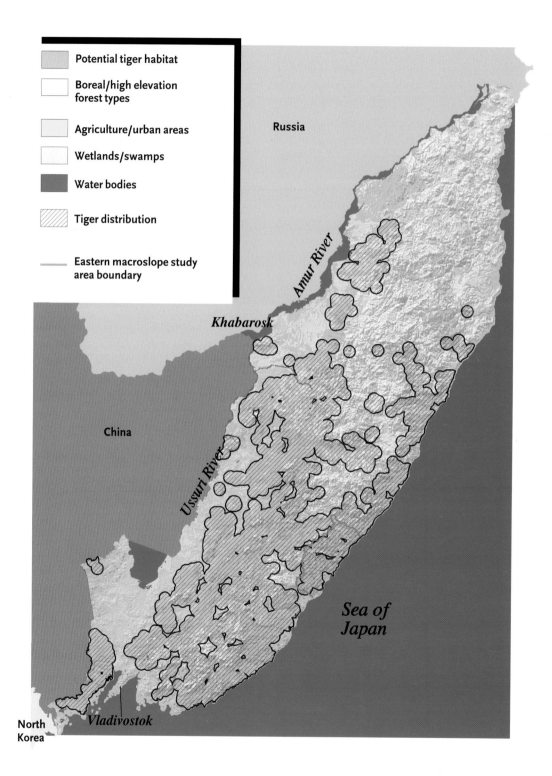

FIGURE 6.2

Distribution of major habitat types and distribution of Amur tigers in the Russian Far East. Tiger distribution was based on a 1996 range-wide survey (Matyushkin *et al.* 1996). The eastern macroslope study area is delineated in red. Scale: 1cm = 58km.

complex of the region is represented by a mixture of Asian and boreal life forms. The ungulate complex is represented by seven species: red deer, wild boar, sika deer, roe deer, Manchurian moose, musk deer and ghoral. Red deer and wild boar are the most common; both are found throughout the Sikhote-Alin and eastern Manchurian Mountains, but are rare in higher altitude spruce-fir forests. Manchurian moose are near the southern limits of their distribution in the central Sikhote-Alin Mountains, and are distributed sparsely in the inland boreal forests. Sika deer are near their northern limits at Sikhote-Alin Zapovednik. Musk deer are associated with the upper elevation conifer forests, and roe deer are confined to regions of limited snow depth. Ghoral, an endangered wild goat, are mostly restricted to coastal cliffs and some coastal mountains.

Sikhote-Alin Zapovednik and associated lands on the eastern 'macroslope' of the Sikhote-Alin Mountains

We confined the analysis of second order selection to those habitats within east-flowing drainages with some portion of their basins within Sikhote-Alin Zapovednik (Fig. 6.3). Although the crest of the Sikhote-Alin does not form a barrier to tiger movements, major ridges often form home range boundaries (Miquelle *et al.* in prep.), and the eastern 'macroslope' represents a discrete set of habitat parameters at a spatial scale within which tigers are likely to select home range location and associated habitat parameters. On the eastern coastal slopes of the Sikhote-Alin Mountains, temperatures and snow depth are moderated by the Sea of Japan, resulting in greater representation of southern forest types. Secondary Mongolian oak and white birch upland forests are most common nearer the coast. More inland, and at slightly higher elevations, Korean pine forest types are dominant and contain a mixture of deciduous and coniferous species, including birches, basswood, fir and larch. Spruce-fir and larch forests make up the remainder of the higher elevation, cooler forest types. Riverine forests are most often comprised of a variety of deciduous species, including willow, elm, chosenia,

cottonwood and ash, or a mixture of these deciduous species with Korean pine.

All radio-collared tigers included in this analysis had home ranges at least partially within the boundaries of Sikhote-Alin Zapovednik on the eastern side of the Sikhote-Alin Mountains. A total of 554 805 ha were included for second order analyses, representing 265 758 ha within the Zapovednik (approximately 66% of the Zapovednik) and 289 047 ha outside.

Methods

We defined two seasons for analyses: winter, from November to March (those months when snow was usually on the ground); and summer, extending from April to October (snow-free months). This division allows comparison of winter results to existing Russian studies, which have relied totally on traditional snow-tracking methods (Abramov 1962; Yudakov & Nikolaev 1987), and is relevant when considering tiger habitat selection in relation to ungulate distributions.

Radio-collaring and radio locations

Between January 1992 and November 1994, tigers were captured with modified Aldrich foothold snares, anaesthetised with a mixture of ketamine and Rompun and fitted with radio-collars. Of 11 animals captured and monitored, five (one male and four adult females) were monitored long enough (more than 15 months) to obtain sufficient sample sizes for analyses. Of the four females, one animal was followed from approximately 12 months old through her first litter; we include data starting 11 months after family break-up, when it was apparent that she had settled into her own home range. Two females appeared to be young animals establishing new territories at the time of capture (only one produced cubs during the study period), and one tigress was an adult already with cubs.

Locations from radio-collared tigers were obtained by three methods: aerial locations, triangulation and homing (White & Garrot 1990). Locations were plotted on 1:25 000 topographic

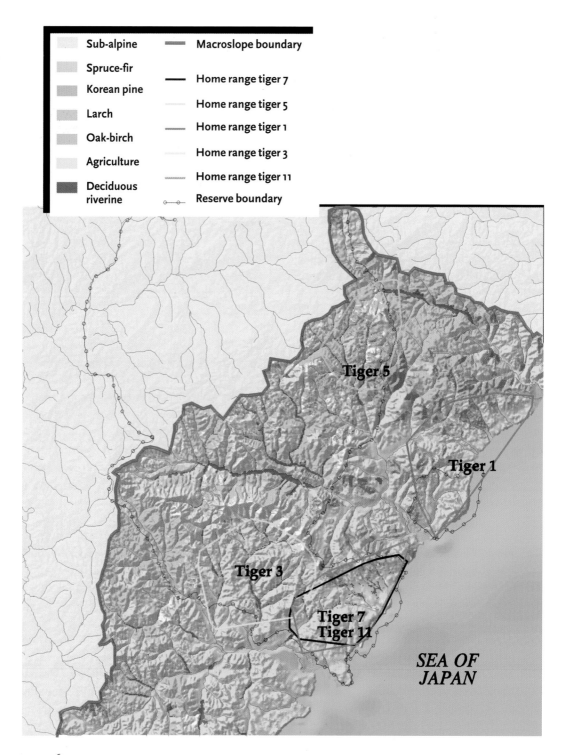

FIGURE 6.3

Study area on the eastern slopes of the Sikhote-Alin Mountain Range (eastern macroslope) used for second order habitat selection analysis, and home range boundaries (95% minimum convex polygons) of five tigers used for third order analyses (home range of male 11 overlaps that of female 7). Scale: 1cm = 8.26km.

maps. Many of our telemetry data were collected opportunistically and, except for aerial locations, a certain percentage of locations was biased by proximity of animals to roads, i.e. there was a greater probability of obtaining locations when tigers were close to roads. This bias affected methods used for home range analyses. For analyses of habitat selection, locations with a linear error estimated at less than 500 m were included.

To reduce the probability of serial dependence of locations, we used only one location per day for all analyses, unless subsequent same-day locations were separated by at least 5 km (Miquelle et al. unpubl. data). Sequential daily locations at den sites were eliminated, as were all locations within two days of capture. For third order analyses, boundaries of home ranges were defined by 95% minimum convex polygons of radio-collared tigers.

Spatial data

We developed a Geographic Information System (GIS) spatial data base at several levels of resolution for analyses. For first order analyses, landscape and forest-cover maps were developed at resolutions of 1:100 000 to 1:500 000, and generalised to a resolution of approximately 1:500 000. A forest-cover map was developed from the most recent Russian Forest Service inventory data (3–15 years old, depending on location) that categorised forest stands (averaging 30 ha) into an hierarchical classification scheme that could be resolved into 65 forest types (V. A. Rosenburg, unpubl. data). To ensure compatibility of the region-wide forest-cover map and the Sikhote-Alin forest-cover map, which were based on a different forest classification schemes, it was necessary to merge ('cross-walk') the 54 forest/habitat cover types of the Zapovednik into the 65 forest cover types of the region-wide map (S. M. Krasnepeev, unpubl. data). A landscape map for the entire region, developed from Russian satellite imagery (Murzin et al. unpubl. data) and existing 1:500 000 habitat maps, were used to define non-forest habitat. Sixty-five forest cover types from the forest-cover map and 42 landscape types were collapsed into 18 general habitat types used in first order analyses. For analysis of habitat selection at

both the second and third order, these general habitat types were further collapsed into seven forest types: upland oak, upland birch/aspen, riverine, Korean pine, spruce-fir, larch, and 'other' (including rare forest types and human-influenced landscapes such as villages, agriculture and grazing lands).

Range-wide distribution of Amur tigers was based on the results of a 1995–1996 winter census (Matyushkin et al. 1996). Tracks of tigers were reported on 652 count units distributed throughout approximately 90% of potential tiger habitat. Field counters plotted out the location of each track on 1:100 000 maps. To develop a distribution map of tigers from this data set, each track was encompassed by a 10 km-radius circle that represents a potential radius of travel. Because each 10-km circle is less than the estimated home range size of adult female tigers (Miquelle et al. in prep.), and some tigers were likely missed during the survey, we believe the distribution map represents a conservative estimate of tiger distribution.

Distribution maps of large ungulate species, including red deer, wild boar, roe deer, sika deer, and moose and were developed in a two-stage process. First, general distribution and relative density maps were developed at a small scale of resolution (approximately 1:2 500 000) based on existing literature (e.g. Bromley & Kucherenko 1983) and unpublished information. Ungulate distributions were plotted at three relative densities (high, medium and low) that were species specific (e.g. high densities of moose were not equivalent, in absolute terms, to high densities of red deer). Secondly, predicted distributions of ungulates within this range, were then mapped 1:500 000 based on habitat preferences defined by local specialists (V. A. Nesterenko et al. unpubl. data) and the forest-cover map (based on 18 habitat types). Mapping predicted distributions based on habitat associations provides a much higher level of resolution than the original distribution maps (Scott et al. 1993; Csuti 1996). For our analyses comparing ungulate and tiger distribution, we excluded low-density contour distributions for each ungulate species, assuming that areas with low prey densities

Table 6.1. *Summary of hypotheses tested, methods of analysis, and results of analysis assessing the relationship of Amur tiger distribution to prey and habitat parameters at three spatial scales in the Russian Far East*

Order of selection	Scale		Hypothesis	Methods of analysis	Results
First	Geographical distribution	1	Tiger distribution associated with a single key prey species.	3 overlap measurements.	Closest association with red deer distribution.
		2	Tiger distribution associated with two prey species.	3 overlap measurements.	Closest distribution with red deer & any other prey.
		3	Tiger distribution associated with habitat of key prey species.	3 overlap measurements.	68% weighted mean overlap with wild boar and red deer habitat.
		4	Tiger distribution associated with Korean pine forests.	3 overlap measurements.	Weak association (36%).
Second	Eastern macroslope Sikhote-Alin Mountains	1	Tigers select for specific habitat types within eastern macroslope.	Friedman test Fisher's LSD.	Riverine most preferred, spruce-fir and Korean pine avoided.
		2	Habitat composition of home ranges differs from eastern macroslope.	Chi-square goodness-of-fit.	Significant differences ($P<0.001$) for 4 home ranges.
		3a	Tigers select for specific elevations.	ANOVA Tukey comparisons.	Home ranges lower than macroslope average for 4 of 5 tigers.
		3b	Tigers select for distance to water.	ANOVA.	No significant differences.
		3c	Tigers select for specific slopes.	1-sample t-test.	Steep slopes avoided by all 5 tigers.
		3d	Tigers select for specific aspects.	Chi-square.	1 of 5 tigers demonstrated a preference for aspect.
		4	Habitat selection of tigers mimics habitat selection of ungulates in winter.	Spearman's rank correlation.	$r = 0.712$, $P <0.05$.
		5	Home range of tigers seasonally shifts to track ungulate distribution.	Overlap of seasonal home ranges.	Overlap of seasonal home ranges <70% confounded by sample size.
Third	Within home range	1	Tigers select for specific habitat types within home ranges.	Friedman test Fisher's LSD.	Preferences exist ($P <0.05$) but tendency for neutral selection.
		2	Composition of habitat types within home ranges varies among tigers.	Friedman test Fisher's LSD.	Composition of tiger home ranges varies ($P <0.01$).
		3a	Tigers select for specific elevations.	ANOVA Tukey comparisons.	Only 1 tigress preferred lower elevations in winter.
		3b	Tigers select for distance to water.	ANOVA.	No significant differences.
		3c	Tigers select for specific slopes.	1-sample t-test.	Steep slopes avoided.
		3d	Tigers select for specific aspect.	Chi-square.	1 of 5 tigers demonstrated a preference for aspect.
		4	Habitat selection of tigers mimics habitat selection of ungulates in winter.	Median difference of ranks.	No signficant differences (weak test).
		5	Seasonal shift in central tendency within home range.	t-test for changes in mean x and y.	Significant shifts in 5 of 10 tests.
		6	Habitat preference varies between summer and winter.	Friedman test Fisher's LSD.	No differences found between summer and winter ($P >0.25$).

Table 6.1. (*cont.*)

Order of selection	Scale	Hypothesis	Methods of analysis	Results
		7 Tigers select lower elevations in summer than winter.	ANOVA.	No significant differences between summer and winter.
		8 Tigers select south slopes more in winter than summer.	Chi-square.	1 of 5 tigers preferred south and east slopes in winter.

are less likely to support tigers and would be poorer predictors of tiger distribution.

Values for habitat parameters within home ranges and the eastern macroslope (elevation, slope and aspect) were developed from a digital elevation model constructed by interpolating 100-m contour intervals from a 1:100 000-scale topographical map, supplemented with elevation points of major peaks and guided by flow direction of streams and rivers. Measurements of distance from water are based on a digital map of streams, rivers and lakes.

First order habitat selection (geographical distribution)

To assess parameters potentially related to distribution of Amur tigers, we tested the following hypotheses with first order spatial data (summarised in Table 6.1):

1 Amur tiger distribution in the Russian Far East is associated with the presence of red deer, wild boar, roe deer, sika deer, or moose (each species assessed separately).

2 Amur tiger distribution in the Russian Far East is associated with the aggregate presence of two key prey species.

3 Amur tiger distribution can be predicted by mapping the distribution of preferred habitat of key prey species.

4 Amur tiger distribution in the Russian Far East is associated with the presence of key habitat types (Korean pine forests).

Korean pine forests have been identified as key tiger habitat (Kucherenko 1985), primarily because they are considered key habitat for wild boar and, to

a lesser extent, for red deer. We therefore compared tiger distribution to this habitat type. Because large ungulates (red deer, roe deer, sika deer, moose and wild boar) comprise 96% of tiger prey (Miquelle *et al.* 1996b), we measured coincidence of tiger distribution to each of these species separately. However, it has also been suggested that two species (namely, red deer and wild boar) are the key combination of prey (Kucherenko 1985; Matyushkin 1992; Miquelle *et al.* 1996b). Therefore, we compared aggregate distributions of all combinations of two prey species with tiger distribution. Finally, we compared the predicted distribution of red deer and wild boar based on habitat preferences to tiger distribution.

We used three measurements to compare the relationship of tiger distribution to various prey and habitat distributions: (1) percentage overlap of prey species (or habitat type) distribution with tiger distribution; (2) percentage overlap of tiger distribution with prey/habitat distribution; and (3) percentage of the total area of tiger and prey/habitat combined that is shared by both. High percentage overlap of the first measure suggests that this parameter (specific prey or habitat distribution) occurs over much of the tiger range; high percentage overlap of the second indicates that tigers inhabit much of the area where prey/habitat occurs (but could occur over a much larger area), and a high percentage of total area shared (third measure) indicates a good association between tigers and prey/habitat. High percentage overlap of all three measures indicates a high degree of concordance between tiger distribution and prey (or habitat) distribution; however, high percentage overlap of prey/habitat distribution with tigers (the second measure) was not as strong a measure because tigers could occur widely outside the range of any particular prey or habitat. Therefore, we used a weighted mean percentage for these three measures (weighting the second overlap measure by 0.5) as a general indicator of the level of association.

Moose distribution extends far to the north of our defined study area, thereby biasing our analyses because the full distribution of this species was not included. Despite this potential positive bias, we eliminated moose from the two-species comparisons because single species comparisons indicated a poor association with tigers.

Second order habitat selection (within the eastern macroslope)

We used some of the same hypotheses for tests of second order and third order relationships (Table 6.1). However, for clarity, we present them separately here. We tested the following hypotheses on second order (within the eastern macroslope) habitat selection:

1 Tigers demonstrate preference for specific habitat types (based on locations of radio-collared tigers) within the eastern macroslope study site.

2 Composition of habitat types within home ranges is different than composition of habitat types within the eastern macroslope.

3 Tigers select for some component of the following habitat parameters on a seasonal basis within the eastern macroslope:
   a.   elevation;
   b.   distance to water;
   c.   slope;
   d.   aspect.

4 Habitat selection by tigers should mimic habitat selection by ungulates within the eastern macroslope (tested for winter only).

5 Because ungulate distributions shift seasonally, the geographical location of tiger home ranges should shift to 'track' areas with high ungulate densities. Therefore, home range locations in summer and winter may

not overlap, i.e. low site fidelity (at the second order spatial scale) on a seasonal basis may occur.

Hypotheses 1–3 imply that tigers demonstrate preferences for habitat types, or features of the environment, at the second order of selection. Hypotheses 4 and 5 imply that habitat selection and location of home ranges is driven, at the second order of selection, by the distribution of ungulates on a seasonal basis.

Third order habitat selection (within home ranges)

We tested the following hypotheses on third order habitat selection (summarised in Table 6.1):

1 Tigers demonstrate preference for specific habitat types within their home ranges.

2 Composition of habitat types within home ranges varies among tigers.

3 Tigers select for some component of the following habitat parameters on a seasonal basis within home ranges:
   a.   elevation;
   b.   distance to water;
   c.   slope;
   d.   aspect.

4 Habitat selection of tigers should mimic habitat selection of ungulates within home ranges (tested for winter only).

5 As an alternative to hypothesis 5 in second order selection, if a single year-round home range provides an adequate prey base, there may still be seasonal shifts within a home range as tigers track localised shifts in prey distribution (i.e. a shift in central tendency of home range).

6 Habitat preference within the home ranges of tigers varies between summer and winter season, reflecting seasonal changes in prey distribution.

7 Because ungulates usually move to lower elevation habitats in winter (Myslenkov unpubl. data), average elevation selected by tigers within their home range (third order selection) should be lower in winter than summer.

8 Because ungulates tend to use south slopes more in winter (A. E. Myslenkov unpubl. data), tigers should select south slopes in winter more than summer, and more than the percentage available within the home range (third order selection) in winter.

Hypotheses 1–5 at the second and third order of selection examine the same variables, but at different spatial scales. Because it is likely that a variety of 'combinations' of habitat types can meet the needs of tigers, hypothesis 2 (third order spatial scale) suggests that the composition of habitat types within home ranges of tigers is likely to vary. Hypotheses 2 and 3 imply that tigers demonstrate preference for habitat types, or features of the environment on a seasonal basis within their home ranges. Hypotheses 4 (at second and third order spatial scales) both state that tigers should demonstrate the same preference for habitat types as their prey, but at different spatial scales. Hypotheses 5–8 test various ways in which tigers may 'track' distribution of prey seasonally, which may affect use of the home range. As in most seasonal environments, the distribution of ungulates in Sikhote-Alin Zapovednik shifts seasonally, reflecting the effect of snow depth, food availability and shelter from extreme weather conditions (A. E. Myslenkov unpubl. data). Without specifying the exact distributions within each tiger home range, hypotheses 5–8 test for evidence of tigers 'tracking' changes in assumed ungulate distribution through seasonal changes in preference for habitat parameters. Changes in either home range location (hypothesis 5, second order scale), or focal use areas within home ranges (hypothesis 5, third order) are expected due to changes in ungulate distribution.

Habitat selection was compared seasonally at both the second and third order spatial scale. The most commonly used test for habitat preference, the χ^2 goodness-of-fit test (Neu *et al.* 1974; Byers *et al.* 1984) was inappropriate because low expected cell frequencies violated basic assumptions of this statistic. We used the Friedman test, in which habitats represented 'treatments' and animals were 'blocks' (Conover 1980; Alldredge & Ratti 1986),

and ranked availability (determined from GIS spatial databases) and use of habitats by tigers (based on radio locations) to compare the following: differences in habitat availability and selection within the eastern macroslope (hypothesis 1, second order); the difference in habitat availability and selection within home ranges (hypothesis 1, third order); variation in the composition of habitat types among four home ranges (tigress 7 and tiger 11 had virtually identical home ranges, so one was deleted for these comparisons) (hypothesis 2, third order); and seasonal differences (winter and summer) in habitat preferences of all five tigers (hypothesis 6, third order). Where significant differences were found, Fisher's least significant difference (LSD) method was used to determine which habitats were different in terms of selection versus availability (Conover 1980; Alldredge & Ratti 1986). To compare composition of habitat types within home ranges to the eastern macroslope (hypothesis 2, second order selection), individual χ^2 goodness-of-fit tests were conducted for each home range, for which expected values were represented by composition of the eastern macroslope.

We assessed the importance of habitat parameters that may influence site selection by tigers at both the second and third order spatial scale, including elevation, distance to water, slope and aspect (hypotheses 3a–d at both scales). We assessed the relationship of tiger selection to availability of habitat parameters by comparing an equal number of tiger locations with randomly selected 200-m-wide grid cells across the eastern macroslope (second order selection), and within home ranges (third order selection); this level of resolution approximately equalling the average accuracy of radio locations. For two continuous variable parameters (elevation and distance to water), '*a posteriori*' Tukey pair-wise comparisons were made if an analysis of variance (ANOVA) test on each individual tiger determined significant differences among four categories of data: summer tiger locations, winter tiger locations, random locations within the eastern macroslope study area, and random locations within home ranges. This analysis of elevation also provided a test of hypothesis 7, third order selection. Winter and summer mean slope selection by tigers were compared to each other with two-sample *t*-tests (hypothesis 8, third order selection), and to mean values within each home range and on the eastern macroslope (derived from a digital elevation model) using one-sample *t*-tests (hypotheses 3c second and third orders). Aspect was converted from a continuous variable of the digital elevation model into categorical data with four elements: 315–45°=north, 46–135°= east, 136–225°=south; 226–315°=west. Because the digital elevation model delineated few areas as truly flat, this category was deleted from analyses. Overall and seasonal use of aspect by each tiger was tested independently with χ^2 goodness-of-fit tests based on expected distributions for each home range and for the eastern macroslope (derived from the digital elevation model).

We used data developed by Matyushkin (1992) to assess the correlation between winter habitat selection by prey species (red deer, roe deer and wild boar) and tigers across the eastern macroslope (hypothesis 4, second order selection). These data were based on relative track densities of prey and tigers along prescribed transect routes that included 10 habitat types. We used a Spearman's rank correlation test to assess the relationship between tigers and each of the three main prey species. Definitions of habitat types and study site boundaries by Matyushkin (1992) are slightly different from those described above, although still focused on the eastern macroslope. Despite some differences, the data provide an opportunity to test this hypothesis within the same general study site at the appropriate level of resolution.

Because data on summer habitat selection of prey were unavailable, we compared a ranking of total winter prey density in habitat types within each home range to winter habitat preference of tigers (hypothesis 4, third order scale). Prey density in winter was based on 1994 aerial surveys within home ranges of four radio-collared tigers (A. E. Myslenkov & D. G. Miquelle unpubl. data) where conifer habitat types were rare and visibility was equal among the dominant habitat types. For the fifth home range, we used relative ungulate track

density in habitat types. We believe that these two methods are comparable in our analysis because, in all cases, density counts were reduced to a ranking by habitat type. Confidence intervals for the median of the absolute difference in ranks of habitats selected by tigers and habitats selected by prey in winter were constructed to determine if selection varied between tigers and prey (i.e. median difference greater than 0 representing variation in selection).

We tested tiger home range fidelity at two spatial scales: (1) overlap of winter and summer home ranges, based on 95% minimum convex polygons (hypothesis 5 second order selection) estimated using the software programme 'CALHOME' (US Forest Service Pacific Southwest Research Station and California Department of Fish and Game); and (2) changes in mean x and y coordinates of tiger locations within a home range (hypothesis 5, third order selection). High overlap of summer and winter home ranges should be an indicator of high seasonal fidelity to home range location (second order selection). Insignificant changes of mean x and y coordinate values between seasons would suggest no change in central tendency of year-round home range location (third order selection).

Null hypotheses of statistical tests were rejected when significance levels exceeded 5%.

Results

First order habitat selection (geographical distribution)
Tiger distribution was more closely associated with distribution of red deer (61% overlap) than any other prey species (Table 6.2, column 1). Distributions of wild boar, red deer and roe deer overlapped more than 70% with tiger distribution (Table 6.2, column 2), indicating that tigers occurred over most of the region where these prey species occurred within the region defined by our study. The high overlap of tiger distribution to red deer in comparison to other single-species distributions (Table 6.2, column 1) and the percentage of total area shared (Table 6.2, column 3) suggest a strong correlation between tiger distribution and red deer (Fig. 6.4a). Although wild

Table 6.2. *Overlap of Amur tiger distribution over their entire range in Russia with prey and habitat distributions, overlap of prey and habitat distributions with tiger distribution, and percentage of total area of combined distributions shared by both tiger and prey or habitat. Distribution of prey based on moderate to high distribution contours (see text)*

Species/habitat	Overlap of tiger distribution with prey/ habitat distribution (%)	Overlap of prey/ habitat distribution with tiger distribution (%)	Percentage of total area shared with tiger (%)	Weighted mean[a] (%)
One species				
Red deer	61	73	50	59.0
Wild boar	37	84	34	45.0
Sika deer	9	67	9	21.0
Roe deer	31	72	27	38.0
Moose[b]	19	29	3	15.0
Two species				
Red deer and wild boar	74	73	58	67.0
Red deer and roe deer	71	78	59	68.0
Red deer and sika deer	72	67	53	63.0
Boar and roe deer	76	46	41	56.0
Boar and sika deer	80	41	37	55.0
Roe and sika deer	72	32	28	46.0
Habitat				
Korean pine forests	24	85	23	36.0
Red deer and boar habitat	93	52	50	68.0

[a] Weighted mean of 3 measures of percentage overlap, with overlap of prey/habitat distribution with tiger weighted by 0.5, and the other two weighted as 1.
[b] Moose were not included in two-way comparisons (see text).

boar have been considered key prey for tigers, their distribution was more patchy than that of red deer (Fig. 6.4b), and consequently, although tigers occurred almost everywhere boar were found at high and medium densities (84% overlap), boar distribution was overall not as strong a predictor of tiger distribution (Table 6.2, column 4). Although tigers prey on both roe deer and sika deer, overlap of these two species with tigers was low (Table 6.2; Figs. 6.4c, d). Distribution of moose was poorly associated with tiger distribution (Table 6.2; Fig. 6.4e).

The combination of two prey species improved the relationship between tiger and prey distribution (Table 6.2), but not dramatically. Red deer plus any other prey species provided consistently high measures of weighted mean overlap (63–68%) (e.g. Fig. 6.4f), while boar and other species (except red deer) provided slightly lower measures (Table 6.2). As expected based on the single-species analysis, the combination of roe deer and sika deer was relatively poorly correlated with tiger distribution.

The distribution of Korean pine forests did not relate well to tiger distribution (Table 6.2; Fig. 6.4g). Although tigers occurred throughout Korean pine forests (Table 6.2, column 2), tigers occurred in a variety of other habitats, and consequently the overall association was low. In contrast, the aggregate set of habitats preferred by red deer and wild boar (Fig. 6.4h) provided as strong a relationship to tiger distribution as the distributions of the species themselves. (Fig. 6.4f; Table 6.2).

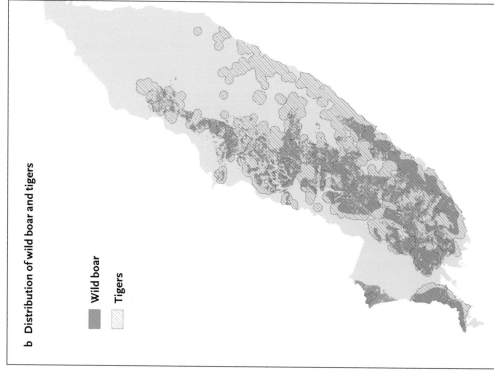

b Distribution of wild boar and tigers

Wild boar

Tigers

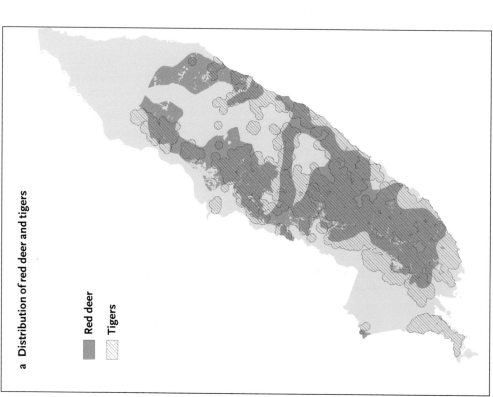

a Distribution of red deer and tigers

Red deer

Tigers

FIGURE 6.4

Overlap of tiger distribution with distribution of five prey species and two habitat complexes: **a.** red deer; **b.** wild boar; **c.** roe deer; **d.** sika deer; **e.** moose;
f. red deer and wild boar distribution; **g.** Korean pine forests; **h.** habitats preferred by red deer and wild boar.

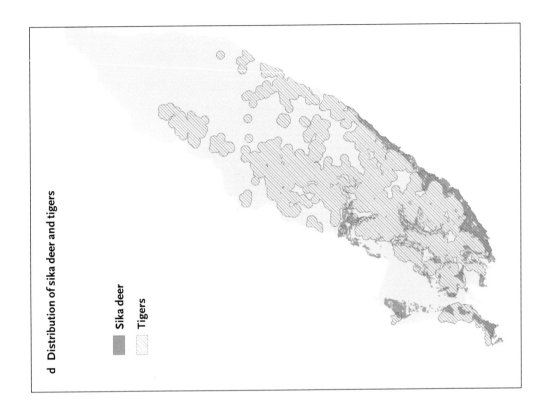

c Distribution of roe deer and tigers

Roe deer

Tigers

d Distribution of sika deer and tigers

Sika deer

Tigers

f Distribution of wild boar, red deer and tigers

Wild boar or red deer

Wild boar and red deer

Tigers

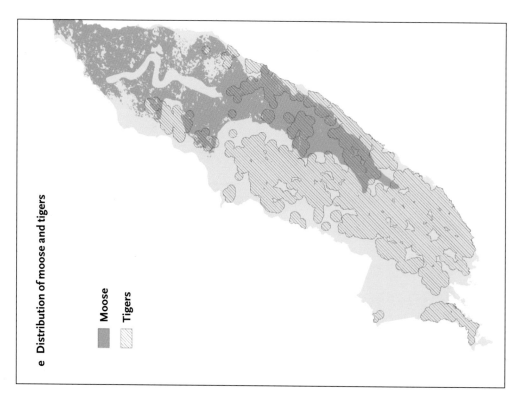

e Distribution of moose and tigers

Moose

Tigers

h Distribution of wild boar and red deer habitat and tigers

Wild boar or red deer habitat

Wild boar and red deer habitat

Tigers

g Distribution of Korean pine and tigers

Korean pine

Tigers

Table 6.3. *Habitat availability and use based on locations of radio-collared Amur tigers on the eastern slope of the Sikhote-Alin Mountains (second-order selection), and within home ranges (third-order selection), 1992–1995, and results of multiple comparisons of habitat preferences[a]*

| Forest type description | Eastern macroslope (2nd order) | | Home range (3rd order) | | | | | | | | | | Multiple comparisons[a] | |
| | | | Tigress 1 | | Tigress 3 | | Tigress 5 | | Tigress 7 | | Tiger 11 | | | |
	Area (m)	Percent (%)	Avail. (%)	Used (%)	Avail. (%)	Used (%)	Avail. (%)	Used (%)	Avail. (%)	Used (%)	Avail. (%)	Used (%)	2nd order	3rd order
Riverine	12 680	2.3	0.1	0	2.7	9.9	3.8	10.9	2.5	4.4	2.5	4.2	a	a
Upland white birch/aspen	73 168	13.2	15.3	10.3	35.0	48.7	5.5	10.9	7.9	8.7	7.9	10.8	abc	ab
Upland oak	174 632	31.5	51.8	70.4	42.2	31.5	0.7	1.5	74.4	79.6	74.4	68.8	ab	abc
Larch	48 813	8.8	19.9	18.6	2.3	3.6	16.7	8.8	2.6	2.4	2.6	7.8	ab	abc
Spruce-fir	46 729	8.4	0.3	0	6.4	0.9	7.8	4.4	0.3	1.0	0.3	1.2	c	abc
Korean pine	154 611	27.9	6.0	0.7	4.0	2.7	61.5	62.0	1.4	0	1.4	0.6	c	bc
Other forests/ human landscapes	44 172	8.0	6.6	0	7.4	2.7	4.0	1.5	10.9	3.9	10.9	6.6	bc	c

[a] Tigers demonstrated significantly different ($P < 0.05$) preferences (a=most preferred) for habitat types with different letters based on Fisher's least significant difference multiple comparison test (after a Friedman's test demonstrated overall significance).

Second and third order habitat selection (within the eastern macroslope and within home ranges)

Within the eastern slope of the Sikhote-Alin study area, the dominant habitat types were upland oak (31.5%) and Korean pine forests (27.9%) (Table 6.3). Upland birch/aspen forests were also fairly common (13.2%), whereas riverine forests were the rarest habitat type (2.3%).

The composition of habitat types within home ranges of tigers differed from availability on the eastern macroslope (four goodness-of-fit tests for each of four home ranges, $P<0.001$) (Table 6.3), and habitat composition of individual home ranges also varied significantly from each other ($T=5.7$, $P<0.01$). Tigers 1, 3, 7 and 11 lived close to the coast, where upland oak was the dominant forest type and upland birch/aspen was common. Tigress 5 lived more inland and at a higher elevation, where Korean pine forests were dominant (Table 6.3).

We found no seasonal differences in habitat selection by tigers ($T=0.58$, $P>0.25$) so data from both seasons were combined for further analyses. Tigers selected habitat types non-randomly in comparison to availability on the eastern macroslope ($T=3.25$, $P<0.01$). Although riverine forests made up only a small percentage of the available habitat, they were strongly preferred at the second order of selection (Table 6.3). Spruce-fir, Korean pine and human landscapes tended to be selected against, while upland oak, upland birch/aspen and larch habitats were used roughly in accordance with their abundance (Table 6.3).

Tigers were selective at the third order of habitat selection (within home ranges) as well, although the level of significance indicated a less robust association ($T=2.73$, $P<0.05$), and there was a greater tendency for neutral selection. Within home ranges, riverine forests appeared to be most strongly preferred, although they comprised only a small percentage of the available habitat (Table 6.3).

Table 6.4. *Seasonal variation in mean elevation of Amur tiger locations in comparison to random locations within the eastern macroslope (second-order spatial scale) and within home ranges (third-order spatial scale) of the Sikhote-Alin Mountains, 1992–95*

	Elevation (m)											
	Tiger locations						Random points					
							Eastern macroslope (second-order)			Within home range (third-order)		
	Summer			Winter								
	n	mean	SE	n	mean	SE	n	mean	SE	n	mean	SE
Female 1	77	255[a]	18	74	210[a]	17	77	456[b]	29	77	278[a]	22
Female 3	73	369[a]	20	40	371[ab]	26	73	451[b]	30	73	346[a]	14
Female 5	63	590[a]	23	76	433[ab]	15	76	451[b]	29	76	616[a]	17
Female 7	86	194[a]	14	124	212[a]	12	124	414[b]	21	124	219[a]	14
Male 11	58	248[a]	26	115	262[a]	16	115	416[b]	24	115	276[a]	25

[ab] For each tiger, means with different letters are significantly different, based on *a posteriori* Tukey pair-wise comparisons after an ANOVA determined overall significant differences ($P < 0.05$).

Human-impacted landscapes tended to be selected against (Table 6.3), whereas tigers used upland oak, upland birch/aspen, larch and spruce-fir roughly in accordance with their abundance. Korean pine and spruce-fir forests appeared to be avoided at the second order, but were used approximately in amounts equal to availability within home ranges, especially by tigress 5, which was the only animal with substantial tracts of Korean pine habitat within her home range.

Home ranges of four of five tigers were significantly lower than the average elevation within the eastern macroslope, and summer and winter tiger locations were significantly lower than mean elevation within the macroslope study area in six of eight comparisons (Table 6.4). However, the elevation of summer locations for all five tigers and elevation of four out of five tigers' winter locations did not vary from random locations within their respective home ranges (Table 6.4). The mean elevation of winter and summer tiger locations was not significantly different (i.e. tigers did not move to lower elevations in winter to avoid deep snow) except for female 5 (Table 6.4), whose home range was situated high, abutting the crest of the Sikhote-Alin. Mean elevation within her home range was greater than average along the eastern macroslope,

and mean elevation of her summer locations was also greater. In winter, however, she selected lower elevations.

Distance from water appeared to be an unimportant parameter determining tiger locations; there were no significant relationships at any level of analysis (Table 6.5).

Tigers selected slopes that were significantly less steep than the average slope over the entire eastern macroslope (Table 6.6). Slope selection by tigers differed from the average within home ranges only when mean home range slope approached that for the eastern macroslope (i.e. for tigers 1, 3, and 5) (Table 6.6). Selection of slopes by tigers did not vary between seasons but, in general, tigers avoided steep slopes.

No tiger demonstrated preferences for specific aspects in summer or winter except female 7, who preferred east and south slopes and avoided north slopes in winter (Table 6.7). Winter selection of aspect by female 7 was also significantly different from available aspects within the eastern macroslope (but not within her home range) with the same pattern (avoidance of north slopes and preference for east and south slopes). No other tigers showed any such trend (Table 6.7).

Across the eastern macroslope, there was a

Table 6.5. *Seasonal variation in distance from water of Amur tigers in comparison to random locations within the eastern macroslope (second-order spatial scale) and home ranges (third-order spatial scale) of the Sikhote-Alin Mountains, 1992–95*

	Distance to water (m)											
	Tiger locations						Random points					
							Eastern macroslope (second-order)			Within home range (third-order)		
	Summer			Winter								
	n	mean	SE	n	mean	SE	n	mean	SE	n	mean	SE
Female 1	77	361	29	74	349	29	77	357	32	77	381	31
Female 3	73	351	36	40	258	44	73	358	33	73	334	32
Female 5	63	325	32	76	252	28	76	357	32	76	329	30
Female 7	86	373	31	124	363	24	124	369	25	124	366	24
Male 11	58	359	35	115	330	27	115	367	25	115	337	25

[a,b] For each tiger, means with different letters are significantly different, based on *a posteriori* Tukey pair-wise comparisons after an ANOVA determined overall significant differences ($P < 0.05$). Rows with no letters indicate no overall significant differences.

Table 6.6. *Seasonal selection of slope by Amur tigers in comparison to mean slope within the eastern macroslope (second-order spatial scale) and within home ranges (third-order spatial scale) of the Sikhote-Alin Mountains, 1992–95*

	Slope (degrees)									
	Tiger locations						Area (spatial scale)[c]			
							Eastern macroslope (second-order)		Within home range (third-order)	
	Summer			Winter						
	n	mean	SE	n	mean	SE	mean	SD	mean	SD
Female 1	77	10.6[ab]	0.9	74	10.0[ab]	0.9	15.0	10.0	13.6	9.4
Female 3	73	12.6[ab]	0.9	40	10.2[ab]	1.3	15.0	10.0	16.0	9.8
Female 5	63	11.7[ab]	1.0	76	12.3[ab]	1.0	15.0	10.0	15.5	10.0
Female 7	86	9.7[b]	0.9	124	10.4[b]	0.8	15.0	10.0	10.4	9.0
Male 11	58	11.3[b]	1.6	115	10.3[b]	0.9	15.0	10.0	10.4	9.0

[a] Tiger locations significantly differ from the average slope within the home range, based on one-sample *t*-tests.
[b] Tiger locations significantly differ from the average slope on the eastern macroslope, based on one-sample *t*-tests.
[c] Summary statistics from digital elevation model for home ranges and eastern macroslope.

significant correlation between habitat selection by tigers and red deer ($r=0.712$, $P<0.05$) and tigers and wild boar ($r=0.757$, $P<0.05$), but not between tigers and roe deer ($r=0.388$, $P>0.20$).

Estimates of ungulate use of many habitat types within tiger home ranges were not available because those habitats comprised such a small percentage of tiger home ranges that they were not sampled for ungulate density (Table 6.8). The confidence interval for the median difference in ranks of ungulate habitat use and tiger habitat preference included 0 (1 ± 1) indicating no significant difference in habitat selection by tigers and ungulates. Missing values (i.e. habitats not sampled) precluded a more

Table 6.7. *Seasonal selection of aspect by 5 radio-collared Amur tigers in comparison to available aspects within home ranges and within the eastern slopes (macroslope) of the Sikhote-Alin Mountains, 1992–95*

Tiger home range/ eastern macroslope	Aspect	% used summer[c]	% used winter[c]	% available[c]
Female 1	(n)	(77)	(74)	(151)
	N	27.3	16.2	20.1
	E	28.6	25.7	33.5
	S	26.0	28.4	24.6
	W	18.2	29.7	21.8
Female 3	(n)	(73)	(40)	(113)
	N	16.4	15.0	17.6
	E	23.3	27.5	24.4
	S	31.5	27.5	31.1
	W	28.8	30.0	26.9
Female 5	(n)	(62)	(76)	(138)
	N	24.2	18.4	25.7
	E	22.6	35.5	22.3
	S	32.3	23.7	24.3
	W	21.0	22.4	27.7
Female 7	(n)	(86)	(124)	(210)
	N	26.7[a]	12.1[b]	16.7[a]
	E	26.7[a]	36.3[b]	31.5[ab]
	S	30.2	35.5	26.6
	W	16.3[ab]	16.1[a]	25.2[b]
Male 11	(n)	(58)	(115)	(173)
	N	19.0	18.3	16.9
	E	37.9	27.0	29.5
	S	29.3	28.7	33.9
	W	13.8	26.1	19.7
Eastern macroslope[d]	N	—	—	24.6
	E	—	—	28.8
	S	—	—	24.1
	W	—	—	22.5

[ab]For each tiger, values with different letters within a row indicate where largest cell chi-square values contributed to significant overall chi-square goodness-of-fit values.
[c]No sample size because values are based on the sum of values generated by the digital elevation model.

powerful test, but results suggest only moderate concordance in winter habitat use by tigers and prey within home ranges (Table 6.8).

Although tigers demonstrated strong year-round fidelity to home ranges, all tigers demonstrated some seasonal shift in home range boundaries, and shifts in mean locations. Although tigers remained within the same basic home range, overlap of summer and winter home ranges did not exceed 70% for any of the five tigers (Table 6.9). These relatively low measures of overlap are partially due to biases associated with small sample sizes. The minimum convex polygon method is sensitive to sample size, and our initial analyses suggest that greater sample sizes may be necessary before the size of minimum convex polygons begin to

Table 6.8. *Ranks of prey abundance and tiger preference (use versus availability) in winter for 5 habitat types, and the difference in those two ranks within home ranges of 5 Amur tigers on the eastern slope of the Sikhote-Alin Mountains, 1992–95*

Relative prey abundance and tiger habitat preferences within home ranges[a]

	Tigress 1			Tigress 3			Tigress 5			Tigress 7			Tiger 11		
	Prey[b] (rank)	Tiger[c] (rank)	Difference[d] (tiger–prey)	Prey (rank)	Tiger (rank)	Difference (tiger–prey)	Prey (rank)	Tiger (rank)	Difference (tiger–prey)	Prey (rank)	Tiger (rank)	Difference (tiger–prey)	Prey (rank)	Tiger (rank)	Difference (tiger–prey)
Riverine	2	2	0	1	2	1	3	1	−2	2	2	0	2	2	0
Birch/aspen	4.5	4	−0.5	3	1	−2	1	2	1						
Upland oak	1	1	0	2	3	1				1	1	0	1	3	2
Larch	4.5	3	−1.5							3	3	0	3	1	−2
Korean pine	3	5	2				2	3	1						

[a] Missing values indicate habitat was not present in home range or was inadequately sampled.
[b] Habitat with highest prey density = 1.
[c] Habitat most preferred by tiger = 1.
[d] Zero differences indicate concordance of habitat selection by tigers and relative prey density.

Table 6.9. *Seasonal home range size (95% minimum convex polygon estimates) and fidelity to location (% overlap of summer and winter home ranges) of five radio-collared Amur tigers in Sikhote-Alin Zapovednik, 1992–95*

Animal	Summer home range		Winter home range		Overlap[a]	
	n	(km²)	n	(km²)	(km²)	(%)
Female 1	103	353	76	302	225	53
Female 3	77	413	40	203	155	34
Female 5	65	224	81	172	128	48
Female 7	88	245	145	227	190	68
Male 11	59	235	129	288	208	66

[a] Total area (summer and winter home ranges combined) that includes both summer and winter home ranges.

Table 6.10. *Differences in mean coordinates of locations of five radio-collared tigers in summer and winter in Sikhote-Alin Zapovednik, 1992–95*

Tiger	df	Seasonal difference in mean coordinates			
		x coordinate		y coordinate	
		Test statistic F	Mean difference (km)	Test statistic F	Mean difference (km)
Female 1	177	33.19	6.2***	0.04	0.2
Female 3	116	1.27	1.4	7.67	3.0**
Female 5	145	30.62	3.5***	51.60	6.6***
Female 7	231	0.23	0.4	0.01	0.1
Male 11	187	14.14	3.3***	9.85	2.2**

* Significant difference in seasonal mean coordinates, based on t-tests ($P < 0.05$).
** ($P < 0.01$).
*** ($P < 0.001$).

asymptote (Miquelle *et al.* unpubl. data). Even with five animals, there was a strong correlation between total sample size and degree of seasonal overlap (r^2=0.87, P<0.02). No doubt, with larger sample sizes, overlap of summer and winter home ranges would increase, reinforcing the argument of strong seasonal fidelity to a single home range.

There were significant shifts in the mean x and/or y coordinates for summer and winter locations for four of five tigers (Table 6.10). The largest differences (for tigresses 1 and 5) were largely artefacts of seasonal localisation during denning. The largest linear shift of mean locations (6.6 km) was still relatively small in comparison to total home range dimensions. Nonetheless, tigers did show small but significant shifts in mean coordinates of locations between summer and winter.

Discussion

First order habitat selection (geographical distribution)

In assessing the four hypotheses associated with first order selection (Table 6.1), we found:

1   Amur tiger distribution was associated closely with red deer distribution, but there

was poor association between tigers and wild boar, another prey species considered 'key'.

2   A combination of two prey species, especially red deer and any other prey, did increase the strength of the relationship between prey and tiger distribution.

3   Distribution of preferred habitat for key prey species was an accurate predictor of tiger distribution.

4   Amur tiger distribution was poorly related to what has been considered a 'key' habitat type, namely Korean pine.

Amur tiger distribution is closely associated with prey distributions. There has been a long-standing debate on the relative importance of red deer versus wild boar in the diet of Amur tigers (Abramov 1962; Pikunov 1981; Kucherenko 1985; Yudakov & Nikolaev 1987). Our results concur with those of Matyushkin (1992) in that red deer distribution appears to be the strongest predictor of tiger distribution in Russia. However any combination of red deer and a second prey species improves the overall relationship (Table 6.2). In the case of roe deer and red deer, the strength of the relationship is likely to be an artefact of excluding low-density prey areas from the analysis. Density of roe deer is largely inversely related to that of red deer and wild boar; roe deer reach highest densities in fragmented landscapes that include agricultural production areas, whereas red deer and boar reach highest densities in predominately forested areas. The cumulative result, in terms of our analysis, is that roe deer 'fill in' the distribution map where tigers occur and red deer and boar are found at low density. The same is true, although to a lesser extent, for sika deer. Therefore, although the weighted means of red deer and boar versus red deer and roe deer are nearly equivalent (67 and 68%), we believe that the first value may be more meaningful biologically. Roe deer make up a relatively small percentage of the diet of tigers throughout their range in Russia (Miquelle *et al.* 1996b). Although the strength of the relationship between elk, boar and tiger distributions is likely to be greater if low density contours were included in the analysis, exclusion of low-

density areas provided a more powerful means of assessing the relative value of each prey species in predicting tiger distribution.

Moose distribution was a poor predictor of tiger distribution, even within our limited (for moose) study area. Moose distribution extends far to the north of our defined study area, and extension of the area of study to include all moose habitat would show a very low, and in fact inverse, correlation between tiger and moose distribution.

Our finding that habitat preferences of prey can be used with the same reliability as prey distributions themselves should be viewed with caution for two reasons. First, elimination of low-density contours in the analysis provided greater discriminative powers in discerning differences between prey species, but suggested weaker associations than actually existed. Because it is a multi-prey system, tigers will occur in regions even if one species is at a lower density. Secondly, because our definition of prey distribution is dependent on distribution of prey habitat, there is some redundancy in the comparison of prey distribution versus prey habitat distribution, and the results have an inherent bias towards stronger associations with habitats. Mapping preferred prey habitat can give a meaningful picture of 'potential' tiger habitat, but it may not be a good predictor of tiger presence because the presence of prey habitat does not always equate with presence of prey. Therefore, it is dangerous to make predictions on tiger distribution based solely on existence of prey habitat; some assessment of the habitat must be made to determine the status of prey populations and human disturbance in those regions.

Contrary to earlier assessments, Amur tigers do not appear to be closely tied to Korean pine forests *per se*. Distribution of Korean pine habitat has decreased dramatically due to repeated fires over the past century and intensive harvesting in the last half century (Budzan 1996; Kolosova & Kondrashov 1996; Petropavlovski 1996). Tiger distribution was probably more closely related to Korean pine forests prior to these disturbances, but the fact that tigers have thrived while Korean pine forests have decreased in distribution and quality is evidence that tigers are not dependent on this forest type. Red deer and tigers thrive in a variety of forest types; wild boar may be more dependent on pine mast crops for winter survival (Bromley 1964).

The northernmost distribution of tigers is tied closely to the northern distribution of red deer and wild boar (Kucherenko 1985; this analysis). Red deer and wild boar are in turn linked clearly to temperate forest complexes in the Russian Far East: where spruce-fir and larch forests become dominant in the north and along the crest of the Sikhote-Alin Range (Fig. 6.2), these ungulates are rare or absent, especially in winter. Thus the distribution of this complex of parameters – temperate forests, red deer and wild boar – appear to define the northern limits of tiger distribution.

Second and third order selection (habitat selection within the eastern macroslope and within home ranges)

Our results demonstrate that the spatial scale of analysis has important consequences in interpreting habitat selection data. Spruce-fir and larch forests appeared to be selected in direct relation to abundance (neutral selection) within home ranges (third order spatial scale) (Table 6.3). Across the eastern macroslope landscape (second order spatial scale), larch still appeared to be used in relation to abundance, but spruce-fir forests were avoided. Finally, over the whole tiger range (first order spatial scale), distribution of both larch and spruce-fir forests is clearly inversely related to distribution of tigers and prey (Fig. 6.2). Although this pattern seems intuitively contradictory, it is not unlikely. Patches of larch or spruce-fir within a home range may represent unique habitats that provide parameters selected by a particular animal when they occur in limited amounts (e.g. tigress 1 gave birth to her first litter in a larch stand) but, at another scale, that forest type may be associated negatively with overall distribution. Relative abundance of habitats, at the landscape level and within home ranges, also no doubt affects use-availability analyses: Korean pine forests appeared to be avoided by radio-collared tigers when availability was low, but tigress 5 showed a slight preference for that

forest type when it represented 61% of available habitat (Table 6.3).

Hypotheses pertaining to second and third order selection relate either to selection for habitat parameters *per se* (hypotheses 1–3 second and third order, hypothesis 6 third order), or selection for habitat parameters that relate to prey distribution (hypotheses 4–5 second and third order, hypotheses 7–8 third order) (Table 6.1). We review hypotheses within these two broad categories to compare the relative importance of factors driving selection at these two spatial scales.

At the second order spatial scale, an animal should attempt to include within its home range those parameters that increase the potential for reproduction success. Indeed, tigers selected home range locations with a composition of habitat types (hypothesis 2) and habitat parameters (hypothesis 3) different from that of random points within the eastern macroslope. Habitat type (hypothesis 1, second order), elevation (hypothesis 3a), and slope (hypothesis 3c) appeared to be important variables that were incorporated into the 'decision-making process' of home range selection.

Evidence for selection of habitat parameters within home ranges (at the third level of analysis) was not as strong as that for second order selection. For instance, habitat preferences were not as pronounced at the third order, i.e. there was a stronger tendency for habitat types to be used in amounts equal to their availability within home ranges. Where selection occurred at the third order, it was often manifested in winter, or when values of parameters within specific home ranges approached those across the eastern macroslope. For example, although tigers selected for lower than average elevations across the eastern macroslope (second order selection), only one tigress preferred lower than average elevations within her home range, in winter only. Tigers avoided steep slopes in comparison to the average slope within the eastern macroslope, but preference at the third order was demonstrated only when the average slope within home ranges approached that of the entire macroslope.

The strength of the relationship between habitat

selection by tigers and habitat selection by their prey varied with the scale of analysis. At the second order of selection, there was a significant correlation in habitat selection between tigers and red deer, and between tigers and wild boar, but no correlation between tigers and roe deer. This relationship is expected given the results of the first order analysis, and the knowledge that red deer and boar are the dominant prey species (Matyushkin 1992; Miquelle *et al.* 1996b). Tigers should select home range locations that correlate with presence of prey.

At the third order of selection, while there were no significant differences between tigers and prey, absence of data for many habitat types precluded a more powerful test. We consider this analysis only a preliminary assessment, and believe that what data exist suggest that the relationship between tiger and prey habitat selection within home ranges is not strong.

It is not known whether the observed shifts in location of seasonal home ranges (hypothesis 5, second order) are a true representation of tiger movements, or an artefact of inadequate sampling. A longer term data base in the study areas (J. M. Goodrich *et al.* unpubl. data) suggests the later, and that home ranges are in most cases stable over years and through seasons. Similarly, observed shifts in central tendency within home ranges (hypothesis 5, third order) may relate to factors other than prey distribution (e.g. birthing and associated localisations). Perhaps these hypotheses are ill-defined, because it is not known whether seasonal ungulate movements occurred within tiger home ranges, or whether ungulates migrate in and out of tiger home ranges, resulting in a seasonally fluctuating density of prey for an individual tiger. Available evidence suggests that wild boar can move far greater distances than the diameter of tiger home ranges in search of winter mast crops (Bromley 1964). Less is know about seasonal movements of red deer in this region. Despite the potential for fluctuating prey abundance, available evidence suggests that Amur tigers within the study area show high site fidelity.

With the exception of one animal, tigers did not move to lower elevations in winter (hypothesis 7, third order), and did not use south slopes more in winter (hypothesis 8), as would be expected if tigers were 'tracking' distribution of prey.

Some caution is required in interpreting some of our results, especially with second order analyses. In many cases, our analyses are hampered by inadequate sample size. The five tigers incorporated in this analysis are obviously not representative of all tigers within Amur tiger range, and are unlikely to be representative of all tigers within the eastern macroslope study area. Tigers are distributed over a broader range of habitats than indicated in our analysis. Nonetheless, given that larger sample sizes for such an analysis are unlikely to be available in the near future, we believe that the attempt to understand habitat selection criteria of tigers at several spatial scales has value and important conservation implications.

We propose that prey density and distribution, and habitat parameters associated with prey, are the key factors driving first and second order habitat selection, but have only an indirect influence on third order selection. That is, geographical distribution of tigers and the locations of tiger home ranges within the landscape are determined by a set of habitat types and habitat parameters that are linked closely to distribution of prey, but within a home range there is relatively little selection for specific parameters. We believe this generality may extend across much of tiger habitat. Exceptions will occur where there are extreme environmental challenges other than prey density. For Amur tigers, selectivity within the home range is likely to be more pronounced on the western macroslope of Sikhote-Alin Mountains, especially in the northern limits of Amur tiger distribution. In these places, deeper snow and colder weather may force tigers to seek habitat parameters that mitigate those factors (Kucherenko 1985). On the eastern macroslope, moderate temperatures and relatively low snow depths probably do not impose serious constraints on tigers. In the arid Panna region of India, tigers are apparently closely tied to water (R. Chundawat *et al.* this volume), a factor that likely affects home range location (second order selection) and selection of sites within home ranges (third order selection). Therefore, while third order selection of habitat parameters within home ranges may be important in extreme conditions, we predict that under most conditions tigers will show little selection at this spatial scale. Selectivity will be reduced within home ranges by other mitigating factors.

In the absence of human-induced mortality and disturbance, the geographical distribution of all subspecies as well as the distribution and relative abundance of tigers within their range are driven by prey distribution and density; but within an individual's home range, habitat selection is driven, for females, by the need to maintain a home range that ensures cub production and survival, and, for males, by the need to obtain access to females. Neither of these constraints is necessarily related to structural components of the habitat, habitat types, or directly to prey density (see below).

Both female and male tigers in Nepal and the Russian Far East have demonstrated site fidelity and territoriality (McDougal 1977; Sunquist 1981; Smith *et al.* 1987a; this chapter, D. G. Miquelle *et al.* in prep. and J. M. Goodrich *et al.* unpubl. data). Maintenance of an exclusive home range (territory) appears to be an important component of tiger social structure (although it is unclear if territoriality is a characteristic of all tiger populations, e.g. N. Franklin *et al.* this volume), and may be a defining element of reproductive success. If this hypothesis is true (or for populations where this is true) we predict that selection of preferred habitats and habitat parameters within a home range will be tempered by the cost of maintaining the integrity of the territory. Scent-marking and 'patrolling' are costs associated with maintaining a territory (Yudakov & Nikolaev 1987; Smith *et al.* 1989; Matyushkin 1992) and may require use of areas that do not necessarily contain high prey density or include preferred habitat. We predict that this pattern should be most consistently demonstrated by females. Males may also attempt to maintain exclusive access to females by maintaining territories (e.g. Smith *et al.* 1987a), but male competition for access to mates could be expressed in a variety of ways.

If prey biomass is a key variable driving second order selection by tigers, then home range size

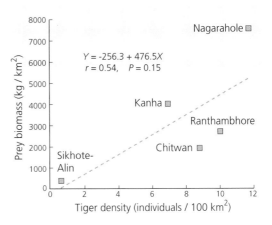

FIGURE 6.5
Relationship between prey biomass and tiger density across tiger range. Data from Karanth (1991), and Miquelle & Myslenkov (unpubl. data).

should be directly related to prey density. Across tiger range, there is insufficient information to assess this relationship. Only two 'data points' exist: in Chitwan National Park, Nepal, where ungulate biomass is nearly 2000 kg/km² (Tamang 1982), home ranges of female tigers average 20.7 km² (Smith *et al.* 1987a); and in Sikhote-Alin Zapovednik, Russia, where ungulate biomass is less than 400 kg/km², female home ranges are an order of magnitude larger (D. G. Miquelle *et al.* this volume, chapter 19). We predict that while prey density fluctuates greatly across tiger range, the total available prey biomass within the female home range size should be fairly consistent.

A related measurement does exist, indicating a clear relationship between prey biomass and tiger density across tiger subspecies (Fig. 6.5). Although this relationship is not statistically significant (r=0.54, P=0.15), given the small sample size, biases and methodological variations inherent in these datasets the relationship appears biologically meaningful.

The difference between these two relationships, tiger home range size versus prey biomass and tiger density versus prey biomass, is slight but important. An understanding of the response of a tiger population to long-term changes in prey biomass – in terms of home range size and social structure – could have important conservation implications; can home range sizes be reduced by increasing prey density (and what is the mechanism for this change), thus increasing the density of the reproductive segment of the tiger population? Karanth (1991) suggests that by increasing prey density, more tigers can be 'packed' into a given space, but an important question remains; what percentage of that increase represents a resident breeding population, versus greater numbers of transients and young (survivorship of both these population segments would presumably be higher with higher prey density)? Changes in home range size and the number of resident animals within a 'tiger conservation unit' (Dinerstein *et al.* 1997) relate directly to changes in effective population size, while a change in density has an unknown relationship with changes in effective population size.

Management implications

What are the conservation implications of this spatial hierarchical analysis? First, distribution of tigers across their geographical range appears tied to a complex of ungulate species, as has been suggested by Seidensticker (1986). But the key prey species, i.e. where management efforts should focus, may be limited; our results are in agreement with Matyushkin (1992), suggesting that red deer are the key component in the Russian Far East, contrary to earlier assessments (Abramov *et al.* 1978; Kucherenko 1985). Secondly, 'micro-management' of habitat parameters within a tiger's home range to increase suitability is probably unnecessary. While tigers may select some components of a home range over others, the important selection process occurs at the landscape level (second order spatial scale). The one exception is where there exist extreme environmental challenges which management

The formula for the future of wild Amur tigers is straightforward: protect large blocks of habitat so that tiger populations are demographically and genetically viable; give local people a reason not to poach tigers; and give local people an incentive to support higher populations of key prey species.

actions could mitigate, e.g. development of water sources in arid environments (Karanth 1991).

Riverine forests appeared to be highly preferred by Amur tigers at both the second and third order spatial scales. While riverine forests make up only a small proportion of the available habitat, their importance to tiger conservation is high; prey densities are often high in these habitats, especially in winter (Matyushkin 1992), and tigers use river valleys as travel corridors (Matyushkin 1977). Just as importantly, because riverine habitats are the first disturbed by human intrusion, they act as potential fragmentation points. Trails and roads are built along river bottoms, and forests are cleared for agricultural production and human settlements. Most of the large riverine forest complexes are gone in the Russian Far East. Continued clearing of riverine complexes may result in fragmentation of the Sikhote-Alin tiger population.

In addition to riverine forests, Korean pine and oak forests appear to be important for Amur tigers. Although not always 'preferred', these habitats are common (in conjunction with riverine forests they total over 50% of tiger habitat in Primorski Krai), they provide winter mast for prey species (pine nuts and acorns), and prey species can be maintained at relatively high densities in these habitats.

Although many of the details are still lacking, in general the conservation implications are clear – the higher the prey biomass, the more tigers can fit into a unit area (Fig. 6.5), and the less land is required for a given population size. Land requirements of Amur tigers – for individuals and populations – are vast because prey densities are low. Other subspecies of tigers could theoretically live at the same low densities as Amur tigers, but in most range countries the land base does not exist to support demographically stable populations at such low densities.

Although the hierarchical approach suggested by Johnson (1980) provides a valuable construct for understanding what variables drive habitat selection at varying levels of resolution, realisation of the importance of the prey base for a predator is certainly not new. However, the importance of incorporating the prey into a conservation strategy for tigers is only now being recognised (K. U. Karanth & B. M. Stith this volume; M. Sunquist et al. this volume). Ungulate management is an integral component of tiger conservation. Prey density is one of the critical issues in the Russian Far East today due to intensive legal and illegal harvest by humans. Similar situations exist elsewhere (Rabinowitz 1989; K. U. Karanth & B. M. Stith this volume). Biologists, conservation organisations, funding organisations, and policy makers need to focus on this issue. We need to protect habitat at the local level for ungulates as much as for tigers; we need to manage habitat for ungulates, not tigers; and we need to make responsible ungulate management worthwhile to local hunters and the local communities. It is not essential to define tiger habitat except in terms of defining it for the key prey species. The formula is straightforward: (1) protect large blocks of habitat so that tiger populations are demographically stable and genetically viable; (2) give local people a reason not to poach tigers; and (3) give the local people an incentive to support higher populations of key prey species. If the prey species are there, and poaching is minimised, the land will support tigers.

7

Prey depletion as a critical determinant of tiger population viability

K. Ullas Karanth and Bradley M. Stith

Introduction

The decline of tiger populations: perceptions and reality

Earlier optimism among conservationists about 'saving' the tiger (Mountfort 1981; Jackson 1990) has now given way to a more realistic perception that the species continues to decline over most of its range. Recent predictions that wild tigers will be 'virtually extinct' by the turn of the century (e.g. Jackson 1995) have generated broad media coverage. Despite wide public interest in tigers, causes of the tiger's decline have been poorly studied. Poaching of tigers (Jackson & Kempf 1994), loss of genetic variability resulting from insularisation of populations (Seal et al. 1994; Wiese et al. 1994), and habitat loss from agricultural expansion and development projects (Mountfort 1981; Panwar 1987; Thapar 1992) are usually cited as the major factors driving the tiger's decline. Unfortunately, the covert nature of poaching and the difficulties of monitoring tigers in the wild (Karanth 1987) make it extremely difficult to evaluate the relative contribution of different factors.

Habitat shrinkage has been a historically well-known factor responsible for tiger population declines (Schaller 1967; Mountfort 1981; Panwar 1987; Thapar 1992). However, recent assessments (Wikramanayake et al. 1998) based on forest-cover maps show that extensive stretches of potentially suitable tiger habitats still exist in most range countries. For instance, potential tiger habitat in the Indian subcontinent alone is estimated to exceed 300 000 km². Moreover, tigers are adapted to occur in a diverse range of tropical and temperate habitats (Sunquist 1981; Seidensticker & McDougal 1993). Until recently, there have been few reliable data on

the size of wild tiger populations (Karanth 1987, 1995). Nevertheless, presence–absence data suggest that tigers are either extinct or in decline even where suitable forests still exist (Karanth 1991; WCS 1995d; Wikramanayake et al. 1998). Clearly, additional causal factors are at work.

Poaching of tigers for the oriental medicine trade is widely perceived to be the chief causal factor driving the current decline of tigers (Jackson & Kempf 1994). Kenney et al. (1995) used a model of a tiger population in Nepal to argue that, in the long term, tigers are at high risk of extinction from poaching. Consequently, to reverse the tiger's decline, conservationists have advocated a primary strategy of reducing the demand for tiger body parts (Jackson & Kempf 1994). This strategy has generated considerable media attention and funding support. However, there are few data on tiger population dynamics, or on rates of illegal poaching, to evaluate this threat. In this context, we believe that a risk assessment model based on available data, which evaluates the factors driving the tiger's decline, may be more useful than dire prophecies about the big cat's imminent extinction.

Several workers involved in captive breeding (Seal et al. 1993; Wiese et al. 1994) have hypothesised that potential genetic problems arising from population insularisation may pose serious threats to tigers. Unfortunately, there are no relevant genetic data for wild tigers. Furthermore, the relative importance of genetic factors in driving demographic declines is debatable (Lande 1988; Boyce 1992; Caro & Laurenson 1994; Caughley 1994). Therefore, here we have considered a purely demographic framework of risk assessment (Ginzburg et al. 1982; Akcakaya 1991; Burgman et al. 1993).

Prey depletion is a critical determinant of tiger population viability. Where densities of ungulate prey are depressed by uncontrolled shooting and snaring, tiger populations will continue to decline, even if tiger poaching is curbed.

We thank S. Ferson for advice on the use of RAMAS/STAGE software, and J. Seidensticker, A. R. Rabinowitz, G. B. Schaller, J. D. Nichols, M. Sunquist and F. Sunquist for useful discussions about some of the ideas in this chapter; and G. Mace and P. Beier for useful comments on the manuscript.

Prey depletion: a neglected factor

Earlier studies (Schaller 1967; Sunquist 1981; Seidensticker & McDougal 1993) qualitatively described a positive correlation between tiger and prey densities. Karanth (1995) has quantitatively demonstrated this correlation in a high-density tiger population in Nagarahole, India. Recent studies (Karanth & Sunquist 1995; Miquelle *et al.* 1996b;

Karanth & Nichols 1998) suggest that abundances of tigers and other similar predators are largely mediated by densities of different-sized ungulate prey. There is ample evidence that over-hunting by humans has depressed the densities of ungulate prey over the tiger's range (Karanth 1991; Rabinowitz 1993; WCS 1995d; Miquelle *et al.* 1996b; Madhusudan & Karanth, in press). Based on above empirical evidence, we propose that prey depletion is a major factor driving the current decline of wild tigers. Clearly, if depressed prey density is a significant negative factor, reducing poaching pressure on tigers alone is not an adequate conservation response.

In the past three decades, some quantitative data have been generated on population density and survival rates in wild tiger populations (Schaller 1967; Sunquist 1981; Smith *et al.* 1987a; Smith & McDougal 1991; Rabinowitz 1993; Seidensticker & McDougal 1993; Smith 1993; Karanth 1995, Karanth & Nichols 1998) and in other large, solitary felids, such as leopards (Martin & de Meulenaer 1988; Bailey 1993) and cougars (Seidensticker *et al.* 1973; Lindzey *et al.* 1992; Laing & Lindzey 1993; Lindzey *et al.* 1994). Using these data, and insights from field studies (Karanth 1995; Karanth & Sunquist 1995, Karanth & Nichols 1998), we developed a stochastic demographic model (Boyce 1992; Burgman *et al.* 1993) of 'typical' wild tiger populations that now survive in Asia. Our model simulates the effect of two important environmental factors: poaching directed specifically at tigers, and prey depletion. Although survival and fecundity rates for tigers in relatively healthy populations are reasonably known, changes to these rates due to poaching and prey depletion are poorly known. Demographic modelling provides a useful means of evaluating the relative importance of different factors in driving the decline of tiger populations. Based on insights gained from our results, we recommend modifications to the tiger conservation strategies being presently promoted.

Methods

Modelling philosophy

Caswell (1976) distinguished two primary uses for models: gaining understanding and making predictions. Forecasting the future is perhaps the most difficult and important use of models, yet models with high predictive accuracy are quite rare. More commonly, models are used as 'assumption analysers' (R. Holthausen quoted in Bart 1995). The goal of such models is to gain qualitative understanding about the way a system works, given a certain set of assumptions. This use of models is valid and important in conservation, even in situations where key types of data are missing or poorly known (Starfield & Bleloch 1986; Rykiel 1996). By varying the unknown parameters across a plausible range, modelling can provide a better understanding of the importance of different parameters to system behaviour.

Thus, the goal of our model is to understand the relative importance of several factors that influence the demographics of tiger populations. The rates these factors obtain in the wild and their corresponding effects on demographics are extremely difficult – if not impossible – to measure. Nevertheless, we can estimate the range that these factors are likely to have in the field. Modelling provides a way to analyse how sensitive demographic performance is to these different factors, given the reasonableness of the model assumptions and the plausibility of the parameter values. Our model seeks understanding rather than predictive accuracy, and we acknowledge that this understanding may need to be revised in the face of new data or better models. Nevertheless, our models provide quantifiable and repeatable means of explaining and evaluating assumptions, generating new hypotheses and guiding future research.

Model assumptions: population size

We simulated a range of population sizes that are characteristic of wild tiger populations that now survive in increasingly insular forest patches. Tiger population densities (excluding cubs <1 year old) ranged from 4.1 to 16.8 animals/100 km² at some protected sites in India (Karanth & Nichols 1998). We assume that breeding tigresses have stable, exclusive home ranges (Sunquist 1981; Smith *et al.* 1987a; Smith 1993), and that their densities are positively correlated to densities of ungulate prey (Seidensticker & McDougal 1993; Karanth 1995).

Breeding female range sizes vary from 10 to 15 km² in prime habitats with prey densities of 25–75 ungulates/km² (Sunquist 1981; Smith *et al.* 1987a; Karanth & Sunquist 1992; K.U. Karanth unpubl. data). Although home ranges of breeding tigresses are likely to be much larger in tropical evergreen, mangrove and temperate forests because of lower prey densities (Seidensticker & McDougal 1993), larger blocks (5000–10 000 km²) of these habitat types remain (Wikramanayake *et al.* 1998). The density and home range estimates shown above suggest a carrying capacity (*K*) of 66–100 breeding females for a 1000 km² area of high-prey-density habitat in the Indian subcontinent. Assessments by Wikramanayake *et al.* (1998) show that protected habitats that contain tigers range in size from 488 to 7013 km² in the Indian region. However, only smaller patches of high prey density now remain within such larger tiger habitat matrices (Karanth 1991; Wikramanayake *et al.* 1998).

Based on the above considerations, we modelled a range of typical wild tiger populations with carrying capacities set between K=24 breeding females (to represent a fairly large population), down to K=3 (for a very small population). Our 'typical' large population size (K=24) is very similar to the estimate of Kenney *et al.* (1995, p. 1128), '25 female territories represents the approximate size of most tiger populations'. These simulations provided a base set of models of 'normal' tiger populations. We examined the possible effects of tiger poaching and prey depletion on these base models by depressing survival rates of age classes affected by tiger poaching or prey depletion.

Model structure

For large felids, survival and fecundity data are available only for demographic 'stages' (such as breeders, transients, juveniles and cubs), rather than for specific age classes. Therefore, we developed a stage-based population model (Lefkovitch 1965; Caswell 1989) using the program RAMAS/STAGE (version 1.4, Ferson 1991). We developed our own set of equations rather than use the built-in matrix projections provided in RAMAS/STAGE. By incorporating a carrying capacity, K, in our equations, we avoided using the density-dependent functions available in RAMAS. Our model projects tiger population size, structure, mortality and quasi-extinction risk (Burgman *et al.* 1993) in a 100-year time frame, using yearly time steps. Because of lower survival rates assumed for males based on field data (Smith 1993), they had higher extinction risks than females. Thus, males were the limiting sex in our model. We ran 1000 simulations of each scenario.

The model has four stages for both sexes: breeder, transient, juvenile and cub. In our model,

Table 7.1. *Annual survival probabilities used in the model of the normal tiger population*

Demographic stage	Age (years)	Survival rate
Breeding females	3+	0.90
Breeding males	3+	0.80
Transient females	2+	0.70
Transient males	2+	0.65
Juveniles	1–2	0.90
Cubs	0–1	0.60

tiger cubs either die in their first year or transit to the juvenile stage. At two years, juveniles either die or become transients. Transients may die in a particular yearly time step, or survive and continue as transients, or transit into the breeding stage. Breeders either survive a particular time step, or die creating vacancies that some of the transients can fill. We parametrised our model using survival rates (Table 7.1) based on field data on tigers (Sunquist 1981; Smith *et al.* 1987a; Smith & McDougal 1991; Smith 1993), leopards (Martin & de Meulenaer 1988; Bailey 1993) and cougars (Lindzey *et al.* 1992; Laing & Lindzey 1993; Lindzey *et al.* 1994).

Female ranges are stable in wild tiger populations, and vacancies among breeders are rapidly filled by new tigresses capable of quick reproduction (Smith *et al.* 1987a; Smith 1993; K. U. Karanth unpubl. data). Therefore, K is treated as an invariant in our model. Our tiger population model includes animals of both sexes in four demographic stages: cubs (younger than one year), pre-dispersal juveniles (between one and two years), post-dispersal, non-breeding transients (older than two years) and breeding residents. Although the male:female ratio appears to be 1:3 among breeders (Sunquist 1981), the number of males fluctuates due to intraspecific competition (Smith & McDougal 1991; Kenney *et al.* 1995). Therefore, we assume the number of breeding males to be a Poisson variate in our models, with a mean value of K/3. Based on field data (Sunquist 1981; Smith & McDougal 1991; Kenney *et al.* 1995; K. U. Karanth unpubl. data), we assumed an equal sex ratio at birth, mean litter

Box 7.1   Counting the tiger's prey, reliably
       K. Ullas Karanth

The decline of tiger populations can be primarily driven by the loss of their prey base due to over-hunting by humans (K. U. Karanth & B. M. Stith this volume). Therefore, enhancing and monitoring the tiger's prey base is perhaps the single most important task facing wildlife managers across Asia. Although tigers kill a wide range of prey types (Karanth & Sunquist 1995), most of their biomass needs are met by large (>20 kg) ungulate species. Cervids, bovids and suids form the principal wild prey of tigers (M. Sunquist et al. this volume).

Although the issue of counting tigers has received considerable attention, counting their prey has been almost neglected. Most of the conceptual issues of population estimation (Lancia et al. 1994) outlined in relation to counting tigers (K. U. Karanth this volume Appendix 5) apply to counting prey species also. Even in this case, we are interested in estimating the sampling fraction p (see Appendix 5), so that we can estimate population size N from the sample count C using the equation:

$$N = C/p$$

Once we have estimated N within the effectively sampled area (A), then the population density D (e.g. number of sambar deer/km²) can be estimated. Using the estimated density, population size over regions larger than the sampled area can be estimated.

However, there is one major difference. Tigers are individually 'marked' animals. Although we cannot see them easily, we can photographically 'catch' them through camera traps and estimate capture probabilities. Prey species are not individually identifiable and, therefore, not amenable to capture-recapture sampling. Fortunately, because the vision and hearing of ungulates are relatively poor, we can see and count them. However, because of the screening effect of the vegetation in forested habitats, it is not appropriate to count tiger prey using 'block counts', 'total counts', or other methods that assume that all animals in the sampled area are seen (Lancia et al.

1994). Consequently, we still face the task of estimating the sampling fraction p, but using sighting probabilities rather than capture probabilities.

Line transect survey is the preferred formal population estimation method based on visual detection of animals (Buckland et al. 1993; Lancia et al. 1994). The method involves observers moving along straight trails (transects), counting animals seen on either side. Additionally, using range finders and compasses, the observer measures the distance from the transect to the animals which are seen. During analysis, these count and distance data are used to estimate sighting probabilities. Transect data can be best analysed using computer programs that apply stochastic models to estimate animal densities based on these detection probabilities. Although the mathematical complexity underlying distance sampling models may appear formidable, essentially these models compute the sampling fraction (p) to estimate N (in the sampled strip) from the field counts (C).

Table 7.2 shows results from line transect density estimation carried out at Nagarahole, India (Karanth & Sunquist 1995). The field data came from six teams, of two observers each, who walked on six permanent transects in the mornings and evenings, sampling a total distance of 468 km over a 13-day period. They recorded, at each animal sighting (detection), the species, number of animals, sighting distance and sighting angle (Buckland et al. 1993). The number of sightings (n), the estimated densities D and their standard errors (SE[D]) were derived using the computer program DISTANCE (Laake et al. 1993) for six prey species.

If it is not practical to walk on the transects (as in the tall grasslands of Kaziranga, India or Chitwan, Nepal), riding elephants can be used as mobile platforms. Generally, it is not appropriate to treat roads as transect lines to count ungulates. Roads are not straight lines, and animals may be non-randomly distributed along roads for ecological reasons. These factors lead to violations of fundamental assumptions underlying probabilistic line transect estimation (Buckland *et al.* 1993).

Table 7.2. *Densities of ungulate prey estimated using line transect surveys in Nagarahole, India, during 1989 (Karanth & Sunquist 1995)*

Prey species	Sample size (*n*)	Animal density *D* (SE[*D*])
Chital	356	52.3 (5.24)
Sambar	60	3.4 (0.68)
Gaur	55	4.4 (0.89)
Wild pig	64	3.0 (0.54)
Muntjac	68	3.9 (0.59)

Sample size *n* is the number of sightings. Mean densities and their standard errors are reported in number of animals/km².

Line transect methods have some limitations. For efficient density estimation, more than 40 separate sightings (not individual animals) of each species may be needed (Buckland *et al.* 1993). The effort required to obtain such samples will be high, if animal densities are low. Distance estimation requires equipment such as range-finders and compasses, and reasonably skilled persons to use them. How do we count prey reliably if these requirements are not met? Once again, rather than getting density estimates, we have to settle

for indices of abundance based on signs of prey animals. Counts of standing crops of pellets or dung (dung density, e.g. number of sambar pellet groups/100 m²) deposited by prey species are useful indices of the relative abundance of prey species. Although dung density can also be estimated from transects using probabilistic distance sampling, if the plots are linear and narrow (1–2 m wide) we can ensure total counts of dung within the strip and obtain densities through simpler computations.

If dung density indices are to be used to compare prey abundance at a site over time, or among different sites, it is necessary to standardise sampling procedures to achieve constancy of the sampled fraction (Lancia *et al.* 1994). The dung/pellet plots or transects must representatively sample the habitat. The dung identities (species), and definition of what constitutes a pellet group or dung pile, must be established at the outset. Counts must be made at the same time of the year, so that dung decay rates do not vary greatly. Indices based on dung counts are useful to monitor population trends over time, or to compare different sites. However, to do so, ensuring that dung decay rates are also comparable is critically important. Although many investigators measure dung decay rates after doing the counts, Hiby & Lovell (1991) argue convincingly that it is better to calibrate decay rates before the counts are done, or through replicated counts during each dung survey.

size of three and an interbirth interval of 2.5 years. We treated the litter size as a Poisson variate to simulate demographic stochasticity (Akcakaya 1991; Burgman *et al.* 1993).

Three important variables in our model are; number of breeding tigresses (K), tiger poaching factor (P_1) and prey depletion factor (P_2). We modified the basic survival rates (Table 7.1) of the relevant demographic stages by multiplying these rates by the values set for factors P_1 and P_2. For example, a value of P_1 or P_2 set at 0.90 results in a 10% depression in survival rates of the affected

stages. We first investigated the effect of different values of K (24, 12, 6 and 3) on population persistence in the absence of either tiger poaching or prey depletion ($P_1=P_2=$ 1.0). These simulations provided a base set of population trajectories and extinction/quasi-extinction risk curves (Burgman *et al.* 1993), for tiger populations of different sizes.

Mortality sources and survival rates

Three factors affecting tiger survival rates were considered in our model. These are background mortality, poaching mortality and prey-depletion-related mortality. We refer to tiger deaths resulting from natural violence, disease, starvation and dispersal into unsuitable habitat as background mortality. Such mortality occurs among all normal, healthy tiger populations, even when poaching is absent (Kotwal 1984; Smith 1993; K. U. Karanth unpubl. data). These normal background survival rates used in the basic model (Table 7.1) were modified by varying P_1 or P_2, to simulate the effects of tiger poaching or prey depletion.

The mortalities of juvenile and adult tigers resulting from systematic poaching (Jackson & Kempf 1994; WCS 1995d) through shooting, poisoning, trapping, snaring and electrocution are distinct from the background mortality. Although cubs are not usually targeted by poachers, killing of breeders may sometimes result in social instability, infanticide and depressed reproduction (Smith 1993; Kenney et al. 1995). However, such effects also arise following natural replacement of resident males in unhunted tiger populations (Smith 1993; K. U. Karanth unpubl. data). Consequently, although poaching may increase the frequency of episodic social instability as argued by Kenney et al. (1995), we point out that it may simultaneously reduce intraspecific competition, thus counteracting its own destabilising effects.

Even though poaching mortality is likely to be compensatory to some degree (Martin & de Meulenaer 1988), for modelling purposes we have, conservatively, treated it as being completely additive to background mortalities. Organised poaching targets the juveniles, transients and breeders of both sexes. We depressed the basic survival rates by 7%, 10% and 25% for these stages, respectively, to simulate low, moderate and severe poaching pressures.

Comparison of data from different tiger habitats suggests that home range sizes of breeding females are strongly and negatively correlated with prey densities (Sunquist 1981; Smith et al. 1987a; E. N. Smirnov & D. G. Miquelle this volume; K. U. Karanth unpubl. data). Therefore, prey depletion can be expected to result in lowered carrying capacities (K) for breeding females. How are reduced prey densities likely to affect female home range sizes and K? Prime areas in India and Nepal support prey densities of around 25–50 ungulates/km², with female range sizes of 10–15 km² (Sunquist 1981; K. U Karanth unpubl. data). At the other end of the spectrum, in the Russian Far East, female ranges are around 200 km², at prey densities lower than five ungulates/km². For our normal tiger population in a 1000 km² area (female home range size of 42 km²; $K=24$), we have assumed prey densities as being 20% lower than in prime areas. We have assumed a direct relationship between prey depletion and K. With a moderately depleted prey base (50% lower than at $K=24$), we have assumed a female home range of 84 km² and $K=7$. To model severe prey depletion (75% lower than at $K=24$), we have assumed a female home range size of 168 km² and $K=6$.

Because prey depletion lowers the tiger's encounter rates with prey, we expect lowered hunting success, greater energy expenditure per kill, and increased foraging movements. These ecological consequences will result in poorer nutrition, increased intraspecific competition at kills and reduced attention towards cubs. Therefore, prey depletion is likely to affect survival rates of tigers in all demographic stages. A tigress with three cubs has to kill about 60–70 ungulate prey within her home range every year to successfully raise her cubs (Sunquist 1981). Clearly, lowered prey densities will depress cub survival rates substantially. For modelling purposes, conservatively, we have assumed that prey depletion lowers survival rates of only the tiger cubs, not of other stages. Since no data are available on the effect of prey depletion on cub survival, we simulated a range of potential effects, respectively, by lowering cub survival rates by 33% ($P_2=0.67$), 50% ($P_2=0.50$) and 67% ($P_2=0.33$).

Results

We recognise some limitations of the demographic model we have developed. This model deals with a single species (tiger), while a more realistic approach would be to build a multi-species (tiger–other large predators–ungulates) model. Simulation of the potential density-dependent responses of tiger fecundity and survival rates to poaching and prey depletion awaits further refinements of the model, although the absence of this refinement probably overemphasises the impact of tiger poaching and underemphasises the impact of prey depletion in our current model. We are currently working on incorporating better parameter estimates for prey density, tiger density and survival rates, using data now being generated from the ongoing field studies of the first author with the more refined models being developed by the second author. However, we believe that these developments are unlikely to radically change the thrust of our arguments presented in the following parts of this chapter, because of the rather conservative assumptions underlying our current 'first cut' model.

Extinction risks

Figure 7.1 shows quasi-extinction risk curves (Ginzburg *et al.* 1982; Burgman *et al.* 1993) for normal tiger populations with survival rates unaffected by poaching or prey depletion, but with different carrying capacities of breeding females ($K=3$, 6, 12, 24). The quasi-extinction probability curves are shown only for male tigers, which are the limiting sex according to the model assumptions. Populations with $K=24$, $K=12$ and $K=6$ breeding females show no extinction risk, while those starting with $K=3$ tigresses show a 29% probability of extinction (Fig. 7.1).

Population trajectories

Figure 7.2 shows population trajectories for six modelling scenarios, all starting with $K=24$ (i.e. no reduction in carrying capacity). These trajectories depict the total number of tigers in the age categories targeted by poachers (adults and juveniles). The simulations include the basic model of tiger populations unaffected by poaching or prey depletion ($P_1=1.0$; $P_2=1.0$), as well as of the populations subjected to three levels of tiger poaching ($P_1=0.93$, 0.90 and 0.75). Additionally, trajectories

for populations subjected to two levels of prey depletion ($P_2=0.50$ and 0.33) are shown.

The basic population, starting with 64 tigers ($K=24$), shows a stable mean trajectory for 100 years. Under low levels of tiger poaching ($P_1=0.93$), the population shows a similar stable trajectory. A fairly small increase in poaching pressure ($P_1=0.90$) produces a moderately declining mean trajectory. Under severe, continued poaching pressure ($P_1=0.75$), the same initial population declines rapidly to extinction. Under the scenarios simulating prey depletion, but only through reduced cub survival rates, a moderate level of depletion ($P_2=0.50$) produces a slightly declining mean trajectory, while a higher level ($P_2=0.33$) produces a moderately declining mean trajectory.

The trajectories in Figure 7.2 are for simulations that do not include reduced carrying capacities. Figure 7.3 shows mean trajectories for the more likely scenarios of combined effects of reduced number of breeding females (K) and lowered cub survival, resulting from prey depletion. At $K=12$, the projections show intermediate and strongly declining trajectories, for $P_2=0.50$ and $P_2=0.33$, respectively. At $K=6$, the trajectories decline at an intermediate rate for both $P_2=0.50$ and $P_2=0.33$. The base runs ($P_2=1.0$) for $K=12$ and $K=6$, which are not shown to avoid clutter, both show stable population trajectories.

In Figure 7.4, we have summarised probabilities of extinction as a function of a range of prey depletion factor values (P_2). Figure 7.4 summarises the extinction probabilities for all runs involving depletion (P_2), at different carrying capacities. At $K=24$, the extinction risk is significant only when prey depletion is very severe ($P_2=0.33$; cub survival rates depressed by 67%). However, this is an unlikely scenario, because prey depletion is also accompanied by reduction in K. When $K=12$, extinction risks are significant even at $P_2=0.50$ and quite high at $P_2=0.33$. When carrying capacity is reduced to six and three breeding females, the extinction risks are quite high even for low levels of prey depletion ($P_2=0.67$) and rise very sharply as prey depletion effects become more severe (cub survival is depressed further).

Discussion

Tiger population size and viability

The quasi-extinction probabilities generated using field-derived vital rates for normal tiger populations (Fig. 7.1) suggest that even small, insular populations (e.g. with six breeding tigresses) have a low probability of extinction. This capacity for population persistence is basically a function of the high reproductive potential of tigers, which is a pattern typical of large felids (Martin & de Meulenaer 1988; Lindzey et al. 1992; Laing & Lindzey 1993; Lindzey et al. 1994). We caution that our model does not include several other factors that could affect tiger population dynamics. Such factors, for which no data exist, include environmental stochasticity, catastrophes, disease and genetic deterioration, some of which may substantially increase extinction probabilities projected in Figure 7.1.

Effect of tiger poaching

There is empirical evidence that hunting mortalities among big cats may not depress their densities if hunters remove less than 10–25% of the total population annually (Lindzey et al. 1992, 1994). Martin & de Meulenaer (1988) argue that hunting can drive big cat populations into rapid extinction only if it exceeds threshold levels set by habitat quality and reproductive potential of the species. The work of Lindzey et al. (1992) suggests that typical cougar populations can sustain annual removal of 10–20% of all animals over one year of age. Our results suggest that these rates might also be relevant to tigers, unless hunting mortality is completely additive to background mortalities.

Martin & de Meulenaer (1988) argued that the effect of hunting mortality in leopards is likely to be largely compensatory rather than additive. Estimates from our basic model ($K=24$) show that, normally, of all tigers older than a year, about 20–25% die annually as a result of intraspecific aggression, hunting injuries, starvation and accidental death during attempted dispersal (background mortalities). Consequently, specifically directed tiger poaching may substitute some of these deaths, thus having a partially compensatory rather than completely additive effect. Also, poaching of any tiger from a population in which there is intense competition for limited slots is likely to improve survival probabilities of other tigers. Therefore, tiger poaching mortalities are unlikely to be totally additive to background mortalities.

In reality, as tigers become scarce, poaching mortalities may decline due to decreased hunter success (even with increased effort), or even due to reduced hunting effort in response to diminishing returns. Realistically, tiger survival rates under the poaching scenarios may fluctuate rather than stay depressed continuously. This argument is supported by historical evidence (Schaller 1967; Rangarajan 1996) of large-scale hunting mortalities repeatedly inflicted on tiger populations in several parts of India and Nepal. In such cases, it appears that when hunting pressure slackened, tiger survival improved, and remaining small populations of breeders quickly rebounded from episodic slaughters.

Our model makes the conservative assumption that tiger poaching mortality is completely additive rather than compensatory. Nevertheless, the results suggest that modest-sized tiger populations ($K=24$) can sustain low levels of poaching indefinitely (Fig. 7.2). Although higher levels of poaching may be sustained for short periods of time, our model does not directly address this question.

Effect of prey depletion

When the effect of prey depletion is simulated (Fig. 7.3), the carrying capacity for breeding females (K) is depressed, cub survival is reduced and the tiger population size declines rapidly. Extinction risks are substantial for nearly all the prey depletion scenarios (Fig. 7.4). This sensitivity is due to the additional effect of lowered cub survival rates, and not simply due to lowered K. We emphasise here that, although we have conservatively assumed that prey depletion has no effect on survival rates of juveniles and adults, this is almost certainly not true for wild tiger populations. Juvenile tigers are nutritionally dependent on their mother to a significant extent (Smith 1993). Even among the adults,

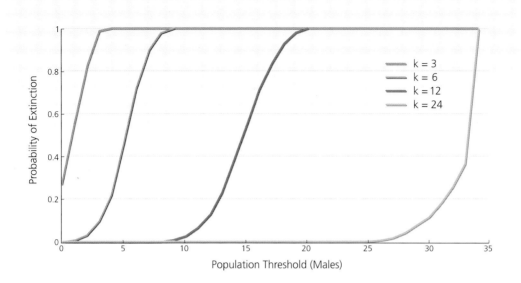

FIGURE 7.1
Probabilities of extinction for tiger populations unaffected by either poaching or prey depletion, at different carrying capacities. Threshold population sizes are shown for male tigers, which are the limiting sex in the model.

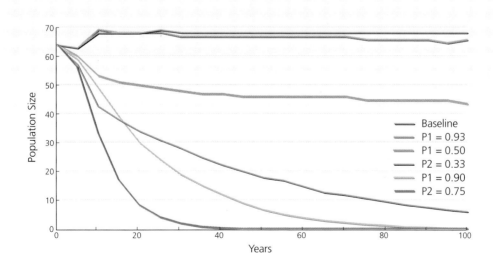

FIGURE 7.2
Trajectories of tiger population size (excluding cubs) for simulations that show the basic population (K=24 breeding females), as well as the effects of different levels of tiger poaching and prey depletion on the basic population.

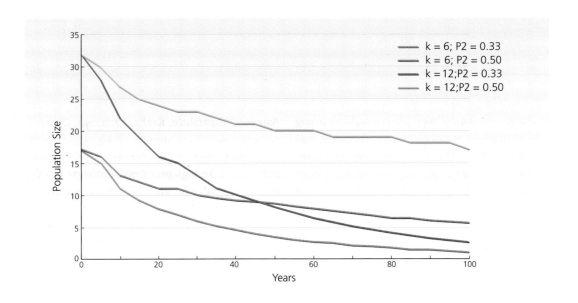

FIGURE 7.3
Trajectories of tiger population size (excluding cubs) for simulations that show the combined effects of reduced carrying capacities (*K*=12, *K*=6) and reduced cub survival rates, resulting from different levels of prey depletion.

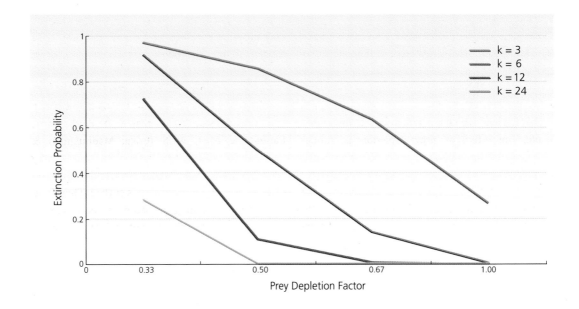

FIGURE 7.4
Probabilities of extinction for tiger populations at different carrying capacities (*K*=24 to *K*=3), affected by different degrees of prey depletion.

prey scarcity will affect nutrition and force movement or dispersal into unsuitable habitats with increased mortality risks. Therefore, in reality, the effects of prey depletion are likely to be much more severe than those depicted by our conservative modelling scenarios.

Our results show that prey depletion has a strong effect on tiger population dynamics. Therefore, prey depletion may be a major factor driving the current decline of wild tiger populations and hence a significant constraint on their recovery. The absence or scarcity of tigers over extensive forest areas over several decades (Karanth 1991; Rabinowitz 1993; WCS 1995d), even where organised tiger poaching was not common, may be largely due to lowered prey densities. Therefore, prey depletion needs to be explicitly recognised as a threat distinct from physical shrinkage or structural degradation of tiger habitat. Our results suggest that Geographic Information System (GIS) analyses, which rely on forest-cover maps to predict the distribution and abundance of wild tigers, may have poor predictive value if they do not include specific data overlays of prey abundance.

Conservation implications

If prey depletion is such a critical determinant of tiger population viability, the question arises as to why this factor has been largely overlooked by conservationists until recently. There are several possible explanations. First, unlike tiger poaching and habitat loss, which are both dramatic and highly visible, the effect of prey depletion is subtle. Depression in prey densities goes unnoticed in the absence of annually replicated prey density surveys. Appropriate 'distance sampling' methods are available for such surveys (Burnham *et al.* 1980; Buckland *et al.* 1993) and have been used by researchers to estimate densities of tiger's prey (Karanth & Sunquist 1992, 1995; Srikosamatara 1993). However, wildlife managers and policy makers in Asia have generally ignored such systematic estimation of prey densities, possibly because the work is tedious and unglamorous. Instead they have favoured direct 'censusing' of tigers, using demonstrably failure-prone techniques (Karanth

1987, 1995, this volume Appendix 5).

Our results suggest that tiger populations can persist in relatively small reserves (300–3000 km²), even in the presence of low-level poaching, provided their prey base is protected and maintained at adequate density. There appear to be several areas in Asia (e.g. Nagarahole, Kanha, Kaziranga, Chitwan, Gunung Leuser, Taman Negara, Huai Kha Khaeng) that could support demographically viable tiger populations. In high-prey-density habitats like the alluvial grasslands and deciduous forests of southern Asia (Karanth & Sunquist 1992; Seidensticker & McDougal 1993), the necessary protected areas could be as small as 300 km², whereas in prey-poor habitats such as mangrove, evergreen or temperate forests, they may exceed 3000 km².

The potential for sustaining small but productive tiger populations depends primarily on maintaining high prey densities. Large ungulates (50–1000 kg), which are the principal prey of tigers (Schaller 1967; Sunquist 1981; Seidensticker & McDougal 1993; Karanth & Sunquist 1995) attain high densities even in successional and disturbed tropical forests (Eisenberg & Seidensticker 1976; Karanth 1991; Karanth & Sunquist 1992). These data suggest that physical shortage of habitat may not be a constraint on tiger population recovery in most areas.

Densities of ungulate prey are depressed primarily because of uncontrolled shooting, snaring and trapping by local people (Karanth 1991; Rabinowitz 1993; WCS 1995d; Madhusudan & Karanth in press). In such areas, unless prey densities increase, tiger populations will continue to decline, even if tiger poaching is curbed. Therefore, reduction of hunting pressure on prey species must be the central component of tiger population recovery plans. Specifically, in critical tiger conservation areas, it is essential to legislate and implement a ban on hunting of principal prey species. Although strategies to reduce illegal killing of prey species may vary among locations, armed patrols to deter hunters and to eliminate snares, traps, pitfalls, baits and other devices set by them may be necessary almost everywhere. Because tigers

are also often killed using similar techniques, such anti-prey-hunting activity will simultaneously reduce poaching of tigers. Even if some tigers continue to be poached on the edges of such small but productive populations, the probabilities of population persistence will be reasonably high.

Establishing and maintaining small, high-prey-density tiger reserves appears feasible as demonstrated by Indian and Nepalese wildlife managers during the 1974–1990 period (Mishra *et al.* 1987;

Panwar 1987; Karanth 1991; WCS 1995d). Although such a strict preservationist approach will be impossible to implement over the entire tiger range, protecting several reserves of 300–3000 km² in size still appears to be feasible. Identifying, protecting, maintaining and monitoring such prey-rich enclaves, embedded within larger landscape matrices under multiple uses, should be the central concerns of any future tiger population recovery efforts.

8

Long-term monitoring of tigers: lessons from Nagarahole

K. Ullas Karanth, Mel Sunquist and K. M. Chinnappa

Introduction

The last of the Javan tigers disappeared in the 1970s. Seidensticker (1987) bore witness to the death throes of the subspecies, which was down to three to four individuals, and found it difficult to isolate any single factor that pushed it over the brink. Once tigers are down to such small numbers, environmental factors, demographic stochasticity, or genetic deterioration (Burgman *et al.* 1993) will surely push them over the edge. Therefore, the definitive answer to the question of what finished off the last Javan tigers may be largely irrelevant. Of greater concern is the question of whether we can effectively recover some of the small wild tiger populations now surviving in different parts of Asia before they reach such low levels. If so, what are the critical components of a recovery strategy for such tiger populations?

Within the international conservation arena, during the last decade, the earlier ethos of preservation has been replaced by a new paradigm that promotes 'sustainable' human uses of wildlife landscapes. A strong component of biological resource extraction by local people is seen almost as a prerequisite for conserving biodiversity (IUCN *et al.* 1991; UNEP 1996). Some scholars argue that this paradigm of 'sustainable use everywhere' is an appropriate framework for recovering wild tiger populations also (Saberwal 1997). However, they offer little empirical or theoretical evidence to support this position (Karanth & Madhusudan 1997).

Between 1970 and 1990, wildlife managers of India and Nepal implemented a preservationist policy, establishing many effectively protected nature reserves (Karanth 1987, 1991; Mishra *et al.*

1987; Panwar 1987). Their management focus was on reducing human and livestock population densities within protected areas, and on stopping or regulating biomass removal in the form of hunting, grazing, wood harvesting, burning, etc. Although it is recognised that in many reserves in India tiger populations recovered significantly following such interventions (Mountfort 1981; Panwar 1987), the process of this recovery was not examined carefully in an objective ecological framework (Karanth 1987). Consequently, real world lessons from these large-scale tiger population recovery 'experiments' carried out in India and Nepal are not clearly understood even within the conservation community.

In this chapter, we have tried to document critical factors underlying the actual recovery of a small tiger population witnessed in Nagarahole, India, during 1970–96. As expected from observations from an 'unplanned experiment' extending over three decades, the quality of our data is variable. We support our arguments based on observations of the third author who managed a part of Nagarahole from 1969 to 1992, periodic visits by the first author between 1967 and 1996, and interviews with informants. We also substantiate our observations through *post-hoc* extrapolations of quantitative ecological data on tigers and their principal prey species (Karanth & Sunquist 1992, 1995), work which is currently being continued by the first author.

We first present an overview of the historical factors that affected tigers and their habitats in the Nagarahole region. Secondly, we chronicle conservation interventions that appeared to significantly influence the tiger population recovery. Thirdly, we present data on the status of tigers and their principal prey in Nagarahole after 1986, and, through these, provide our best guess at the trajectory of the recovery process. Finally, we discuss what we believe to be the lessons from the Nagarahole experience, relevant to the task of recovering and maintaining free-ranging tiger populations in the overcrowded, market-driven, fragmented landscapes of twenty-first century Asia.

The first and second authors would like to acknowledge the funding support received for their long-term studies in Nagarahole from Wildlife Conservation Society, New York; US Fish & Wildlife Service, Washington DC; and US–India Fund, New Delhi. We are grateful to the Karnataka State Forest Department and the Ministry of Environment and Forests, Government of India, for facilitating our work in Nagarahole. We have benefited greatly from discussions with the late K. P. Achaiah, the late A. Chengappa, A. C. Lakshmana, S. Shyam Sundar and M. D. Madhusudan. We thank Sriyanie Miththapala, Sarah Christie and John Seidensticker for useful suggestions. We are indebted to all of them.

The ecological and social setting

Land, vegetation and the tiger-prey community

The following brief summary is based on earlier detailed accounts (Pascal *et al.* 1982; Karanth & Sunquist 1992; Madhusudan & Karanth, in press). Nagarahole National Park (644 km² area; 11° 50' to 12° 15' N. latitude and 76° 0' to 76° 15' E. longitude) is located in Karnataka State, India, at an elevation of 700–960 m. It has a tropical climate with mean monthly temperatures between 20°C and 27°C and an annual monsoon rainfall of 1000–1500 mm. Many streams and two rivers drain the tract. The natural vegetation of the region (Pascal *et al.* 1982) consists of tropical moist-deciduous forests where rainfall is higher, and tropical dry deciduous forests where it is lower. Between 1870 and 1980, about 14% of the area was clear-cut to raise monocultures of teak. Wherever such teak plantations have failed, dense secondary forests now occur. Both the moist and dry deciduous forests have been selectively logged until recently. Low-lying areas contain grassy swamps called *hadlu*. Outside the park boundaries,

the landscape is covered by a vegetation matrix consisting of government-owned forests and private lands under coffee or rice in the moist zone, and a variety of dry crops in the low rainfall zone (Pascal *et al.* 1982).

Nagarahole is one of the richest wildlife areas in India, being particularly noted for its intact assemblage of seven large ungulate species (muntjac, chital, sambar, chousingha, gaur, wild pig and elephant) and three large predatory carnivores – tiger, leopard and dhole (Karanth & Sunquist 1992, 1995).

Local people and hunting patterns

Historical evidence shows that hunter-gatherer tribes and agricultural enclaves have been present within the boundaries of the present park for several centuries. British colonial administration established hegemony over the area in the 1850s (Stebbing 1929), after subduing the local king. At present about 1500 tribal people live in the interiors of the park, and 5000 more on its fringes. These tribal people are *Jenu Kuruba*, *Betta Kuruba* and *Yerava*; many more of them live outside the park as labourers or tenants of land-owning castes. The tribal groups were originally hunter-gatherers, later switching to slash and burn farming and, subsequently, to collecting non-timber forest products for sale to urban and industrial commercial markets. Their diet and life styles now fully reflect this integration with the market economy. The tribals are mostly poor, landless and socially dominated by other caste groups. Currently, their livelihood is derived from wage labour in coffee plantations or farms outside Nagarahole and, seasonally, from work provided by the park management. Many tribal men are skilled in the use of an array of traditional hunting techniques, as well as shotguns. Because of their familiarity with the interior of the park, tribals are preferred as guides by other hunters (Madhusudan & Karanth, in press).

Outside Nagarahole, the land-owning *Kodava*, being former warriors, have a strong cultural tradition of hunting. Colonial rulers exempted loyal *Kodava* from the stringent Indian gun control laws, resulting in a high density of legal guns around

Nagarahole. Since 'game' hunting became illegal in 1974, most literate and wealthy *Kodava* have stopped hunting. However, some of them lend guns to illegal hunters to obtain wild meat. People belonging to other land-owning castes also sometimes illegally hunt wildlife using their 'crop protection guns' (Madhusudan & Karanth, in press).

A majority of the population outside the park are workers in the coffee estates or farms. Workers are a heterogeneous group of poor people, comprising tribals, other 'lower castes', and migrant labourers from the neighbouring districts. Those who hunt among them either borrow guns or use wire snares. The enterprising petty tradesmen who own small eateries, shops, liquor stores, etc. also influence local hunting patterns. Their legitimate businesses sometimes cover up ancillary illegal trade in wildlife or non-timber forest products. Although a variety of animals are hunted by local people, large ungulate species (except elephants) are preferentially killed (Madhusudan & Karanth, in press).

Management history

From the 1870s onwards, forests of the Nagarahole region were gradually demarcated and notified as government-owned Reserved Forests by the British. The Forest Department was set up in the 1860s to manage these Reserved Forests, with the authority to prohibit slash and burn agriculture by tribal groups, to control encroachments by settled farmers, and to sustainably harvest valuable timber trees. The long-term management goal was to replace natural forests with the more profitable teak monocultures (Stebbing 1929; Somaiah 1953). The colonial foresters pursued these goals until Indian independence in 1947. Between 1947 and 1955 the new Indian government's policy was to maximise production of timber and food from the land. This policy encouraged large-scale immigration of tribal and non-tribal people from surrounding areas into Nagarahole's *hadlus*, to cultivate rice and to provide cheap logging labour. Throughout this period there were effectively no wildlife protection laws, and bounty hunting of predators (tiger, leopard and dhole) was actively encouraged. The intensity of hunting increased with the availability of better

guns, electric flashlights and motor vehicles, and with the liberalised issue of crop-protection guns.

According to anecdotal accounts (K. P. Achaiah & A. Chengappa, pers. comm.), the wildlife populations declined rapidly. A single bounty hunter, who lived about 30 km away from the present boundaries of the park, killed 27 tigers between 1948 and 1964 without ever entering the park. According to several informants, by the early 1950s the abundance of ungulates and tigers in Nagarahole was 'very much lower' than at the time of writing.

In 1955, following new government initiatives to preserve wildlife, the northern part of the present park was notified as a 285 km^2 'Game Sanctuary' in which hunting of large mammals became illegal. Although logging and other habitat perturbations continued, licensed hunting was stopped. After 1967 a separate wildlife protection wing was gradually established within the Forest Department. However, the anti-hunting laws were weak and poorly enforced against illegal hunters. Following the introduction of tough new wildlife protection laws in 1974, anti-poaching efforts became more effective. Special wildlife protection staff were deployed and anti-poaching camps and patrolling systems were introduced.

In 1974, Nagarahole was enlarged and upgraded as a National Park. Following this, considerable effort was made to remove illegal agricultural encroachments, and to curb poaching and livestock grazing. Between 1970 and 1980 many squatters (>1000 people) were moved out of the park into resettlements. However, this resettlement process was not handled with the required degree of compassion, sensitivity and planning.

After the notification of the National Park, forest product exploitation was increasingly regulated in response to lobbying by wildlife conservationists. A core zone of 200 km^2, from which all forestry activities and even tourism were excluded, was established in 1982. These conservation measures led to a significant decrease in livelihood opportunities for tribal people living in the interior of the park. However, employment opportunities in the coffee plantations on the western side of Nagarahole increased significantly consequent to an economic boom. Movement of people out of the park in search of employment during daytime, and even permanent emigration, began. During the period of observations, serious conflicts periodically erupted between the park management and local communities, who resented the restrictions imposed on them against hunting, cattle grazing and forest product collection. Some tribal people have demanded proper resettlement outside the park, while others have asked for provision of agricultural land, livestock and forest product collection rights within it.

From 1970 onwards, population densities of ungulate species appeared to increase significantly, following the on-ground anti-hunting and habitat protection measures initiated between 1967 and 1970. For the purpose of this chapter, we have taken 1970 as the beginning of the observed recovery. However, evidence for this increase (observed by the first and third authors), is largely anecdotal. By 1984 tigers (and other predators) were being seen by tourists. At least two instances of tigresses raising their full litters to adulthood were observed in 1984–85. Evidence for rising tiger densities also came from increased reports of intraspecific aggression, and of tigers dispersing out of the park and being killed. However, in the Reserved Forests adjacent to Nagarahole, which are ecologically very similar but lack strong protective measures, densities of ungulates were very much lower and tiger reproduction was not reported. The tiger and prey density increases were restricted to better protected parts of the park.

Tiger and prey populations: status and recovery

From 1986 onwards, the first author studied the dynamics of tiger and prey populations in association with other collaborators through a series of projects (Karanth & Sunquist 1992, 1995; K. U. Karanth & B. M. Stith this volume; Karanth & Nichols 1998). These studies generated estimates of population density and related parameters for wild tigers and their prey populations, derived using theoretically well-founded population estimation techniques (Lancia *et al.* 1994; K. U. Karanth this volume Appendix 5). Some preliminary results from these studies are summarised later in this chapter to portray the current population status of tigers and their principal prey (chital, sambar, muntjac, gaur and wild pig), and to reconstruct the trajectory of their recovery since the 1970s.

Densities of ungulate prey

Ecological densities of prey species vary greatly over the park, depending on two primary factors: the habitat type and the degree of protection from illegal hunting. Calculations based on Karanth & Sunquist (1992) show that well-protected moist zone forests support much higher biomass densities of ungulate prey (8954 kg/km²) than the dry zone forests (2078 kg/km²). It must be noted that these estimates are for an area within the park that is almost free from illegal hunting. In 1996, Madhusudan & Karanth (in press) compared two 70-km² forest blocks that were ecologically similar but had different levels of hunting pressure. They found considerable differences in ungulate densities, attributable to illegal hunting. The ecological densities (animals/km²) were substantially depressed by hunting pressure for chital (9.1<66.2), sambar (3.9<4.4), gaur (1.7<5.8) and wild pig (0.5<6.2). The combined biomass density of prey species was 74% lower at the site prone to illegal hunting. Considering that even this hunting-prone site had some anti-hunting protection in 1996, and that in the early 1970s several additional adverse habitat-related pressures prevailed, we estimate that the average densities of ungulate prey all over Nagarahole were about 80% lower than the densities that now prevail in well-protected parts of the park. Based on this premise, the initial prey biomass densities can be put at 1791 kg/km² for the moist zone forests and 416 kg/km² for the dry zone. Considering that about 44% of the park area is in the dry zone and 56% in the moist zone, the average park-wide biomass density of principal tiger prey at the beginning of the population recovery works out at 1186 kg/km².

Box 8.1   How volunteer naturalists can monitor tigers and prey
K. Ullas Karanth

A major constraint in reliably monitoring tigers and prey populations is that the appropriate sampling methods require substantial manpower (K. U. Karanth this volume Appendix 5). There are many problems in using lower-level reserve staff or local labour for such monitoring work, particularly in countries like India, where education and hence instrument-use skills are often inadequate. Pressures of routine duties, lack of motivation, and a vested interest in inflating animal numbers compound these problems. Although Project Tiger authorities in India sometimes allow volunteers to participate in official 'censuses', there are no attempts to screen or train participants for long-term, sampling-based monitoring. As a part of my long-term studies in Nagarahole, India, from 1989 onwards, I tried to train and deploy amateur naturalists to reliably monitor populations of ungulates and tigers.

Nagarahole attracts many visitors. From among them, physically fit young persons who repeatedly visited the park, showing a keen interest in wildlife, were targeted. Initially I took the volunteers on walks in the forests to assess their ability to spot and identify animals, and to recognise tiger sign. I also judged their fitness and ability to work safely in elephant habitats. The selected candidates (usually six to eight per batch) were trained to do line transect surveys, dung/pellet surveys and tiger sign encounter rate surveys. The training usually lasted for six days, and involved identification of animals and their signs, field counts along transects, and the use of instruments such as range finders and compasses. In the initial stages, trainees who could not dependably read instruments, or spot animals from transects, were segregated and used only for dung and scat encounter rate surveys. Only those who had requisite skills actually collected line transect data.

Volunteers checking the height on a detector beam for a 'camera-trap'.

Generally, I found that after about three days of field work the trainees were capable of collecting reliable data. To ensure safety from wild elephants, and help with spotting and counting of animals, trainees were initially accompanied by instructors, and later by helpers with good field skills. Every year, during summer, such combined training and survey camps were held in Nagarahole for both newcomers and those trained earlier. The trainees came from a variety of backgrounds: students, tribal youth, businessmen, professionals, entrepreneurs and executives; they often scheduled their summer vacations to participate in these training camps.

Ullas Karanth teaches volunteers to distinguish tiger sign from that of other predators.

In Pench and Kanha Tiger Reserves, where such a network of trained para-naturalists did not exist, I used another approach to increase the sample size (and hence precision) of my line transect estimates. Trained research assistants conducted line transect surveys in the normal manner, to estimate the effective strip width and sighting probabilities. Additional line transect surveys were carried out by tribal trackers who had good spotting and counting abilities but could not read range finders or compasses. During analysis, I merged these two datasets, using counts from the pooled dataset in combination with the sighting probabilities derived from the distance data collected by my research assistants.

Based on this experience, I suggest that there is substantial scope to involve local naturalists in tiger and prey monitoring, provided the necessary time and energy can be invested in short-term, rigorous training programmes. Through such volunteer participation, the local community's involvement with tiger conservation, as well as the reliability of tiger population monitoring, can be increased.

Size and dynamics of tiger populations

Karanth & Nichols (1998), using photographic capture-recaptures (Karanth 1995), estimated that density of tigers (excluding cubs) in an effectively sampled area of 243 km^2 was 11.5 tigers/100 km^2. Their sampled area was a well-protected part of the park, covering both the dry and moist zone habitats. Based on our judgements, conservatively assuming that the remaining park area of 401 km^2, on average, supports 50% lower densities of tigers (5.8 animals/100 km^2), the entire park may now support a density of 8.0 tigers/100 km^2 for a population of 52 tigers excluding cubs. If we assume that about one-third of these (18) are breeding females, using a 2.5-year interbirth interval and a mean litter size of three (K. U. Karanth & B. M. Stith, this volume), these tigresses would produce about 22 cubs per year on average. With a cub survival rate of 60% (based on Kenney *et al.* 1995; K. U. Karanth & B. M. Stith this volume), the number of tigers being recruited into the juvenile class would be around 13 per year. Clearly, if the park's tiger population had reached saturation levels by 1986, since then, every year, on average, 13 juvenile or adult tigers may have been dispersing out of or perishing within the park. As K. U. Karanth & B. M. Stith (this volume) suggest, Nagarahole appears to have a demographically viable population, if current prey density levels can be maintained and the habitat integrity is preserved.

What was the tiger population size at the beginning of the recovery period? With the estimated prey biomass density of 1186 kg/km^2, the total standing biomass of prey was about 763 784 kg for the entire park area. Assuming an 8% annual cropping by tigers, and an average annual prey requirement of 3000 kg/tiger (Sunquist 1981), we estimate an initial total population size of 20 tigers, including cubs. Based on the age distribution generated by the demographic model of K. U. Karanth & B. M. Stith (this volume), this tiger population would have 15 adults and juveniles, including 5 breeding resident females, at the beginning of the population recovery. The conservation interventions since the 1970s appear to have increased the carrying capacity of the park for breeding females from 5 to 18, densities of tigers (excluding cubs) from 2.3 to 8 tigers/100 km^2, and their population size from 15 to 52. Clearly, factors which caused this dramatic recovery (an increase of 347% over 15 years at an annual rate of 8.3%) of a small and vulnerable tiger population to the verge of demographic viability over a short period deserve closer scrutiny.

Effect of conservation interventions

The conservation interventions in Nagarahole that led to increased densities of prey acted in two ways. First, they increased survival rates of ungulates through depression of hunting pressure. Secondly, they increased ungulate reproduction rates due to improved habitat conditions. In Nagarahole, these two consequences appear to have resulted both from planned managerial interventions and from unforeseen consequences of the pursuit of altogether different goals.

Regulation of hunting pressure

Madhusudan & Karanth (in press) documented the patterns of hunting practised by different groups of local people and showed that such hunting significantly lowered ungulate densities. Ungulate survival rates are depressed as a result of the following factors:

1   Active searching on foot at night and shooting using spotlights.
2   Active searching and shooting or netting during daytime, with the help of dogs.
3   Night shooting at spots that attract ungulates (e.g. fruiting trees, water holes, crops).
4   Setting wire snares or baited explosives (for pigs) on forest trails.
5   Driving along public access roads bordering the park to spotlight-shoot from vehicles.
6   Stealing of predator kills, leading to increased natural predation rates.
7   Predation by domestic dogs kept as pets by people living in the park.

From the early 1970s several measures were initiated by the park staff to control the above activities. Anti-poaching patrols of four to six men, armed with at least one gun, patrolled specific 'beats' prone to entry by outside hunters. These foot patrols operate in daytime and actively seek out poaching parties. At night, anti-poaching patrols operate from fixed camps in interior forest areas, or from mobile patrol jeeps or elephant-back, responding to gunshot sounds or waiting at travel routes known to be used by poachers. The anti-poaching patrols are legally empowered to fire on poachers in self-defence. The patrols often act on information provided by local informants. Patrols also periodically search neighbourhoods of human settlements inside the park, to detect and remove wire snares and to eliminate domestic dogs.

In addition to active anti-hunting activity to detect or deter poachers, indirect or passive strategies have also played a major role in reducing hunting pressures in Nagarahole. Reducing legalised entry of potential illegal hunters into the park has been the cornerstone of such indirect anti-hunting operations. Hunting, particularly through use of snares and appropriation of predator kills, is often an activity that is carried out together with or under the guise of non-timber forest product collection or forestry operations. Such legalised access makes it impossible for anti-poaching patrols to deal effectively with poachers. Consequently, the stoppage of non-timber forest product collection and the gradual elimination of forestry operations have significantly reduced hunter access and increased the efficiency of anti-poaching patrols.

Movement or resettlement of people away from inside the park in the 1970–90 period led indirectly to a significant reduction in hunting pressures. First, human settlements with crops, which attracted ungulates and provided convenient spots for shooting or snaring, were eliminated from the park interior. Secondly, following resettlements, the base camps from which outside hunters could explore the park interior using skilled local guides were eliminated. Thirdly, by reducing human-wildlife conflict caused by crop-raiding ungulates or livestock-killing predators, the magnitude of retaliatory killings of wild animals was reduced substantially.

Improving habitat conditions for ungulates

Modifications of ungulate habitats in Nagarahole have resulted from logging and silvicultural operations carried out by the foresters who also manage wildlife. Additionally, agriculture, animal husbandry and wood and non-timber forest product removals by local people living in and around the park have considerably altered the habitats. Over the years, gradually all these manipulations (both managerial and by local people) have been reduced in intensity or eliminated altogether. In addition, three specific forms of habitat manipulation were carried out with the aim of 'improving' habitats for ungulates. These were active suppression of forest fires, maintaining short-grass clearings along forest roads and creating artificial water ponds. Earlier studies (Karanth & Sunquist 1992; 1995; K. U. Karanth unpubl. data) have tried to analyse the effects of these interventions on ungulate populations, and it is beyond the scope of this chapter to examine the complexities involved. However, the following broad generalisations can be made:

1   The conversion of natural moist deciduous forests to monocultures of teak has depressed densities of ungulate species, with the possible exception of chital.

2   The secondary moist forests occurring on old clear-cut sites support high densities of preferred tiger prey such as gaur and sambar.

3   The reversion of *hadlus* to natural swamp grass following abandonment of rice cultivation has increased the carrying capacity of the area for all ungulate species.

4   The creation of water holes has benefited most ungulates, at least in the short term.

5   Short-grass clearings have benefited chital, possibly at the expense of browsers.

6   Suppression of manmade forest fires, through direct measures as well as indirectly through control of cattle grazing and non-timber forest product collection, has improved understory food availability for ungulates.

7   The elimination of cattle grazing from the park has increased the carrying capacity for wild ungulates.

Overall, it appears that reduction in the intensity of market-linked exploitation of forest products, elimination of livestock grazing and forest fires and removal of agricultural enclaves from inside the park have resulted in increased densities and reproduction rates for ungulates in the park. However, the maintenance of these high ungulate densities is possible only because of increased survival rates resulting from effective anti-poaching measures.

Tiger conservation: lessons from Nagarahole

The foregoing analysis of the factors responsible for the recovery of tiger and prey populations in Nagarahole between 1970 and 1985 has the following general implications for recovering small populations of wild tigers in other parts of tropical Asia:

1   Even relatively small tiger populations inhabiting disturbed tropical deciduous forest habitats (and perhaps other tropical forest types) can be recovered and demographically stabilised, provided that densities of principal ungulate prey (cervids, bovids and suids) can be increased.

2   Elimination of hunting pressures on ungulates and regulation of market-linked exploitation of biomass from ungulate habitats are necessary to maintain high ungulate densities. Considering the high rates of natural predation on ungulates (Karanth & Sunquist 1995), and the virtual impossibility of regulating human hunting off-takes (Madhusudan & Karanth in press), legalised game harvesting appears to be incompatible with the goal of establishing viable breeding populations of tigers, at least in settings like Nagarahole.

3   Hunting and habitat-related pressures can be reduced using both direct measures (e.g.

anti-poaching patrols, fire suppression) and indirect measures (elimination of incompatible human activities, appropriate resettlement programs) in critical tiger conservation areas. Furthermore, these indirect measures will reduce the potential for human-wildlife conflicts and the need for enforcement measures that arouse antagonism among local communities.

4   In order to eliminate social dependencies, which lead to pressures on tiger populations, biomass needs and livelihood opportunities for people should be generated outside critical tiger habitats through appropriate management of multiple-use landscapes (Panwar 1987; Karanth 1991; WCS 1995d). However, such buffer areas under multiple uses may not be capable of supporting viable breeding populations of tigers, and may even need active intervention to periodically eliminate 'problem tigers'.

5   Ultimately, the success (or failure) of any tiger population recovery programme can be assessed only through monitoring the response of tiger and ungulate prey populations, using scientifically valid population estimation techniques (Lancia *et al.* 1994; K. U. Karanth this volume Appendix 5).

Generally, the above measures necessary to recover small tiger populations seem incompatible with the prevailing conservation paradigm of 'sustainable use everywhere'. The alternative approach of 'sustainable landscapes' proposed by Robinson (1993) appears to be a more appropriate framework on which to build realistic, site-specific, enduring tiger conservation strategies. The social mechanisms necessary to implement such strategies can only evolve from a constructive, data-driven dialogue between two concerned human groups: on one side, the champions of livelihood rights of people currently using tiger habitats; on the other, the advocates of the tiger's right to survive on this planet, together with the rights of future generations to bear witness to their survival.

9

Tigers in Panna: preliminary results from an Indian tropical dry forest

Raghunandan S. Chundawat, Neel Gogate and A. J. T. Johnsingh

Introduction

Quantitative information on movements, activity and food habits of tigers living in the tropical dry forest has lagged behind that now available for tigers living in tropical moist forest and alluvial grassland/subtropical moist deciduous forest types in the Indian subcontinent (Seidensticker & McDougal 1993; Karanth & Nichols 1998). This once extensive tiger habitat type has been increasingly fragmented and degraded. Still today about 45% of the remaining tiger habitat in the Indian subcontinent is in tropical dry forests (E. D. Wikramanayake *et al.* this volume). Tiger density is constrained by the availability of large ungulate prey, cover and water (M. Sunquist *et al.* this volume). In the highly seasonal environment that characterises tropical dry forests, these critical resources are limited in time and space. Surface water is highly restricted, and shade available only around it, from March to June. In the subcontinent, large ungulate densities reach their highest levels in the gallery forests/alluvial plains and their lowest levels in tropical dry forest/savannah (Eisenberg & Seidensticker 1976; Karanth & Nichols 1998). Throughout the dry forest habitats, human disturbance from wood cutters and cattle grazers is high and there is low native ungulate biomass because cattle populations compete with the tiger's natural prey species for food. Cattle are also a source of diseases for the native ungulates. All of these factors place tigers living here at high risk.

This is a preliminary report of an ongoing study of tiger ecology in Panna Tiger Reserve, Madhya Pradesh, from early 1996 to mid-1997. We present data on tiger food habits and on the movements and activities of three tigers established through radio-tracking, and preliminary estimates of prey density. Our research revealed the critical ecological needs of tigers living in Panna specifically and tropical dry forests generally, and we make suggestions on how managers can address these needs so that the tiger can survive in Panna and elsewhere in this forest type.

We thank the Ministry of Environment and Forests, Government of India, and the Madhya Pradesh Forest Department for permission to capture and study tigers. The Ranthambhore Foundation provided transport. Drs P. K. Malik, P. K. Peshin and A. B. Shrivastrava helped us radio-collar tigers. We thank our colleagues at the Wildlife Institute of India for their assistance in our work.

The Panna Tiger Reserve and its environs

The 543 km² Panna Tiger Reserve lies along the Ken River, a tributary of the Yamuna, in Madhya Pradesh, northcentral India (Fig. 9.1). It is composed of three landform units: the upper Talgaon Plateau, the middle Hinauta Plateau, and the Ken River valley. The plateaus are separated from each other and from the river valley by steep, 10- to 80-m high escarpments characterised by rock faces, caves and thick vegetation at their bases. Panna supports a diverse mammalian assemblage including tigers, leopards, hyaenas, sloth bears, dholes and wolves. The Hanuman langur is the most numerous primate. Ungulates include sambar, chital, wild pigs, nilgai and chousingha.

Mean maximum temperature ranges from 6 to 43°C. Annual precipitation depends on the southwest monsoon that lasts from July to September, when an average of 1100 mm of rain

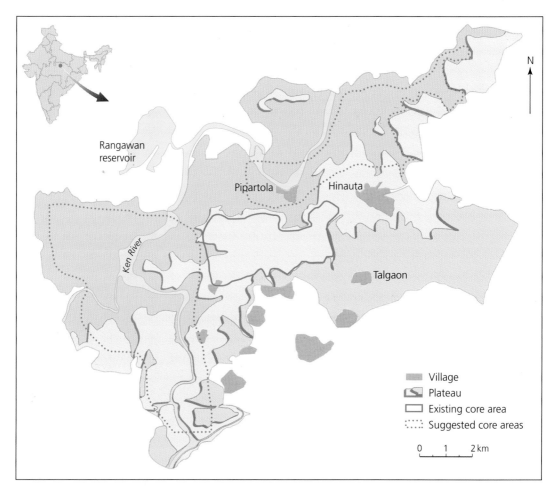

FIGURE 9.1
The Panna Tiger Reserve, India, with existing and suggested core areas.

falls. In the dry season, available water is restricted to the river, seepage at the bases of the escarpments and a few artificial waterholes. Slopes off the plateaus are dominated by *Acacia catechu*. Open woodlands with areas of short grass, the preferred habitat of nilgai and chousingha, are the dominant vegetation on the plateaus, while in the river valleys and along the seasonal drainages are areas of tall grass and closed woodlands that are preferred by chital and sambar.

The central Hinauta Plateau extends over 80 km² and is now relatively undisturbed, following the removal of three villages in the early 1980s. Throughout the remainder of the reserve, however, there is widespread and intense disturbance by

people and cattle. Thirteen villages exist within the Panna Tiger Reserve, with human and cattle populations of about 600 and 9500 respectively. There are 79 villages within 10 km of the boundary outside the reserve. Villages place immense pressure on the reserve, primarily through grazing their livestock there.

Methods

Capturing and radio-tracking tigers

We captured and radio-collared an adult male tiger, an adult tigress with cubs less than one year old, and one subadult female we estimated to be about 16–17

Table 9.1. *Size of home ranges for radio-collared tigers in the Panna Tiger Reserve, India as determined by the minimum convex polygon (MCP) methods of establishing home range sizes*

Tigers	Dates tracked	Number of locations	Home range size (km²)			
			Total	Winter	Summer	Monsoon
Subadult female	4/96–9/96	23	31	—	—	—
Adult female	1/97–7/97	115	27	—	16	27
Adult male	4/96–7/97	134	243	110	200	—

months of age at capture (Table 9.1). We located tigers feeding at kills, approached them on elephant back, and darted them from 15–20 m away using a Telinject gun. Medetomidine (0.05 mg/kg tiger body mass) combined with ketamine (3.5 mg/kg) was used to chemically restrain tigers. Radio-collars from Telonics were attached and standard measurements of body size and condition were made. Atipamezole (Antisedan), an antagonist for medetomidine, was given to revive the tiger. We monitored the newly radio-collared tiger from elephant back until it recovered.

The position of radio-collared tigers was determined by triangulation from known reference points using standard radio-tracking techniques (Sunquist 1981). Tigers were monitored daily for periods of 5–15 days. To calculate home range size, random locations (one every three days for the adult male and one every two days for the adult tigress) were used to ensure independence of sample points. Our index of daily movement was calculated using the distance between the tiger's location on one day and its location on the following day. We used CALHOME to summarise our findings. Home range sizes were estimated using the minimum convex polygon (MCP, Mohr 1947b). Our tiger home range statistics are more complete for the dry season than the wet season because from July to September roads were washed away and we had problem at times in locating tigers.

Estimating tiger diet and prey abundance

Whenever a radio-collared tiger was found in one general location for about eight hours we suspected

that a kill had been made and the area was searched intensively after the tiger moved away. This method was successful for the adult tigress, because she lived in areas that were relatively undisturbed by humans and frequently rested near her kills. Finding kills made by the male was more problematic. In the highly disturbed areas he frequented, he fed largely at night and frequently he fed only once from a kill. We also sought information on tiger kills from the reserve staff and villagers. Tiger scats were collected opportunistically. We randomly selected 15 scats for each season and identified remains based on the micro- and macroscopic characteristics of hair (Mukherjee *et al.* 1994). Ungulate prey densities were estimated by counting ungulates along reserve roads and pre-established transect lines; the data were analysed using the programme TRANSECT (R. Chundawat in prep.).

Results

Movement patterns and tiger home range sizes

The adult male tiger moved on average 4.2±3.3 (1.7–10.5, $n=38$) km between locations on successive days in winter, and 4.1±3.3 (1–13.9; $n=28$) km in summer. The difference was not significant. This is about three times farther than the female (1.4±1, 0.6–2.9, $n=58$). The maximum distances moved between successive days were about three times the average for both adults.

The home range of the adult male (243 km²) was about ten times larger than that of the adult tigress with cubs (Table 9.1, Figs. 9.2 & 9.3). In winter his

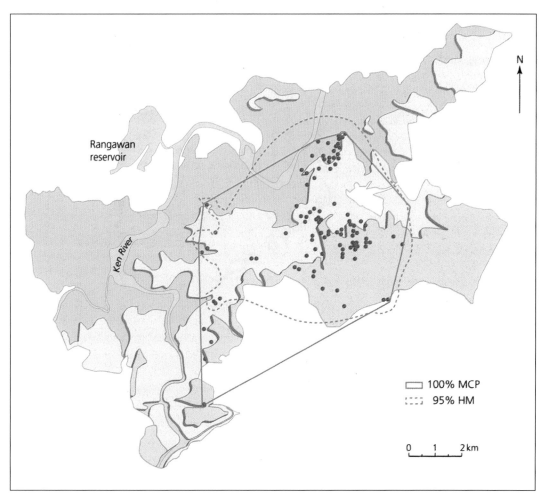

FIGURE 9.2
The home range (territory) of the radio-collared adult male tiger, Panna Tiger Reserve, India. MCP = minimum convex polygon; HM = the harmonic mean method of measuring home range size.

home range was 100 km², nearly half his summer range of 200 km². The adult tigress' home range increased from 16 km² in the pre-monsoon period to 27 km² during the monsoon. The subadult tigress' home range was estimated to be about the size of that of the adult female (Table 9.1, Fig. 9.3). On two occasions she made long movements away when we could not locate her for periods of 7–15 days within an area of about 100 km² around her known home range. She was found dead near a village 10 km away from her home range on September 16, 1996, six months after we started radio-tracking her.

The adult tigress' range covered the central area of the reserve, which was largely undisturbed and supported the highest density of native ungulates in the reserve, and she used this area in a homogeneous manner. No other adult females were living within this area and her home range could be called a territory (Sunquist 1981). The adult male tiger used this same area about 95% of the time and in winter used it in a homogeneous manner. His summer use of the area was more clumped, as he focused his attention on areas along the base of the escarpments where there was dense cover and

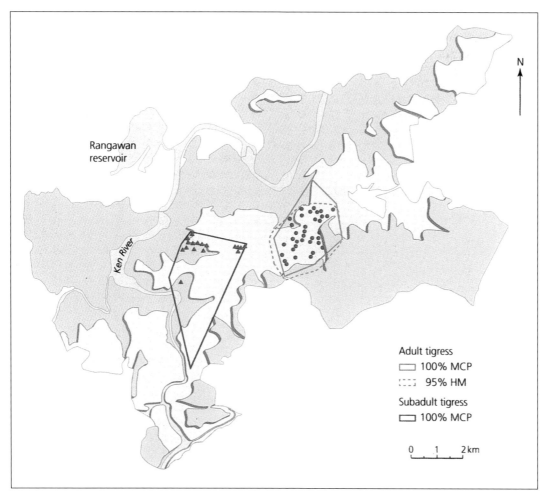

FIGURE 9.3
The home ranges of radio-collared tigresses, Panna Tiger Reserve, India. MCP = minimum convex polygon; HM = the harmonic mean method of measuring home range size.

water. His large home range covered nearly half of the reserve and his movements took him beyond the reserve boundaries. We had no evidence that any other adult male lived within his home range, but we did have evidence that two adult males lived adjacent to his home range.

Prey abundance, food habits and frequency of killing large prey by radio-collared tigers

Our initial estimates of ungulate prey abundance were 1 chital, 1.8 sambar and 2 nilgai/ km² (R. Chundawat unpub. data.) Over a six-month period (January 1997 to July 1997) we located 27 large mammal kills made by the adult tigress and could determine with confidence that she made three additional kills (Table 9.2). On average, she was making a large mammal kill about once every six days while she was accompanied by three cubs less than one year old. We rarely found small-sized kills. Her large prey diet consisted primarily of sambar and nilgai (~80%) in contrast to the diet of the adult male tiger that consisted mostly of cattle (84%). Because of his wide-ranging movements, we were not as successful in finding his sequential kills

Table 9.2. *Tiger kills (% occurrence) found in Panna Tiger Reserve, India*

Prey	Adult male tiger (n=19)	Adult tigress (n=27)
Sambar	5.3	55.6
Chital	0.0	3.7
Nilgai	10.5	25.9
Hanuman langur	0.0	3.7
Wild pig	0.0	3.7
Cattle	84.2	7.4

Table 9.3. *The diet of tigers (% occurrence) based on scats in the Panna Tiger Reserve, India. Each season 15 of all the scats collected were randomly selected for analysis*

Prey	Winter	Summer	Monsoon	Total
Sambar	13.3	46.7	13.3	24.4
Nilgai	40.0	13.3	20.0	25.4
Chital	13.3	0.0	40.0	17.7
Cattle	13.3	20.0	0.0	11.1
Wild pig	13.3	6.7	6.7	8.9
Hanuman langur	0.0	20.0	6.7	8.9
Porcupine	13.3	0.0	6.7	6.7
Chousingha	13.3	0.0	6.7	6.7
Unknown	0.0	6.7	6.7	4.4

and we could not determine the frequency with which he killed large mammals. On five occasions, we found him feeding at sambar kills made by the tigress.

The major components of the tiger's diet in Panna, as revealed by our analysis of scats, were sambar, chital, nilgai and cattle (Table 9.3). While we were unable to find small-sized kills through our radio-tracking, the scats revealed that, based on percentage occurrence, mammals weighing less than 40 kg comprised about one-third of the tiger's diet in Panna (Table 9.3). Our examination of the scats also revealed that >40% of the diet by season was comprised of one large ungulate species: nilgai during winter, sambar during summer and chital during the monsoon. Cattle and Hanamun langur were killed more frequently in summer. Cattle remains were absent from the scats during the monsoon, when there was abundant food for the cattle to graze near villages and they did not have to enter the surrounding forest areas.

Discussion

In this preliminary report, we want to emphasise the important findings that characterise tigers living in Indian tropical dry forests and the risks they face. These habitats support a relatively low large-ungulate biomass and have high human disturbance. Water and cover, key resources for the tiger, are sparse, seasonal, and clumped in their distribution. In Panna, the adult male tiger's home

range was about four times larger and the adult female's home range was about twice as large as comparable tiger home ranges, established through radio-tracking, in the Royal Chitwan National Park, Nepal (Sunquist 1981).

We found just how low the supply of wild ungulate prey is for tigers in Panna. We estimated that there are about five large wild ungulates (chital, nilgai and sambar) per square kilometre, about one-third of the estimated crude density of cattle (17.5/km^2) in the reserve. This compares to ~60 potential wild ungulate prey per square kilometre in the tropical moist forest and alluvial grassland/gallery forest tiger reserves in India and Nepal (Eisenberg & Seidensticker 1976; Seidensticker & McDougal 1993; Karanth & Nichols 1998). In Panna, tigers include a substantial number of small-sized mammals in their diet. They also take about 70 cattle each year. Considering that about 75% of the ungulates in the reserve are cattle, it is surprising that cattle make up only 11% of the tigers' diet (Table 9.3). Tigers in fact take less than 1% of the available cattle each year, but taking cattle on any scale places tigers at risk of poisoning and creates bad feeling towards them.

The extent of disturbance-free habitat with adequate prey in Panna is limited and this limits the area available for females to rear their cubs. The majority of the reserve is heavily disturbed by humans, and the behaviour of ungulates in relation to such areas is a key ecological factor affecting tigers living in tropical dry forests. The key wild ungulate here is sambar. Sambar drink at night, and during the summer they do so every night. We found that the adult tigress focused her hunting to take advantage of this, and killed sambar coming to restricted watering sites. Sambar are less restricted in their wet season habitat use and the tigress nearly doubled the size of her home range during this period. In other areas of India, chital are also an important prey species for tigers, but chital density in Panna is very low. While nilgai are the most numerous wild ungulate in Panna, their availability as prey for tigers is restricted in some seasons by their use of more open habitats.

The single most important finding to emerge from our work to date has been the importance of the watercourses and watering sites in this forest type as critical habitat for tigers. If small reserves such as Panna are to sustain their tiger populations then they must include as much of the watercourse as possible in a disturbance-free zone, so that tigers can hunt wild ungulates and rear their cubs undisturbed. What is urgently needed in Panna is the creation of at least two more disturbance-free 'mini-core' areas in Balaiya Seha and the Ken River Valley (Fig. 9.1). This requires removing about 4000 people from the reserve; without this, the future of the tiger population in Panna is very bleak. The second lesson from our study of tigers in Panna is just how few tigers there are living here. With these low numbers we cannot see how this can be a sustainable tiger population unless habitat connectivity is maintained with other areas where tigers still live, allowing tigers dispersing between areas to survive. Our data show that currently, dispersing tigers leave Panna and die in the hostile environment that surrounds it. We believe that these conditions are the major threat to tigers throughout India's dry forests and place all the tigers living here at very high risk.

10

Last of the Indonesian tigers: a cause for optimism

Neil Franklin, Bastoni, Sriyanto, Dwiatmo Siswomartono, Jansen Manansang and Ronald Tilson

Introduction

The Sumatran tiger is the last of the three tiger sub-species that once inhabited the islands of Indonesia. The World Conservation Union (IUCN) recognises the Sumatran tiger as critically endangered (IUCN 1996). A Population and Habitat Viability Assessment (PHVA) meeting in 1992, conducted by the Indonesian Department of Forest Protection and Nature Conservation (PHPA) and the IUCN/SSC Conservation Breeding Specialist Group (CBSG), suggested that there were only 400 surviving tigers in the five major national parks. A further 100 individuals were proposed to exist in other forested areas (Faust & Tilson 1994; Seal *et al.* 1994; Soemarna *et al.* 1994). This assessment was based on population estimates by PHPA staff over the whole of Sumatra, assuming a tiger density of one to three individuals/100 km² depending on habitat type, and extrapolating over the potential tiger range using a Geographic Information Systems (GIS) analysis of remaining forest habitat.

Of the five national parks, the largest population, 110 tigers, was identified in the Gunung Leuser National Park of northern Sumatra (Griffiths 1994), whilst the remaining populations in the parks were estimated at half this number or fewer. Small populations such as these are extremely vulnerable to environmental catastrophes, as well as to the demographic and genetic problems considered typical of fragmented and isolated population sub-units. Threats also exist from the extraction of single individuals, whether this be from poaching or the official 'removal' of problem tigers, where a dramatic effect on the population's long-term viability is predicted (Wiese *et al.* 1994). However,

though these early estimates were crucial in focusing the attention of conservation agencies, presence/absence data were not available for all potential tiger areas in Sumatra. In addition, ecological parameters used in the viability analysis were, of necessity, derived primarily from the tiger subspecies living in quite different habitat types.

Tigers are under continued threat. Poaching and habitat loss are factors, though little information exists as to the intensity of the problem (Tilson & Traylor-Holzer 1994; Plowden & Bowles 1997). Tigers in Sumatra face other pressures as their populations become increasingly fragmented and isolated, and as the intensity of alternative land use increases. The conservation of the Sumatran tiger calls for a comprehensive management strategy that takes full account of the complex assemblage of social, biological and landscape ecological factors in operation. Above all, this strategy should be firmly based upon the foundation of sound knowledge regarding the tiger's current distribution and status, and should also not ignore the Sumatran tigers' particular ecological parameters and life-history characteristics (N. Franklin unpub. obs.).

A first step in this process must be the amassing of political will. Indonesia has excelled here, with the tiger now recognised as a 'key species' in national biodiversity conservation strategies (BAPPENAS 1993). The Sumatran tiger should be considered not only as a significant single component of Indonesian biodiversity, but also as symbolic of the biodiversity that remains (Tilson *et al.* 1996).

The Sumatran tiger field study, as part of the Sumatran Tiger Project, is grateful to the following collaborating organisations: Tiger Global Animal

Survival Plan (GASP) under the Conservation Breeding Specialist Group (CBSG) of the World Conservation Union (IUCN); the Zoological Society of London; and the following Indonesian institutions – Lembaga Ilmu Pengatahuan Indonesia (LIPI), the Directorate Jeneral Perlindungan Hutan dan Pelestarian Alam (PHPA), the CBSG Indonesia Program at Taman Safari Indonesia (TSI) and the University of Lampung. The rapid evaluation of tigers being carried out by the Tiger Conservation Team is a co-operative effort between the Sumatran Tiger Project, the Zoological Society of London and the Indonesian Department of Forest Protection and Nature Conservation (PHPA).

The Sumatran tiger field study was funded during its first year by Esso UK with subsequent support from the Save the Tiger Fund, a special project of the National Fish and Wildlife Foundation in partnership with Exxon Corporation. Additional support has been provided by the CBSG Indonesia Program at Taman Safari Indonesia, Esso Indonesia, British Airways (Jakarta), the Zoological Society of London and the Federation of Zoos of Great Britain and Ireland.

The Sumatran Tiger Conservation Strategy

The potential crisis that tigers face in Sumatra has been addressed in the Indonesian Sumatran Tiger Conservation Strategy (Ministry of Forestry 1994). In this document the Department of Forest Protection and Nature Conservation recommends a course of action, based on a comprehensive understanding of the tigers' ecology and status, that will ensure the long-term viability of wild Sumatran tigers. Priorities are defined for the development and implementation of a long-term ecological monitoring project for the Sumatran tiger in order to accumulate data on tiger life-history characteristics vital for the management of wild populations. Also recommended as a priority action is the initiation of a parallel programme for community education and awareness of the many villagers living in close proximity to tiger habitat. A primary output of these proposed activities should be the facilitation of

university and PHPA counterparts to become future conservation leaders in Indonesia. In line with these recommendations the Sumatran Tiger Project was initiated in Way Kambas National Park in June of 1995.

The Sumatran Tiger Project

An effective long-term conservation strategy for the Sumatran tiger must be developed through a thorough understanding of the prevailing conditions affecting the subspecies. The Sumatran Tiger Project was initiated with the objective of improving current knowledge about this poorly understood subspecies. This includes the accumulation of accurate data concerning the distribution and status of tigers in Sumatra, both within and outside the protected area system. Also necessary is an understanding of the intensity of threats to tigers from poaching and other disturbances, not forgetting that these same factors may equally affect the tigers' prey species. Ecological parameters and life-history characteristics, particularly those required for the modelling of population viability, are also recognised as high priorities for research.

Other project objectives were defined beyond those of pure ecological research. These included the recognition of the need to provide a long-term tiger monitoring system for the evaluation of tiger protection and management programmes. In addition, it was considered important for the project to initiate means of resolving conflicts between tigers and the forest-edge human communities, as well as to facilitate the dissemination of a strong conservation message to the citizens of Indonesia (Ministry of Forestry 1994).

The field ecology component of the project, investigating the status and distribution of tigers in lowland rainforest, is the subject of this chapter. Wild tigers are extremely difficult to study and census because of their elusive nature and tendency to avoid humans. In the forests of Sumatra, the difficulties in studying tigers are intensified by the vast size and remoteness of the national parks and reserves, and by the thickness and impenetrability

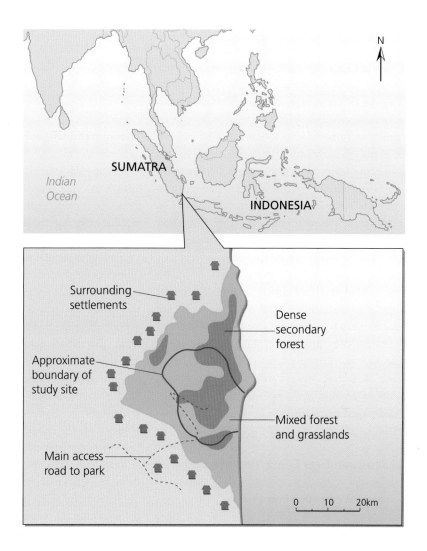

FIGURE 10.1
The location of Way
Kambas National Park,
Sumatra.

of the vegetation itself. Extreme topography and impassable water features also provide further obstacles and logistical difficulties for survey teams. These factors were of significance in the selection of both appropriate research techniques and the study site itself.

Way Kambas National Park – a model study area

Way Kambas National Park was selected as the initial research site, and the ecological study began in August 1995 with remote camera trials and surveys for tiger secondary sign. The park (1300 km²) is located in Lampung Province, on the southeastern coast of lowland Sumatra (Fig. 10.1). This study site was chosen because of the relative ease with which teams can traverse the park, sections of the lowland secondary forest being bisected by navigable waterways. The forests of the park have been logged on more than three occasions during the last 30 years, and this has left behind a mosaic of *Imperata* grassland and secondary forest habitat types. Travel on foot is also facilitated by the continued existence of overgrown trails, once used to transport timber out of the area. During the first months of field work a network of trails was developed, with camps placed at strategic positions within the study site, to allow extended residence in the forest for the field teams. The project's headquarters were sited on the Way Kanan river, at a central enclave some 14 km inside the park's perimeter boundary.

Although Way Kambas is comprised primarily of lowland secondary rainforest and grasslands, quite different to the largest of the Sumatran national parks (Kerinci Seblat and Gunung Leuser), it can be considered to be more representative of Sumatra as a whole. As the park has been logged in the past, the various habitat types represented have been subject to varying degrees of disturbance, and are currently at different stages of regeneration. Such habitat types are very typical of the scrub and secondary forest found wherever human activity has ceased, which category includes abandoned plantations and old settlements and accounts for much of the eastern plains of Sumatra itself. In areas of the park extensive stands of mature, dense lowland forest can also be found, though it is the mosaic of pioneering and successional vegetation types that is of real management interest. Way Kambas National Park could be considered a representative model for much of the potential tiger habitat remaining in Sumatra, and of greater importance still when we consider the latent options for reclaiming tiger habitat in future conservation initiatives.

Despite disturbances from past logging operations, the remote camera monitoring carried out during 1996 has shown that the park's mammalian biodiversity remains intact. The presence of Sumatran rhino (Siswomartono *et al.* 1996), flat-headed cat, clouded leopard, marbled cat, golden cat, sun bear and Asian tapir, amongst many others, were all confirmed in this way.

The Way Kambas tiger population

Remote camera monitoring

Between October 1995 and January 1997 a system of infrared-activated remote cameras was operational within the Way Kambas study site. These cameras, equipped with Trailmaster TM1500 monitors, were placed over an area of approximately 160 km². The 25–28 cameras used in this study were evenly distributed throughout the region, though placement was primarily along the extensive network of animal and old logging trails (Fig. 10.2). Cameras, equipped with built-in flash for low-light conditions, were optimised to photograph all large mammals, with infrared beam height of 50 cm and an infrared 'gate' with average width of 4.5 m. Locations chosen for the placement of sensors were based on extensive ground surveys for tiger secondary signs. Sensitivity of the cameras is adjustable, and this was set to account for the maximum ambient sunlight levels occurring during the day. The time delay between photographs was set at one minute, which eliminates wastage of film due to the passing of large groups of animals such as the wild pig and the long-tailed macaque. All cameras were operational for 24 hours a day, with no break in monitoring except in cases of camera or film malfunction, or theft. In general, cameras were placed alone, as single units, allowing a photograph of one side of the passing animal. However, it was also feasible to place two cameras together as a pair, connected to the same infrared trigger, in order to photograph the passing animal from both left and right sides simultaneously. Camera locations were maintained for the duration of the study, with only minor modifications made in order to accommodate changing local conditions.

The remote camera system was maintained by field personnel of the project and park staff counterparts. Camera batteries and films were changed every 10 days, whilst batteries in the infrared transmitter and data logger required changing only once every 20 days. Monitoring all of the cameras in the study site necessitated five days in the field, with a distance covered on foot of between 85 and 115 km. Maintenance of cameras involved changing batteries, removing and reinstalling photographic film, cleaning of camera seals, reparation of camera housings, and collection and download of the time/date data from the camera data loggers. Films were developed locally. Photographic results of the monitoring periods were recorded in a database, alongside accurate details regarding the time and date of the photo-capture, and the location and habitat characteristics of the camera location itself.

The organisation of the camera data was facilitated by the use of a detailed habitat-based GIS of the park area. This was used for mapping both the location of remote cameras and the mapping of tiger secondary signs, whose locations were 'marked' using handheld global positioning systems and then later downloaded to the GIS map at the base camp. By plotting the locations where individual tigers were photographed it was possible to gain an understanding of the extent and overlap of the individual tiger home ranges and activity patterns.

Analysis of tiger photographs from the remote cameras

Tiger photographs are separated as individuals on the basis of sex, body stripe pattern, obvious morphological distinguishing features, and then by differences in basic body dimensions (see Box 10.1). A reference database of quality tiger photographs was developed, showing identified tigers from both the left and right, and also from the front and rear where possible. This reference collection consisted of photographs where either a tiger had turned round in front of the camera (showing more than one side) or where remote cameras had been set up in opposing pairs across the trail. Once the reference collection had been established, all new photographs could be classified accordingly.

When uncertainty remained as to the identity of a particular tiger, those photographs were removed from the analysis until such time as additional photographic evidence had accumulated. In general it was sufficient to rely upon the individual's sex and the stripe pattern across the individual's flank in this identification process.

Over 16 months a minimum of 21 tigers were

Adult male tiger was 'BL' repeatedly captured on film at 'camera-traps' in Way Kambas National Park until he disappeared in autumn 1997.

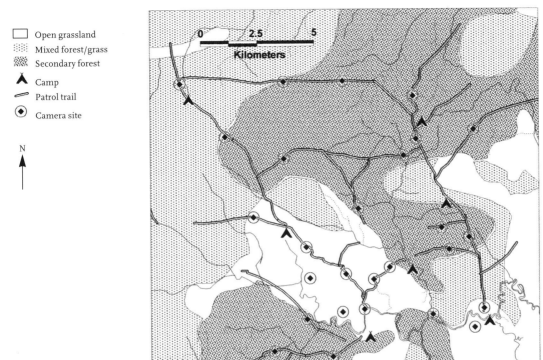

Open grassland
Mixed forest/grass
Secondary forest
Camp
Patrol trail
Camera site

N

FIGURE 10.2
GIS map of remote camera
locations and camps in the
Way Kambas National Park
study site.

identified by the remote camera system, from a database of 223 tiger photographs. Of these, 210 (94%) were confidently categorised in relation to the photographic reference database, whilst the remainder was made up of poor quality and distorted images and photographs of tigers for which insufficient data existed.

Figure 10.3 shows the cumulative numbers of tiger individuals encountered in the study site over time. Despite the consistency of sampling intensity and the uniform spacing of cameras, new tiger individuals were encountered with almost linear regularity, as opposed to declining in encounter frequency over time, as one would expect with a small closed tiger population with no immigrants. Thus, new individuals encountered by the remote camera system were recorded up to the end of the 16 months of monitoring, and the rate of encounter showed no decline at the time of writing. Tiger

passes were recorded from 16 of the 25 camera sites in operation, and the geographical distribution of these photo-events is discussed in more detail in the following section.

Population characteristics of the Way Kambas tigers

Six tigers stand out clearly in terms of the frequency with which they were encountered in the study site. These individuals were photographed on a mean number of 21 occasions, compared to a mean of 5 photo-captures for the other 15 individuals. Figure 10.5 shows, for these 6 tigers, the number of photo-events recorded, and the number of camera locations in which these individuals were encountered.

It is likely that the number of photographs obtained is an indication of the extent to which the tiger frequents the study site. For the six tigers

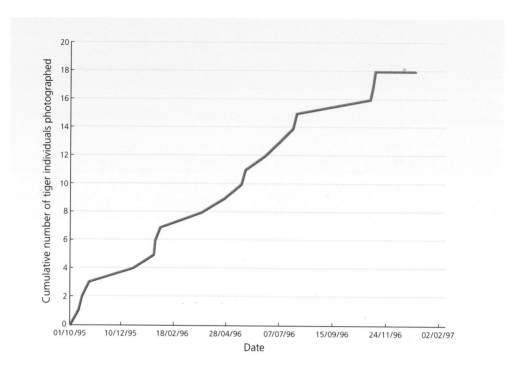

FIGURE 10.3
Cumulative numbers of tiger individuals encountered in the Way Kambas study site by month.

identified here it is suggested that their home ranges are located predominantly within the study site (resident individuals) when compared to the other 15 tigers recorded. The tigers only occasionally observed are considered to be individuals that possess home ranges either adjacent to, or primarily outside of, the study site itself. As such they have little territorial stake within the study site, and are referred to as non-resident individuals. By contrast individual resident tigers were photographed, after the date of first encounter, at least once every five weeks during the period of intensive camera monitoring.

Figure 10.4 also shows the ratio of number of photographs to number of locations (P/L) for the individual resident tigers. We consider that this ratio provides a basic measure of the degree of geographical concentration of individual tiger ranging. As such, female tigers, with greater concentration of activity in a smaller home-range area, should show higher values of the ratio P/L. On the reverse side male tigers, with their larger home ranges, should show more widely distributed activity, and the expected value of P/L observed should be correspondingly lower. This is reflected in Figure 10.4 where male tigers are represented by DJ and GR, and the four females by KP, CE, AY and GK. Female tigers are photographed more frequently, over a smaller geographical range, than the males.

Support for the assumption that the six resident tigers hold relatively stable ranges is provided by Figure 10.6. In contrast to Figure 10.3, where the

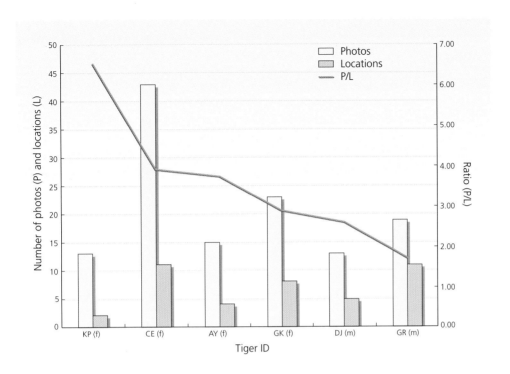

FIGURE 10.4
Resident tigers by number of photo-captures and camera locations encountered.

picture is clouded by the inclusion of large numbers of non-resident tigers, the graph of residents alone shows a levelling off of tiger individual encounter rate over time. This is expected in a small and stable population of tigers, and reflects observations of the stability of territoriality in other tiger subspecies. This figure also shows that two of the resident tigers were not photographed until approximately eight months into the 16-month study period. These two individuals (appearing in June and July 1996 respectively) then maintained a regular presence in the study area until the end of the monitoring period. None of the resident tigers present at the beginning of the study were observed to move out of the area being monitored.

Visibility of external genitalia in the remote camera photographs allows the sex of tiger indi-

viduals to be distinguished. The sex ratio of the entire population of tigers utilising the area of the study site is 8:13 in favour of females. This predominance of females is reflected in both the residents (2:4) and the non-residents (6:9), though whether this is a result of the greater number of females, or of their greater activity around cameras, remains to be tested. Immigration of tigers into the Way Kambas population from areas external to the park is considered unlikely due to dense human settlements and the lack of suitable habitat that could serve as a potential migration corridor.

A preliminary population assessment for Way Kambas National Park

Previous estimates (Tilson *et al.* 1994) have suggested a tiger population in Way Kambas National

Box 10.1 Using tiger stripes to identify individual tigers
Neil Franklin, Bastoni, Sriyanto, Dwiatmo Siswomartono,
Jansen Manansang and Ronald Tilson

Fundamental to a camera-based monitoring programme for wild tigers is the
ability to identify individual tigers from images alone, even when the tiger is
photographed from various perspectives. Previous authors have suggested
potential difficulties of using portraits of tigers as a diagnostic tool (Goyal &
Johnsingh 1996), but not without some dispute (Karanth 1996). Our
experience suggests that many of the identification issues resolve themselves
after time in the field. The key to a good identification scheme is to collate sets
of photographs for every individual tiger, consisting of sharp images taken
from the right, left, front and rear of the animal. Once the reference collection
is established, almost all tigers can be classified with ease, over time. Here we
highlight some diagnostic characteristics of tigers that we use in assigning
identities to individuals. The process is akin to filtering from the general to
the specific, in that young tigers are separated from older tigers, males from
females, and familiar tigers from unfamiliar, until we arrive at an identity.

Size of the tiger's body is one of the first helpful filters. Small cubs, taller
but gracile subadults, and larger, more muscled adults can usually be
separated with ease.

Sex, often established by externally visible genitalia, particularly with
males, is another filter. Also, for older, mature males, their distinctive facial
fur pattern that can be reddish or nearly black sometimes helps to distinguish
sex. Because this character can vary with light conditions, particularly when a
flash is necessary, we are careful not to give too much emphasis to it.

Stripe patterns, especially if they appear strikingly unique, can lead to
misidentification, so caution must be exercised. We have discovered that
establishing a suite of stripe characteristics is the most reliable method for
identifying individual tigers. For these reasons, we are most comfortable
using comparisons from the flank, and shoulder, proximally or distally on the
tail, both inside and outside stripes of the forelegs, and sometimes the cheek
or forehead if it is a frontal image. If no discrepancies are found in
comparison to a reference image, we assign that name in the database.

In general, eight points or more of similarity in stripe pattern are
established between the reference image and the new field image (Fig. 10.5)
before we assign a positive identification to the tiger. We also strive to have at
least two staff familiar with the identification system independently agree
before we confirm the tiger's identity. Finally, because we are often visited by
other field biologists who scrutinise the photographic database, and because
no discrepancies in identities have yet been noted, we are confident that the
system is reasonably effective.

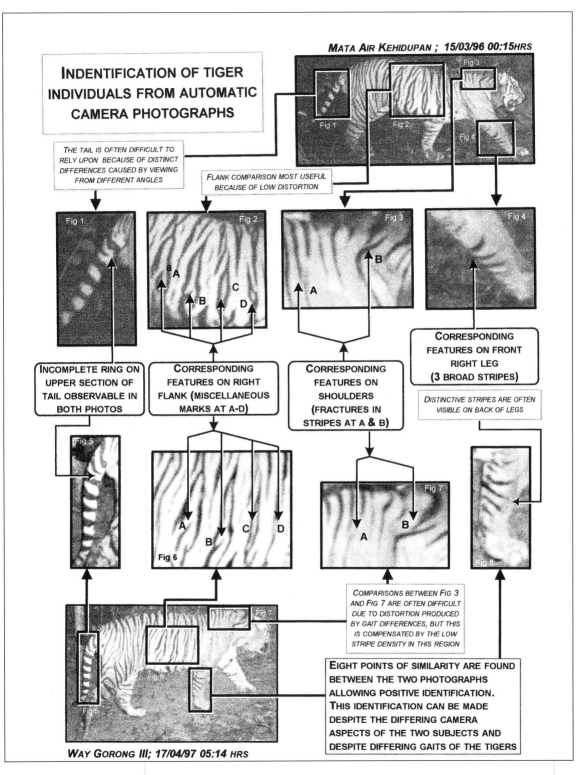

FIGURE 10.5
Using tiger stripes to identify individual tigers.

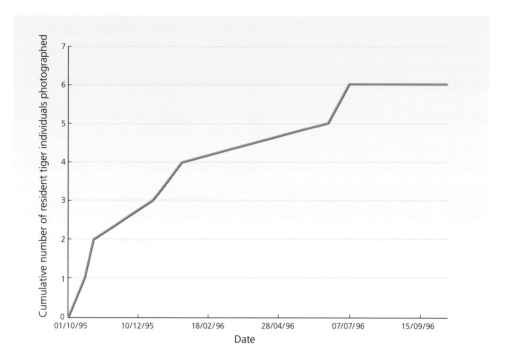

Cumulative number of resident tiger individuals photographed

FIGURE 10.6
Cumulative numbers of resident tiger individuals photographed in the Way Kambas study site by month.

Park of 20 individuals. In the 16 months of this study 21 tigers have been identified within a study area of 160 km². The intensively monitored study site represents 12% of the total park area, and approximately 19% of the more ideal dense secondary forest and mixed secondary forest/grassland habitat types (see below). Six resident tigers were observed (regularly encountered within the area of the study site), and an additional 15 non-resident individuals were observed as having passed through the site during the study period (infrequently recorded).

In any particular month the number of tigers inhabiting the study site is considered to be equal to the sum of the number of permanent residents and the number of non-resident tigers present. Non-resident tigers present in the study site during an average month were calculated from the total number of non-residents observed over the months of camera operation (15 tigers/16 months= 0.94 tigers/month). Thus, the density of tigers in the study site was calculated as 6.94 tigers/160 km², or 4.3 tigers per 100 km².

Using a detailed and ground-truthed (i.e. verified on foot) GIS map of habitat types present in the park, the tiger density obtained above can be extrapolated over the area of all similar habitat found within Way Kambas National Park. In this case only the habitat types most similar to that found within the study site were included (dense secondary forest, mixed secondary forest/grasslands and scrub). This excludes much of the *Imperata cylindrica* grasslands (*alang-alang*), despite evidence from tiger secondary signs indicating that tigers utilise these areas extensively (though perhaps at lower densities than in other, more prey-rich habitats). In a further attempt to remain on the conservative side, the GIS model allowed the removal of a buffer zone representing an area of radius 5 km around all adjacent human settlements. This was considered necessary as a means of compensating for the possible effects of human-induced disturbance on the tiger population.

After the removal of this human-disturbance buffer zone the area of optimal tiger habitat in the park was calculated as 842 km². If the study site tiger density of 4.3 tigers/100 km² is extrapolated over 842 km² of optimal habitat, a total population of 36 tigers is obtained. This estimation does not include cubs, nor does it take into full consideration the high number of non-resident tigers recorded. A population of 36 tigers represents an 80% increase on previous population estimates for Way Kambas National Park. An absolute minimum density of tigers for the entire National Park can also be calculated based on the total number of individuals identified to date, providing a park density of 21 tigers/1300 km² = 1.6 tigers/100 km².

Photographs from camera traps also show that there is some overlap in the home ranges of tiger individuals of both sexes. Complex polygons, representing home ranges, were plotted around the camera sites where the presence of particular individuals was recorded. These range sizes were calculated by incorporating a buffer (or boundary) around the camera sites of 1.16 km in radius, equivalent to the average distance between all neighbouring camera sites throughout the intensively monitored region.

Figure 10.7 shows home ranges for three of the study site resident tigers (two females and one male), where female ranges (70 and 49 km²) are largely overlapped by that of a male (116 km²). Most camera sites have also shown evidence of the regular activity of more than one tiger individual. Three of the central camera locations have been visited by six individuals during the 16 months of monitoring, and a fourth location has been visited by nine individuals during the same period.

Range plotting of all tiger individuals suggests that overlap is common between and amongst both males and females, though the maximum number of resident males encountered at a particular site was only two individuals. Occurrences of multiple females at camera sites accounted for most of the cases of overlap described above.

However, when results from the remote cameras are analysed further it appears that the complex polygons described above are insufficient as an accurate representation of an individual's complete ranging behaviour. For male tigers, though the range polygons suggest considerable overlap, consideration of the occurrence of individuals at neighbouring sites suggests a more complex or micro-organised ranging behaviour (Fig. 10.8). Three pairs of cameras, though separated by a mean distance of only 1.6 km, show domination by either one or the other of two healthy adult males. The overlap of males when their ranges are analysed on a coarse scale, but not on this higher resolution, could have significant implications for tiger density and status assessment in the future, as well as for habitat management focused specifically on tiger conservation.

It is recognised that there are limitations to the methodology of camera monitoring as described above. Over the subsequent months, long-term camera monitoring will be required over a larger area in order to establish more accurate details regarding the size and extent of male and female home ranges. It will be important to establish whether individuals labelled here as non-residents actually have no permanent range of their own, whether they possess ranges primarily outside of the study site, or whether they maintain much smaller ranges within the study site itself. If the latter case is true then there may actually be a greater number of tigers within the study site than reported here. On the other hand, if the resident tigers identified in this study maintain large ranges, primarily outside of the study site, then the density proposed above may be an overestimate.

Radiotelemetry studies, running in parallel to the intensive camera monitoring, would provide more detailed ranging behaviour data for selected individuals, providing a perfect opportunity to compare results of these different techniques. However, a significant advantage of intensive

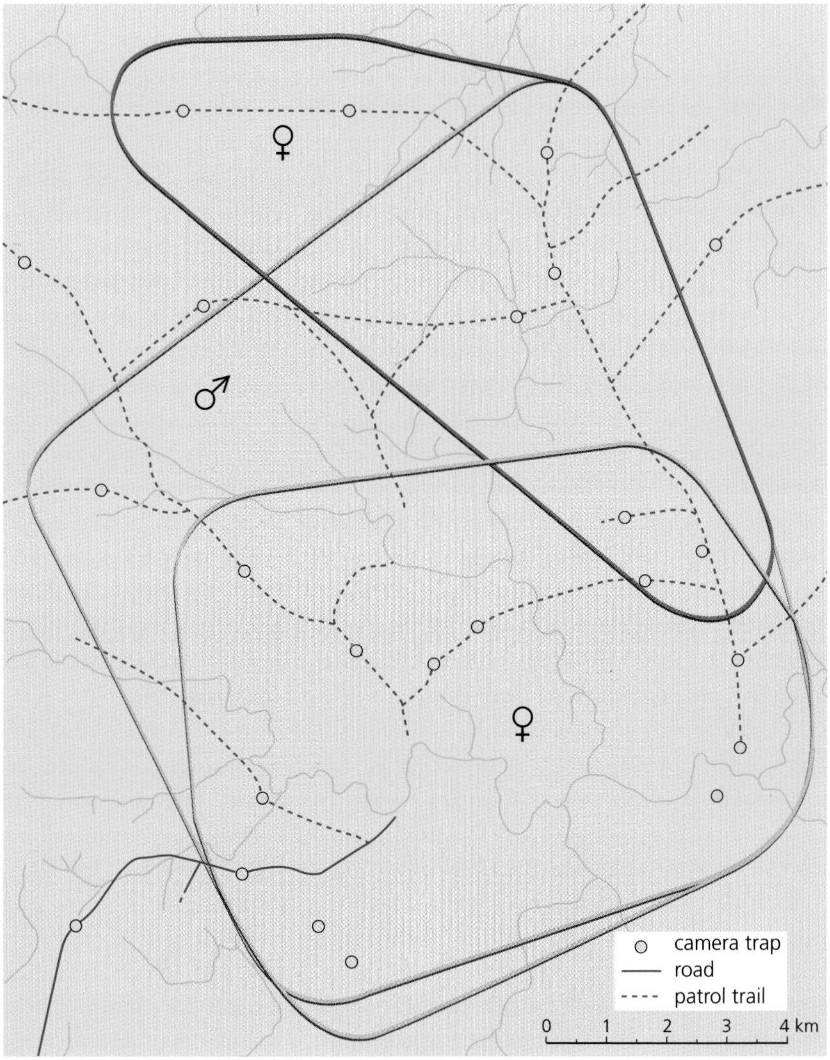

FIGURE 10.7
Three tiger home
ranges (two female
adults and one male
adult) from remote
camera photographs.

camera monitoring over radio-tracking is that all of the tiger individuals inhabiting an area are sampled with equal intensity (including residents, non-residents, recent immigrants and cubs) as opposed to the radio-monitoring of some arbitrarily selected target individuals.

A cause for optimism

It is a fact that tigers are rarely directly observed in Sumatra. When questioned at a tiger PHVA meeting in 1992, only a fraction of Forestry Depart-

ment staff from national parks across Sumatra had actually sighted wild tigers. Even fewer of these observations represented encounters with free-ranging tigers in their natural habitat. In Way Kambas National Park only seven of the 60 forestry field staff had made direct observations, despite having spent considerable proportions of their life working within the tigers' range. Staff from the Sumatran Tiger Project, working with these forest rangers, encountered tigers on only two occasions during more than 950 accumulated team-field days. During 440 days in the vast, mountainous Kerinci Seblat National Park, rhino survey teams reported

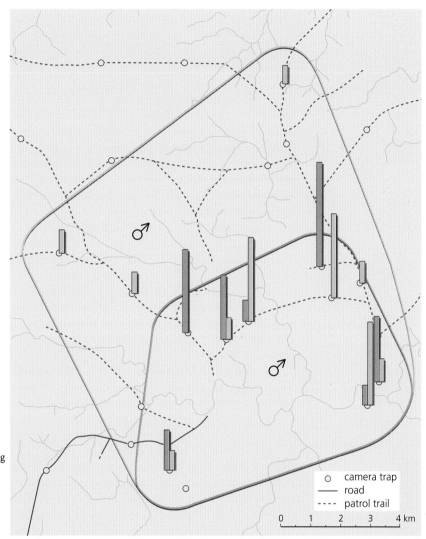

FIGURE 10.8
Two male ranges showing relative frequency of passing of the two individuals at neighbouring camera sites.

○ camera trap
— road
---- patrol trail

0   1   2   3   4 km

no tiger sightings despite covering more than 2800 km of remote and undisturbed forest trail (Franklin & Wells 1994), even though secondary evidence of tiger presence was recorded regularly.

In contrast to other regions of the world, fatalities attributable to tigers (both livestock and human) have been rare in Sumatra, though since the end of 1996 an increased incidence of attacks in several locations has been recorded. In Way Kambas National Park only one attack has been recorded in the last 10 years (Suherti Reddy pers. comm.) despite the park being surrounded by densely populated villages. Villagers living on the perimeter

of forested regions elsewhere in Sumatra seldom encounter tigers, particularly in protected areas where the boundary between park and settlements is distinct (see Box 10.2). Park intruders also report similarly low encounter rates. This lack of contact with tigers is exhibited in villagers' comments about tigers being rare and now only existing in other areas of Sumatra, when in fact tiger populations are often surviving in forests adjacent to their own farmland.

Relying on secondary signs as an indicator of tiger density is also difficult and is prone to lead to underestimation of tiger status. Pug and scrape

Box 10.2   The tiger--human dimension in southeast Sumatra
Philip Nyhus, Sumianto and Ronald Tilson

Before the last Javan tigers quietly disappeared from Meru Betiri National Park, a World Wildlife Fund report (Seidensticker & Suyono 1980) concluded that immediate action was necessary to develop public awareness and sympathy for their plight. Recommendations made in this report were not implemented in time, and the Javan tiger was soon considered extinct. Two decades later and little more than half a century after the death of the last Bali tiger (Nowell & Jackson 1996), a similar scenario is unfolding for the last 500 wild Sumatran tigers – and the challenge of saving these tigers remains just as daunting. For many people in Sumatra, as in much of Asia, tigers are a potent symbol of fear and respect (Wessing 1986; McNeely & Wachtel 1988; Lumpkin 1991). This relationship is so powerful that the tiger's spirit persists long after it has disappeared from the forest (Seidensticker 1987), a concept articulated in the traditional Malay saying, 'the tiger dies, but his stripes remain' (Bakels 1994). It is becoming increasingly clear that the survival of the tiger is ultimately linked to its relationship with people (Seidensticker 1997), and that tiger conservation strategies will fail without understanding these unique interactions. This lesson is particularly applicable to Indonesia, the only country to witness the extinction of two tiger subspecies.

Every year, livestock and people are killed by tigers in Sumatra. Farmers in turn poison tigers to protect their livestock and poachers hunt an undetermined number of tigers to supply the illegal trade in tiger parts (Plowden & Bowles 1997). These conflicts will likely escalate. Indonesian authorities have recognised the need to gather information about these tiger-human interactions in order to develop long-term conservation strategies for tigers and their habitat (Ministry of Forestry 1994). In response, the Sumatran Tiger Project (STP) initiated a community-based conservation programme to describe the tiger-human dimension at Way Kambas National Park, Lampung province, Sumatra. The goal of this programme is to enable conservation authorities to resolve tiger-human conflicts based on a comprehensive database rather than anecdotes and opinion. During its first 15 months, the programme administered more than 900 questionnaires in 15 villages, completed rapid evaluations in 20 villages, monitored human-wildlife conflicts with the assistance of village informants, and collected additional information from non-structured interviews and secondary data from 22 villages representing four major ethnic groups bordering the park.

As Sumatra's southern-most province and the closest to the densely populated island of Java, Lampung has particular relevance to tiger conservation because it provides an opportunity to 'preview' human-caused changes that have occurred, and are likely to occur soon, in many other regions of Sumatra. Just a century ago, the province was sparsely populated and covered by primary lowland forests. As a result of the Indonesian government's transmigration programme, which moved tens of thousands of villagers from the islands of Java, Bali and Madura to Sumatra, Lampung today is one of Sumatra's most densely populated regions. Way Kambas itself is now a habitat 'island' surrounded by intensive agricultural activity. No official buffer zone separates the park from the more than half a million people now living near the border.

Tiger sightings and conflicts were once common in and near Way Kambas. Villagers native to the area, the first transmigrants, and loggers who helped to clear the forest all recount frequent tiger sightings through the 1960s. Villagers report that at least five people were killed by tigers during this time but few serious tiger-human conflicts have occurred outside of the park since that time. The last death attributed to a tiger occurred in 1995 when a villager collecting grass inside the park was reportedly mauled by a tiger. This surprisingly low level of conflict is in stark contrast to figures for the rest of Sumatra. At least eight people were killed by tigers in the rest of Sumatra during 1996-97.

Tigers are rarely observed in or near Way Kambas, either by professional forest protection authorities (PHPA) or by field staff of the Sumatran Tiger Project (N. Franklin *et al.* this volume). Most villagers surveyed can identify tigers correctly from photographs, but few villagers report having actually seen a tiger. Villagers who enter the park to collect non-timber forest products regularly see tiger tracks but few actually encounter tigers. Sporadic but largely unconfirmed reports of tiger signs or sightings outside the park continue. Knowledge about tigers is generally low among villagers surveyed – many respondents believe that tigers are more abundant on Sumatra today than 20 years ago, and some villagers believe that tigers still occur on the island of Java.

Sumatran Tiger Project team members consulting with villagers to identify and seek solutions to human/wildlife conflicts.

Our study suggests that action to raise public awareness for the plight of the tiger is as critical today for the Sumatran tiger as it was for the Javan tiger. In addition, our study highlights the need to address other issues villagers consider important that indirectly threaten the long-term survival of tigers; crop-raiding elephants regularly kill and injure villagers and cause significant agricultural damage and forest resources once freely available are now off-limits to villagers. Villagers perceive few benefits from the park, and these issues create tension between villagers and forest protection authorities. Unless these problems are addressed, conflicts between the park and its economically poor, largely rural neighbours are likely to escalate and undermine efforts to protect southeast Sumatra's last tigers. A new conservation model is needed, one that includes the full participation of local communities, that can ultimately lead to a stronger environmental ethic and support for tiger conservation efforts. Our challenge is to provide the information and the means to enable the people of Indonesia to develop this model, and to ensure that more than just the tiger's stripes remain.

marks are found infrequently in tropical forests, which is primarily a result of the heavy rainfall, thick leaf litter and hard ground substrate. Where they are found they are usually of a quality and clarity that prevents measurement and classification. Faeces decompose rapidly under tropical rainforest conditions, leaving only remnant hairs of consumed prey. These are easily missed by passing field teams.

Remote camera monitoring in Way Kambas has shown that tigers exist in considerable numbers, despite a lack of evidence from either direct observation or the presence of tiger secondary signs. Tigers are observed to be highly mobile during the daylight hours yet they remain undetected, on the whole, by the groups of people working within the park. Evidence from preliminary surveys indicates that this pattern will be reflected in other lowland forest areas of the tiger's range across Sumatra.

Areas outside the protected area system may also hold significant populations, particularly where habitat degradation has produced a mosaic of habitat types similar to Way Kambas. Such areas would include old logging concessions and plantations in various stages of regeneration, but which are typically underrepresented in tiger population status assessments. Anecdotal evidence to support this abounds, even close to Way Kambas, where neighbouring concession holders complain of their reticence to deploy staff in certain areas because of what they say is an unacceptable risk from tigers. Many other areas of Sumatra possess habitat of similar quality and composition, particularly along the plains of the eastern coast.

Even though secondary habitats have been recognised as important for tigers (Seidensticker 1986), previous status assessments have undervalued the contribution of degraded, secondary forest/grassland habitats as potential refuges for tigers. There has also been overemphasis placed on the accuracy of direct observation as an index of tiger presence or absence. Surveys to assess the tiger population in all similar areas are an immediate priority, as they are for the other mountainous, forested regions within the Indonesian protected area system. The methodology used should be efficient in terms of time and personnel required, but should be comparable across the entire range of the Sumatran tiger, even when carried out by independent field teams (see following section).

Assessment of tiger status across Sumatra – Tiger Conservation Team

The Sumatran Tiger Conservation Strategy (Ministry of Forestry 1994) recommends as a high priority the conduction of large-scale and wide-ranging field surveys with the objective of confirming the present range of the Sumatran tiger. Such efforts, crucial in the development of an effective management strategy, will require co-operation on all sides.

A first step in the process of completing this surveying is the development of a ground-truthed GIS map showing all potential tiger habitat, such as that developed recently by the Ministry of Forestry. A second step will be the training of personnel from across the protected area system of Sumatra, which will facilitate the use of standardised survey methods in a wide range of habitat types by these independent teams. A third step in the process will be the initiation of an information network, enabling communication between managers of tiger range regions, assisting with the coordination of the independent survey teams, and providing a channel through which tiger sightings and tiger-human conflicts can be accurately reported.

With operations spread over a huge geographical range, efficient coordination of these survey efforts will be paramount. It is proposed that a specialised Tiger Conservation Team (Ministry of Forestry 1994) should be formed in order to provide support to the Department of Forestry's own forest rangers and park managers. This mobile unit will be equipped with the necessary skills to carry out its own field and local community surveys, using a variety of techniques, but including remote cameras, and will concentrate on covering regions where logistical difficulties are an acutely limiting factor. The Tiger Conservation Team will also take responsibility for collating the resulting tiger presence/absence information from across the

Sumatran subspecies range, and will be available to carry out 'spot-check' field surveys in order to confirm the results of other independent field teams. These efforts will increase the resolution and accuracy of Sumatran tiger distribution, and will provide population status assessments more firmly rooted in regularly updated information from the field. Areas of potential tiger habitat, identified as Tiger Conservation Units (E. Dinerstein *et al.* this volume), provide an ideal starting point for these distribution and status assessments.

Information collected by this mobile unit will include data on habitat type, land protection status and security, tiger presence/absence, prey abundance and other landscape ecology factors of significance to the tigers' long-term survival. Considerations of the characteristics and effects of human communities living adjacent to tiger areas will also be a significant component of the investigation.

11

The status of the Indochinese tiger: separating fact from fiction

Alan Rabinowitz

Introduction

In 1993, the number of tigers that comprised the Indochinese subspecies was listed as between 1050 and 1750 (Jackson 1993a). These numbers were based on little more than speculation. At present, we know less about the tiger in Indochina than we do about tigers throughout any other part of their range. This is due to the animal's secretive, forest-dwelling habits, and the fact that the range of this subspecies mostly encompasses six countries: Myanmar, Thailand, Lao PDR, Cambodia, Malaysia and Vietnam. Since many of these countries have been involved in wars, social unrest and political upheavals over the last several decades, there have been few opportunities to obtain reliable information about tigers in the region.

The use of Geographic Information System (GIS) technology to map existing habitat where tigers are assumed to be present was first proposed by Smith *et al.* (1987b). Only recently has this idea been developed in a more comprehensive manner to identify tiger priority areas and suggest a broader landscape approach to tiger conservation (Dinerstein *et al.* 1997). However, the assumption that tigers exist in all intact forest areas is based more on historical fact than current reality. Since factors such as prey availability, water resources and hunting play a role in tiger distribution and abundance, modelling tiger demographics based on habitat alone can only be a first step. Field research is needed to provide data concerning tiger numbers and the factors affecting their survival.

This chapter takes a look at what we know about tigers throughout the Indochina region. For each country, I consider the history of protection afforded to tigers and their habitats, and the reasons behind

the decline of tigers in that country. The maps for each country show how little recent data there are on known tiger distribution, and that even where tigers are present, they are not uniformly distributed throughout available forest habitat. Finally, the chapter summarises the current status of tiger conservation within the region and recommends some immediate actions to improve tiger protection and management.

Tiger presence data on the maps (red dots) are from field surveys after 1987 that recorded tiger tracks or sightings. Tiger absence data (black dots) may be from pre-1987 surveys if it was assumed that tigers had not come back into the area. Interview surveys (blue dots) and reports that did not cite a source (green dots) are treated as areas where tigers are potentially present, but need further investigation. Points on the maps indicate either locations along survey routes, or represent an entire forest block or protected area where tigers are reported present or absent. The sources for the numbered locations on the maps are given in Appendix 3.

The information presented in this chapter and the maps is derived from the author's search of the scientific and popular literature, unpublished country and trip reports, and from surveys by the author himself. While there is likely to be additional information in other unpublished reports, personal records, or from on-going research, such information was not available to the author during the writing of this chapter. I would like to thank J. Michael Cline and the Cline Family Fund, Gary C. Fink and associates at MCG HealthCare, Inc., Mr. and Mrs. George K. Moss, and the L. X. Bosack and B. M. Kruger Foundation for funding our efforts to study and conserve the earth's last remaining tiger populations. Forest-cover data for the maps were provided by J. MacKinnon, Asian Bureau of Conservation and the World Conservation Monitoring Centre. Base map data are from ArcWorld (Environmental Systems Research Institute). Anthony Lynam, John Robinson, Elizabeth Bennett and Melissa Connor gave critical review of the manuscript. Melissa Connor designed and created the maps.

Country assessments

Cambodia

Tigers have long been known to inhabit Cambodian forests (Ta-Kuan 1993) that, until the middle of this century, covered over 73% of the country's land area (World Bank 1996a). Yet documentation of Cambodia's wildlife, past or present, is meagre. During French occupation (1863–1954) no comprehensive wildlife surveys were reported (Martin & Phipps 1996). During the 1970s and '80s a tumultuous political situation resulted in massive community upheavals, reduced the human population by 10–20%, and virtually closed the country to the outside world. During searches for the elusive kouprey, tigers were noted from some of the more remote forest areas (Wharton 1957; Thouless 1987). Until the removal of the Khmer Rouge from power in 1979 and the withdrawal of Vietnamese troops by 1989, nothing more could be done concerning wildlife in the country.

Despite its violent history, Cambodia still has forests covering nearly 60% of its land area. Although Cambodia was the first country in Southeast Asia to establish a National Park (Angkor Wat), in 1925, there have been no legally protected areas for forests and wildlife until recently. In 1993, a royal decree set the groundwork for establishing a National Protected Area System, creating 23 protected areas, or nearly 20% of the total land area (Lay Khim 1995). Unfortunately, as of 1996, only two of the protected areas had any on-the-ground staff, and only one protected area had a management plan. Most of the remaining forest, however, and probably most tigers, occur outside protected areas. Much of these forests has recently been given as timber

concessions to foreign companies, without any restrictions concerning sustainable harvest levels (World Bank 1996a). Wildlife protection and management is further complicated by the fact that protected areas are under the control of the Ministry of Environment, while forests outside protected areas are under the Ministry of Agriculture, Forestry and Fisheries.

Coinciding with an improved economic situation and increased forest exploitation, commerce in wildlife has also escalated in recent years (Martin & Phipps 1996). Tiger parts and products are openly sold throughout Cambodia and there is increased trade across borders into Thailand and Vietnam (Martin & Phipps 1996). Wholesale consignments of wildlife parts are also shipped to China, Singapore and Taiwan (Stiles & Martin 1994).

Figure 11.1 shows that, despite the extensive remaining forest cover, we know almost nothing about tiger distribution in Cambodia. Even with new legislation that bans the hunting of wildlife, and a Wildlife Protection Office that is empowered to enforce the wildlife laws, large areas of forest are still unsafe. Yet, the potential for priority tiger sites in Cambodia appears great. Reliable survey data are urgently needed to formulate a long-term tiger management strategy for the country.

Lao PDR

Laos is one of the poorest nations in the world, with a per capita income under $200 per year and 85% of the population earning their living from agriculture. Since the communist take-over in 1975, most private businesses have been shut down and the economy has stagnated. Only since 1988 has the government opened its doors to the rest of the world and moved toward a market-oriented economy. As of 1991, Laos still had an estimated 47% forest cover, mostly in the hills and throughout the central and southern part of the country. Much of the north has been deforested from shifting cultivation (Berkmuller *et al.* 1995) and has been more heavily hunted than the rest of the country (Chazee 1990).

The wildlife of Laos has been relatively poorly investigated, with some early work by Delacour & Jabouille (1931) and Deuve (1972). The country's

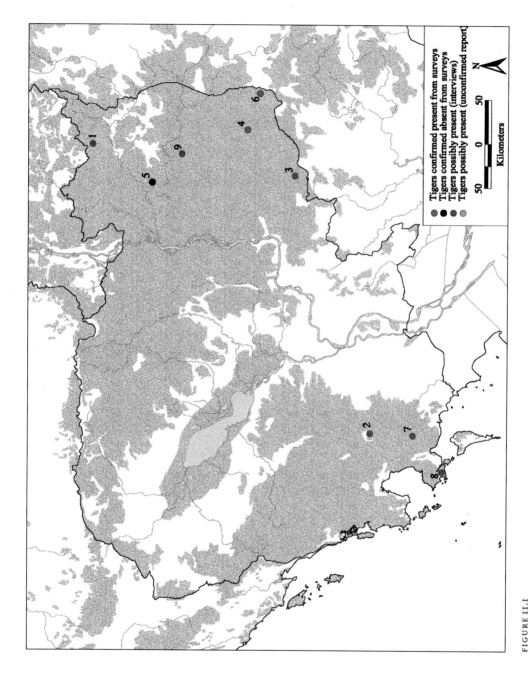

FIGURE 11.1
Tiger distribution data for Cambodia. For key to locations see Appendix 3.

current protected area system was established only in 1993 and contains 10% of the land area (Berkmuller *et al.* 1995). However, these areas were designated under a new protection category – National Biodiversity Conservation Areas (NBCAs) – which allowed for human settlements and subsistence use outside designated core zones. It was anticipated that after three to five years of management, a sizeable core zone meeting the criteria of a national park or nature reserve would be established within these areas (Berkmuller *et al.* 1995). As of 1995, only 5 of 18 protected areas were under some form of management and not a single NBCA had a legally designated 'core area' within its boundaries. Consequently, hunting and land clearing are still widespread within most protected areas.

Hunting of wildlife is a major problem in Laos. After the introduction of modern firearms in the 1940s, the killing of large mammals increased (Salter 1993). Chazee (1990) estimated 1.9 guns/km² in Laos and speculated that most large mammal species would disappear within a decade if nothing was done about the situation. In addition to guns, the use of snares, traps, dogs, crossbows and explosives is also widespread. While some hunting is done for food, much of it is now done to supply the escalating cross-border trade in animal parts.

The hunting of tigers was banned in 1989 when the Government of Laos recognised that they were in a dangerous state of decline (Salter 1993). Today, tiger populations are so depleted that they appear to be at low densities wherever they are reported (Berkmuller *et al.* 1995). Tiger parts are sold openly in the capital city, Vientiane, and in markets along all the borders (Martin 1992b; Srikosamatara *et al.* 1992; TRAFFIC 1993). The three most hunted wildlife species – barking deer, wild pig and sambar deer – are also preferred foods for tigers (Salter 1993). Since tigers need larger prey species to survive (Karanth 1995), depletion of the medium- to large-sized prey base can favour the survival of other species, such as leopards, over tigers (Rabinowitz 1989).

Despite the intense hunting pressures, a rugged landscape and a low human density have allowed Laos to maintain a high species diversity throughout much of the countryside. Recently, several new large mammal species have been discovered in the remote forest pockets of the Annamite Mountains along the Vietnam border (Rabinowitz 1997b). Although recent wildlife surveys have provided some of the only reliable information on tiger presence in the country (see Fig. 11. 2), much more needs to be done. It is particularly disconcerting that recent data point to continuing declines in already low-density tiger populations.

At present there are no management plans for the tiger, nor are there sufficient protected area staff to visit the protected sites. With little else to export for foreign exchange other than timber and hydro-electric power, the Lao government has little choice but to exploit its natural resources. The country is at a crucial crossroads in the battle to balance development with conservation. There is still time to save tigers in Laos if some of the best tiger areas can be protected adequately and maintained, either as separate parks and wildlife sanctuaries or as secure, large core areas within the NBCAs.

Peninsular Malaysia

Within the last century, tigers have been found throughout the extensive forests of Peninsular Malaysia, and have been regularly hunted for sport (Wallace 1869; Whitney 1905; Locke 1954). But from 1957 to 1992, forest clearing accelerated rapidly and the protected area system in Peninsular Malaysia increased from 5.3 to only 6.3% of the land area, with only three protected areas larger than 1000 km² (Aiken 1994). The few new protected areas were in 'economically undesirable areas', usually not the best habitats for wildlife, while some of the best land in the older reserves was excised for agriculture, mining and logging. This loss of optimal habitat, along with the poaching by hunters, caused rapid declines in tiger numbers (Blanchard 1977).

While some significant advances for tigers occurred in Malaysia, such as the improved management of Taman Negara, the country's largest national park, most of the other protected areas continued to suffer from excision, degradation, or simply non-action. When one of the last extensive

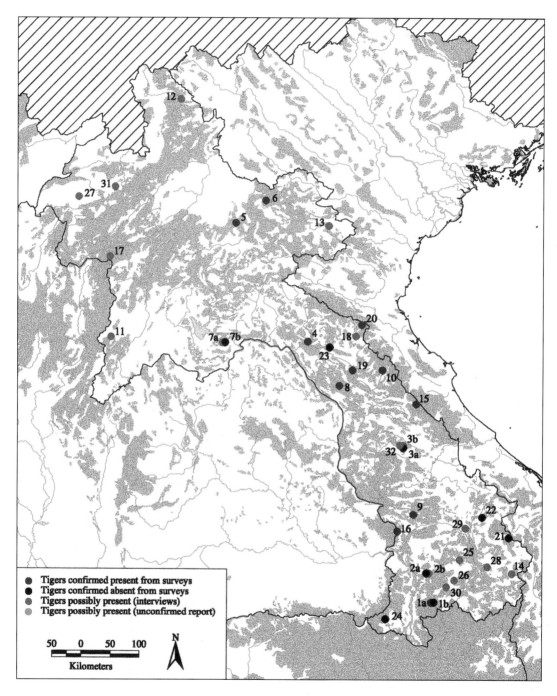

FIGURE 11.2
Tiger distribution data for Lao PDR. For key to locations see Appendix 3.

tracts of rain forest was proposed for protection in the early 1970s as Endau-Rompin National Park, the move was stymied by the long-established constitutional division of powers between the federal and state governments. Finally, in 1992, the proposed 2000 km² area was reduced to a state park of less than 500 km² (Aiken 1993). By this time, the 'effective' protected-area coverage throughout the peninsula was no greater than that of 1940 (Aiken 1994), and much of the wildlife had been lost. It was now the Forest Reserves, set aside mainly for production of timber, that harboured much of the remaining wildlife (Aiken & Leigh 1995).

The tiger was given total protection in 1976, at which time it was estimated that there were only about 300 tigers left in Peninsular Malaysia (Khan 1987) compared with an estimated 500 in 1975, and 3000 in the early 1950s. By 1987, a new government estimate based on 'work carried out over the years' reported 505 tigers, stating that conservation efforts since 1972 had resulted in an increase in the population. In 1988, the penalty for poaching or possession of a tiger or tiger parts was increased to $6000 and/or five years imprisonment. Topani (1990) estimated 482–520 tigers in 1990, based on reported sightings and field surveys. No methodology or justification for the numbers were given in the report. The government then readjusted their estimate of the tiger population in Peninsular Malaysia to 600–650 animals, claiming that Topani's estimate was conservative (Samsudin & Elagupillay 1996).

Government officials state that their country is too small to establish special tiger reserves and they feel that the existing protected area system, along with the Forest Reserves, does a sufficient job in protecting tigers (Samsudin & Elagupillay 1996). However some Forest Reserves that contained tigers in the past no longer do so (Laidlaw 1994), and many protected areas have no detailed management plans (Ngui Siew Kong 1991). Figure 11.3 indicates that there is still a lack of sufficient tiger data for Peninsular Malaysia. If tigers are to survive here in the future, the tiger situation needs to be more closely investigated and tiger management efforts need to be more focused.

Myanmar

Tigers have long been considered abundant in Myanmar (Sangermano 1833; Evans 1911). During the early part of this century, White (1923) reported tigers being killed in the capital city of Rangoon. In the 1929 Burma Game Manual a bounty was placed on the killing of tigers and other carnivores (Tun Yin 1993), resulting in 1100 tigers being killed within a three-year period (Peacock 1933).

Prior to the Anglo-Burmese war in 1852, much of Myanmar was still covered in forest. However, expansion of rice production under British colonial rule led to the massive deforestation of most lowland areas. After independence in 1948, Myanmar was economically devastated, leading eventually to military rule and widespread insurgency among the country's ethnic groups. By 1987, the country was given 'least developed status' by the United Nations, and the government controlled less than 40% of the country.

Today, forest still comprises nearly 40% of the land area, much of it restricted to upland sites. Although the first game reserve was set up in 1918, the current protected area system in Myanmar contains only 21 sites, which cover little more than 1% of the country. Only two protected sites are larger than 1000 km² and few are under any form of active management. A proposal to increase the protected area system to at least 5% of the country (Salter 1994) is currently being implemented by the government.

As seen in Figure 11.4, very little data are available post-1987 on the presence and distribution of tigers in Myanmar. Until recently, government officials could not travel throughout much of the countryside, so protected areas remained 'paper parks' in the truest sense of the term. Now many of these protected areas have been so heavily degraded that they no longer contain the species that they were designated to protect, and most of the tigers are in the forests outside these areas (Tun Yin 1993).

In 1981, the Myanmar Forestry Department used historical records to make an estimate of 3000 tigers in the country. But surveys by Milton & Estes (1963) and by the UNDP/FAO Nature Conservation and National Parks Project in the 1980s reported a

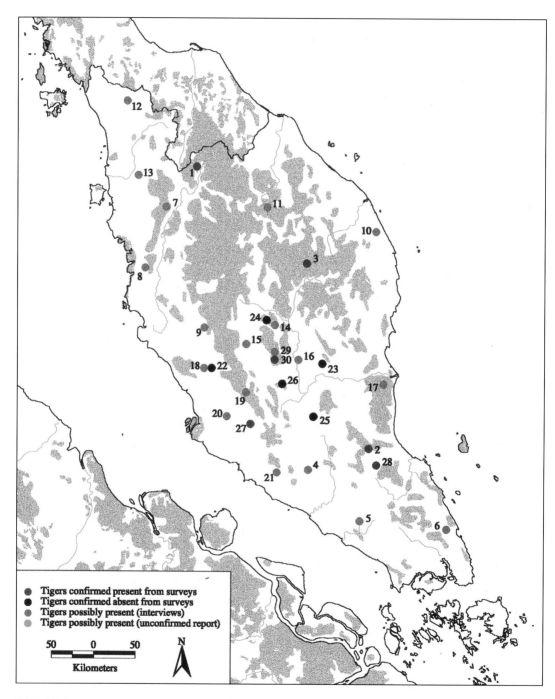

FIGURE II.3
Tiger distribution data for Peninsular Malaysia. For key to locations see Appendix 3.

Tiger parts – skins, penises and other pieces – openly displayed for sale at Tachilek near the Thai/Myanmar border. Tigers will only survive over much of their remaining habitat when demand for tiger parts is significantly reduced.

surprising paucity of tiger evidence in some of the most remote areas. Where tigers occurred, they were subject to hunting and poisoning (UNDP/FAO 1985). During a more recent survey in Tamanthi Wildlife Sanctuary, the country's largest wildlife sanctuary, tiger evidence was not only scarce but Lisu hunters were encountered setting snares for tigers (Rabinowitz *et al.* 1995). When the government devised a National Tiger Action Plan in 1996, they revised their estimate of tiger numbers to 600–1000 for the country. This figure was based on published tiger densities from studies elsewhere, and the supposition that only half of the remaining closed-canopy forests in the country are populated with tigers.

One of the driving forces behind tiger poaching in Myanmar is the well-established trade network for tiger parts. During a recent visit to Mandalay by the author, wildlife shops admitted to sending large shipments of wildlife parts into China. In the city of Myitkyina, tiger skin was selling for $5 per square inch, tiger rib bone for $4.50 per inch, and a piece of a tiger's lower jaw bone for $15. For years the market at Tachilek, along the Thailand border, has sold tiger parts.

With its extensive forests, controlled development and relatively low population pressures in the countryside, Myanmar still has an excellent opportunity to protect and conserve its tigers. An immediate priority is to find out where good tiger

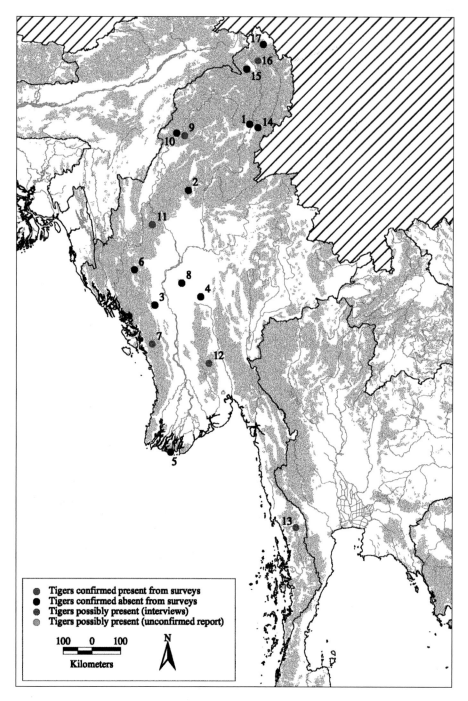

FIGURE 11.4
Tiger distribution data for Myanmar. For key to locations see Appendix 3.

and prey populations still occur and to expand or set up new protected areas that encompass these populations. In addition to these activities, additional Forestry and Wildlife staff need to be recruited and trained in tiger monitoring activities and protected area management.

Thailand

Thailand is the centre of the range of the tiger in Indochina, and early reports of tigers roaming Thailand's forests date back more than three centuries (Gervaise 1688). By the early 1900s, many of the reports of Thailand's remaining tiger populations came from the border areas with Burma (Bock 1884; Gairdner 1915) and Cambodia (Koch-Isenburg 1963). During the last two decades, as Thailand became one of the wealthiest and most economically stable countries in Southeast Asia, forest cover decreased from nearly 60% to 25%. Yet also during this period, Thailand created one of the largest and most extensive protected area systems in the region. Since designating its first national park in 1962, Thailand now has 81 national parks, 39 wildlife sanctuaries, and 49 non-hunting areas (Forest Department statistics 1996). Although most of these protected areas are relatively small and isolated, there are at least 12 national parks and wildlife sanctuaries over 1000 km² in size, and eight protected forest complexes with combined areas greater than 2000 km². In 1989, Thailand became the first country in the region to ban all logging, a move that resulted in increased logging in Myanmar (Hill 1994).

By 1979, guesstimates of tiger numbers in Thailand ranged from 500 to 600 individuals (Leng-EE 1979; Lekagul & McNeely 1988). By the early 1990s, research in one of Thailand's largest wildlife sanctuaries, Huai Kha Khaeng (Rabinowitz 1989), followed by surveys and interviews in other protected areas led to an estimate of no more than 250 tigers throughout the country (Rabinowitz 1993). Hunting was still continuing in most protected areas and the trade in tiger and other wildlife products was still ongoing (Humphrey & Bain 1990; Rabinowitz 1991). Even in Thailand's showpiece protected area, Khao Yai National Park,

protection and management of wildlife was not being effectively carried out (Griffin 1994).

A workshop that resulted in the Indochinese Tiger Masterplan for Thailand (Tilson et al. 1995) concluded that Thailand's Forestry Department was too understaffed and underbudgeted to be effective against the onslaught of hunting, human encroachment, loss of habitat and loss of tiger prey occurring throughout the country. Yet some critics have long maintained that there are more than adequate staff and funds in Thailand's Forestry Department, but that the government needs to concentrate on already existing priority protected areas instead of continually designating new protected areas (Sayer 1982). The 1996 'Status of the Tiger' report (Jackson 1996) cited a range of 250–600 tigers in Thailand. The lower figure was given as a maximum number of tigers as indicated by survey work (Rabinowitz 1993), while the upper figure was an unsubstantiated guess by the Royal Forest Department.

More is known about tigers in Thailand than about tigers in any other country in the region, and new information is still coming forth (see J. L. D. Smith et al. this volume, Chapter 12). Yet, the current available data are still insufficient and mostly unsubstantiated (see Fig. 11.5). However, Thailand is presently in the best position of any country in the region to protect and manage its tiger populations. Most of the country's tigers already occur in established protected areas or large protected forest complexes. The protected area system is relatively well staffed and well funded. The biggest obstacle is that the largest forest blocks, comprised of contiguous protected areas, are not managed as single units. The first stage in a tiger management strategy for Thailand should include more systematic field surveys, and a plan to institute coordinated management of the large protected forest complexes.

Vietnam

With approximately 200 people/km², Vietnam is one of the most densely populated agricultural countries in the world and, at present rates of growth, is expected to double its population in the

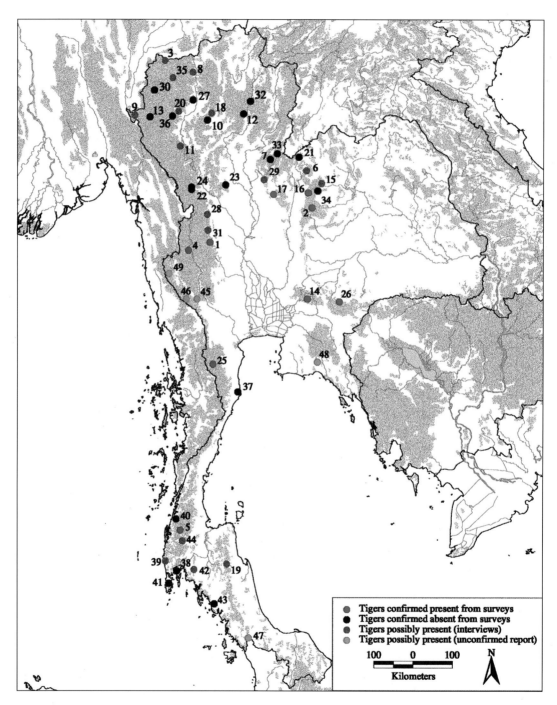

FIGURE II.5
Tiger distribution data for Thailand. For key to locations see Appendix 3.

next quarter century. Since the 1940s, the country has lost 80% of its forest cover, partly due to three decades of uninterrupted warfare (Collins *et al.* 1991). Since the end of the Vietnam War in 1975, slash and burn agriculture, fires and fuelwood collection have caused continued forest loss. Today only 10–15% of the country's original forest remains, much of it in relatively small and fragmented pockets.

Of 59 protected areas listed as of 1992 in Vietnam, only eight were declared before 1986, and only one area is greater than 1000 km². Recent surveys in both protected and unprotected areas have indicated relatively high levels of hunting, illegal logging and collection of forest products. These activities, along with population growth and agricultural expansion, continue to eat away at the remaining forest areas (Eames *et al.* 1994).

A recent checklist of mammals in Vietnam lists the tiger as present throughout much of the north, and most of the border provinces with China, Laos and Cambodia (Dang Huy Huynh *et al.* 1994). Although the tiger has been legally protected since 1960, and a 1989 law prohibits the trade of tigers, tiger parts are still easily available in Vietnamese markets (Martin 1992a). A 1993 survey found a total of 37 tiger skins and 1166 tiger claws on display in five Vietnamese cities (Smith 1993). Despite continuing hunting pressures and loss of forest habitat, the 1996 'Status of the Tiger' report lists tiger numbers in Vietnam as 200–300 individuals (Jackson 1996), the same number given three years earlier (Jackson 1993a). There is no indication how these numbers were derived.

There are few large forest areas left for tigers in Vietnam, and there are not many recent tiger data even from these areas (see Fig. 11.6). With a rapidly expanding economy, increasing population pressures, and much of the remaining forest in very small isolated pockets, Vietnam needs to concentrate its conservation efforts in areas where the forest is still intact and where tigers and their prey still exist in some numbers. If any tigers are to survive in the future, Vietnam needs more field data that can be used to establish a few special 'tiger protection sites'. In conjunction with this, the

government must show initiative in eliminating the wildlife trade.

Discussion

Since 1993, with the establishment of The Delhi Declaration at the Global Tiger Forum in India, there have been numerous regional and international workshops and conferences discussing the status of tigers. At a trans-boundary biodiversity conference for the Indochina region, the Indochinese tiger was declared an important flagship species for promoting regional conservation efforts along border areas (Rabinowitz 1995a). In 1995, the US National Fish and Wildlife Foundation and the Exxon Corporation launched the Save the Tiger Fund, pledging $5 000 000 over a five-year period. All of these actions have helped keep tiger conservation a high-priority issue among government officials, but there have been too few actions 'on-the-ground'.

The establishment and maintenance of a well-managed protected area system is a crucial component of tiger conservation. Local communities, left completely to their own devices, will not save tigers. I concur with the opinion of Karanth & Madhusudan (1997) that while incentive-based community conservation schemes are a laudable future goal for conservation, the current tiger situation is too critical to rely solely on long-term mechanisms. However, the idea that tigers are protected simply by establishing protected areas is no longer a valid concept, if it ever was. Even in the largest remaining forest areas, poaching and the wildlife trade continue to be the most insidious threats to remaining tiger populations. The many small tiger populations in isolated forest blocks are at the greatest risk and may not be salvageable. As poaching in these areas continues over time, the probability of population extinction increases drastically (Kenney *et al.* 1995). And while any one factor alone may not bring a species to extinction, the effects of other factors, e.g. habitat loss and prey reduction, become compounded as extinction mechanisms become synergistic (Pimm 1996).

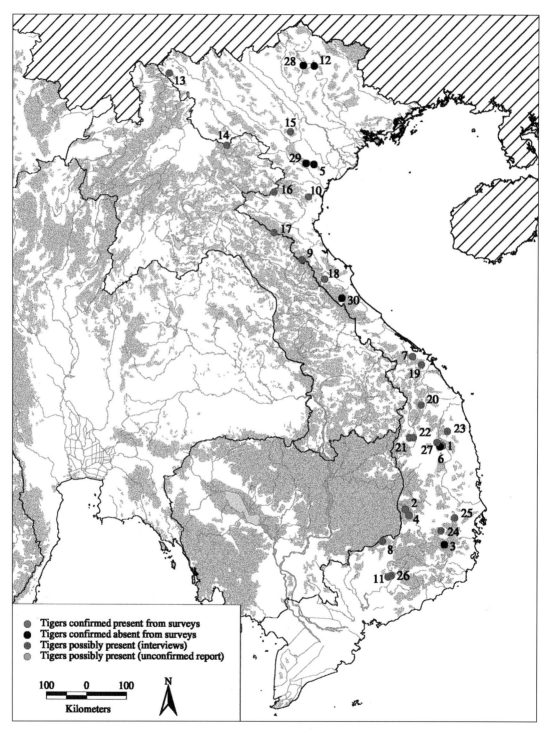

FIGURE II.6
Tiger distribution for Vietnam. For key to locations see Appendix 3.

To protect tigers, two realities must be faced. First, land must be designated, protected and managed for tigers and their prey in perpetuity. High-priority tiger areas cannot be compromised. This will increasingly conflict with the needs and desires of growing numbers of local people. Secondly, not all tigers alive today can be saved for the future. Areas in which small populations may not survive should not drain the limited human and financial resources available for conservation. 'Triage', defined as the allocation of treatment according to a system of priorities that maximises the survivors of a battle or disaster, is how we must now go about our efforts to save the tiger. Otherwise, the tiger may follow the example of the Sumatran rhino (Rabinowitz 1995), whose continued demise has been assisted by years of inaction, misdirected initiatives and political correctness.

The sections that follow summarise some of the most basic issues concerning the conservation of the Indochinese tiger and recommend a few actions that should be immediately implemented if tigers are to be better protected and managed in the region.

The current realities of tiger conservation in Indochina

Data collection

1 There is a startling lack of substantiated, quantitative data on tiger presence and distribution throughout most of the region. Most field survey teams are poorly trained and lack appropriate guidelines for standardised tiger surveys.

2 Interviews and/or questionnaire surveys that attempt to obtain information about tigers are often done in a haphazard, subjective manner that does not provide reliable or accurate information.

3 Field survey data are often not published or made easily available to others involved in tiger conservation. When such data are reported, it is frequently difficult to evaluate or repeat the methodology.

Tiger habitat

4 Much of the prime tiger habitat, consisting of lowland forests with good water resources and a diverse large prey base, has been converted to other land use practices. Many protected areas, particularly those created in the last two decades, are in rugged, often degraded habitat that is suboptimal for the survival of tigers and prey populations.

5 Many areas where tigers live consist of relatively small, isolated forest blocks that may not support viable tiger populations in perpetuity.

6 Many protected areas and other forests with tigers are being degraded by human settlements and activities that destroy or change the forest structure. Access by tigers to reliable water sources is often cut off by such activities.

7 Any landscape or community approach to tiger conservation will not be effective without a core of well-managed protected areas in place.

8 As human populations grow and the demand for land increases, increasing pressures will be brought to bear on remaining forests and protected areas.

Management and protection

9 Most protected areas containing tigers are inadequately staffed and have little or no monitoring and management of wildlife.

10 The staff of most protected areas are poorly trained, and are not given sufficient financial or social incentives to create a feeling of respect for themselves and the importance of their jobs.

11 No effective tiger management policies have yet been designed and implemented in any of the Indochinese tiger range countries.

12 There is a lack of real commitment by many government officials towards any conservation actions that are difficult or controversial.

Hunting

13   Partly due to increased market demands and partly due to the increased political stability of the region, the hunting of tigers and their prey has increased in recent years. Much of the hunting is for animal parts, using non-selective and inhumane hunting methods including snares, jaw traps, poisoned carcasses and traps.

14   Primarily as a result of hunting pressures on both tigers and their prey species, tigers are no longer present in many forest areas where they once occurred.

15   Where tigers do occur, surveys generally point to lower than expected tiger and/or prey densities due to hunting pressures.

Trade

16   The trade in wildlife, particularly tiger parts and their prey, is more extensive today than a decade ago. Such trade occurs at local, regional and international levels.

17   Although the wildlife trade is acknowledged as a major issue in tiger conservation, Cambodia, Laos and Myanmar are still not party to the Convention on International Trade in Endangered Species (CITES). (Subsequent to the Symposium, Cambodia and Myanmar have signed the Convention.)

18   If the trade in tiger and other wildlife parts cannot be effectively controlled, the protection and management of tiger populations will become an almost insurmountable task in most range countries.

Recommended immediate actions to conserve the tiger in Indochina

Data collection

1   Establish simple and inexpensive methods to quickly and accurately assess tiger presence, distribution and relative abundance throughout the different tiger range countries of the region.

2   Publish a manual in each of the range country languages dealing with tiger survey and research techniques, track identification and simple, inexpensive management practices to help existing tiger and prey populations.

3   Create tiger survey units to conduct field surveys and to monitor tiger and prey populations in areas of remaining tiger habitat.

Management and protection

4   Following up on the results of action 3, identify and prioritise the most important tiger areas within each country.

5   Design tiger management action plans for each country in the region, with a timetable for implementation. Progress on implementation should be reviewed by a 'tiger committee' once a year.

6   Ban the use and sale of all equipment used in non-selective and inhumane forms of hunting, particularly snares and jaw traps.

Trade

7   Create a government mechanism within each range country that can deal quickly and effectively with wildlife trade issues. Any range country that has not signed CITES should do so immediately.

Regional

8   Initiate a regional training programme for mid-level and senior government staff. Emphasise cordinated protection and management of tiger populations along international borders.

9   Bring together delegates from all the Indo-chinese tiger range countries once every two years to exchange information, promote trans-boundary conservation, and establish multi-lateral mechanisms for addressing wildlife trade and other conservation issues.

Establishing and maintaining a well-managed protected area system and incentive-based community conservation programs for people who live near tigers is crucial in tiger conservation.

Box 11.1   Moving conservation forward: surveys, training, planning and action must go hand in hand
Alan Rabinowitz

A wildlife biologist radio-collars 10 tigers and follows them for two years. A Peace Corps volunteer teaches maps and compass techniques to forest guards. The chief of a wildlife sanctuary sits down with his staff to draw up a five-year management plan. Rangers distribute wildlife posters in a village and explain about the hunting regulations for the area. All of these different activities appear almost unrelated. But what if the chief of the sanctuary uses the biologist's data for the management plan, and the park guards use their map and compass training to patrol the area that protects the tigers, and the posters teach local hunters about the laws and penalties for killing wildlife in the sanctuary. Now these separate activities create a synergism that goes far beyond the benefits of the individual actions. And it is this synergism that moves conservation forward in a meaningful and lasting manner.

Sharing technical know-how: a powerful conservation toolbox in the hands of local tiger stewards can smooth the path to sustaining tiger populations in the future.

Conservation is a creative art form that involves skills that are acquired by experience, study, observation and thought. As in any creative endeavour, an idea or realisation must initially take shape. In this case, perhaps it dawns on the field scientist that, despite all the apparent efforts, we are still losing tigers! Sightings are a rare event and local villagers no longer regularly find tiger tracks in areas where they were considered abundant only ten years earlier. But why is this happening? The biologist must now search out the recent literature, look at government reports and perhaps discuss the idea with others. What tiger data have been collected over the last decade? What information is lacking? At this point, if the issue of tiger declines is considered urgent enough, perhaps a research project is initiated to collect more data. Surveys are carried out, tiger distributions are mapped, and some long-term research projects, such as radio-tracking, are proposed in some of the best tiger habitats. At the end of such projects, a final report or research paper is produced that might identify priority tiger areas and recommend actions for better protection and management of the species.

The wildlife biologist may feel that he or she has completed the job at this point. But a conservationist realises that the work has just begun. It is now time to step outside the bounds of science and to work with government agencies, non-governmental organisations, community leaders and fellow scientists in designing a realistic plan that will be implemented by the appropriate people. This is perhaps the most difficult of tasks because 'the appropriate people' may involve individuals at the national, state, provincial, district and village level. Furthermore, the objective, such as the protection of management of tigers, must not just be written on paper but must be incorporated into the everyday thinking of the people and government agencies involved. The idea of saving tigers for instance, and the importance of implementing a good management plan, must be considered a social necessity, not a political whim. Very often commitment to an idea must be taught and nurtured. Training programmes for senior and junior level staff that have the flexibility to adapt to different cultures, education levels and perspectives are a crucial part of this process. Decision makers often need to be taught how to make the right decisions, and how better to interact and direct their staff in the field. Field staff need to be given not only a toolbox of field research techniques, but a respect for themselves and their work. There are ways that one can initiate such training, yet what works today may not be appropriate tomorrow. Changing political, social and economic situations may mandate new approaches to protecting and managing wildlife, and the conservationist must be ready to adapt to these changing situations.

A good conservationist is a jack-of-many-trades who participates in and helps expedite the conservation process from start to finish. However, one of the most important and sometimes most difficult parts of this process is to realise the appropriate time to remove yourself from it. There is never closure in conservation, nor will there be a time when we can become lax in our vigilance about protecting forests and wildlife. But in the end, success must be measured by the commitment of the people on the ground, taking care of their own forests and wildlife and acting as mentors to the future generations who will hopefully do the same.

Building conservation partnerships: consultation on needs and problems to be addressed with the people who share their forests with tigers is essential to securing the tiger's future.

12

Metapopulation structure of tigers in Thailand

James L. David Smith, Schwann Tunhikorn, Sompon Tanhan,
Saksit Simcharoen and Budsabong Kanchanasaka

Introduction

In the early 1990s reports of tiger poaching increased dramatically throughout the tiger's range (Kenney *et al.* 1995). This new threat rekindled concern for tigers and in 1993 they were a major topic at regional and international meetings of the Convention on International Trade in Endangered Species (CITES). Also in 1993, an organisational meeting to establish the Global Tiger Forum was held in New Delhi, India. At the Global Tiger Forum there was consensus that tiger poaching had reached a critically high level that required regional action to halt the decline of tigers. In 1994 Thailand's Royal Forest Department (RFD) hosted the First International Conference to Assess the Status of Tigers at Huai Kha Khaeng, Thailand, October 20–24, 1994. This conference was a collaborative effort among the RFD, the IUCN Cat Specialist Group and the World Conservation Monitoring Centre (WCMC), and its purpose was to further Asian cooperation in addressing an Asian conservation problem. At the conference the WCMC provided forest-cover maps of each tiger range state, and preliminary regional tiger distribution maps were made based on the 'expert' (Luscombe 1986) knowledge of the participants. A resolution of this conference called for mapping tiger distribution to the level of individual populations and the convening of two workshops on field assessment techniques and geographic information systems to coordinate mapping efforts.

Wildlife researchers at the RFD realised that rigorous techniques for the evaluation of tiger populations and a comprehensive understanding of the status and threats to those populations were needed. Without baseline data on tiger populations it is very difficult to measure the effect of poaching and habitat loss on the survival of the tiger. Until such measures are developed, it will be difficult to understand the nature of the problems that threaten tiger survival and to take appropriate actions to stem its extinction. To determine the magnitude of threat to the tiger, to prevent its demise as a species and to use the tiger as a symbol for Thailand's biodiversity conservation efforts, the Wildlife Research Division of the RFD launched a tiger conservation project, Saving Thailand's Tigers, in 1995. The goals of this project are to assess the status of tigers, develop an ecosystem approach to managing each remaining tiger population, and formulate a National Tiger Action Plan.

The first step in assessing the status of tigers is to determine the metapopulation structure of this species in Thailand. A metapopulation is a group of geographically related populations; its structure is the number, sizes, spatial configuration and degree of isolation among these populations (Levins 1970; Gilpin & Hanski 1991). To describe the metapopulation structure of tigers requires knowing where tigers live in relation to the size and spatial pattern of forest cover, where barriers exist that separate breeding populations (Smith *et al.* 1987b; Ahearn *et al.*1990; Smith 1993), where habitat is degrading, and where humans have depleted the tiger's prey base to the point that population size declines or tigers become locally extirpated.

This chapter reports the current knowledge of the metapopulation structure of tigers in Thailand. It draws attention to trans-border areas where populations in Thailand extend into neighbouring countries. The methodology to assess the status of tigers and the current actions to implement eco-system management in each of the protected area complexes occupied by tigers are described.

Methods

To determine the geographical distribution of tigers we developed a national Geographic Information System (GIS). The base maps of this spatial database were digital forest cover and the protected area coverages of Thailand. A forest-cover map provided the spatial extent of potential tiger habitat. These forest-cover data were derived from a visual interpretation of 42 1991 Thematic Mapper satellite images (Miller *et al.* 1989; Davis *et al.* 1990; Gagliuso 1991). This manual interpretation was completed in 1995, but like the WCMC forest-cover data used at the 1996 conference and the data used by E. Dinerstein *et al.* (this volume) there are no metadata that document the accuracy of the interpretation of forest-cover data or the spatial characteristics of the resulting maps. We digitised the resulting 1:1 000 000 forest-cover map to create a GIS base map of six forest types (evergreen, mixed deciduous, dry dipterocarp, pine, mangrove and scrub forests). The protected area data layer for the GIS was created by digitising the official RFD boundaries of the protected areas that are demarcated on >700 1:50 000 maps. A higher resolution was used for the protected area coverage because this coverage also serves as the legal demarcation of the protected area boundaries.

The initial step in acquiring tiger distribution data was to hold a series of seven 'expert group' workshops (Luscombe 1986). The 'experts' were RFD rangers and senior staff from protected areas. Using a nominal group process, 562 participants were asked to confirm the presence and absence of tigers and 15 other large mammal species in their respective protected areas and to qualitatively determine the rarity of each species. After rangers from a protected area individually filled out a data card, they were asked to discuss their conclusions to resolve discrepancies in their reports.

At the workshop, participants also received training on how to survey large mammals; the goal of this training was for rangers to conduct field surveys in the vicinity of their substations. They were trained to identify animal sign, determine locations in Universal Transverse Mercator (UTM) coordinates, and record their observations on data cards. They were given data cards to record direct sightings or signs of large mammals and asked to conduct surveys after the workshops. Data we requested included protected area and substation name, the UTM coordinates and type of observations, measurements of tracks and scrapes, habitat type and name of observer. When rangers did not have topographical maps or had difficulty locating their position they were asked to use a generalised UTM location of the substation where the observer resides. At the workshop we mapped the locations of 681 substations distributed in 110 protected areas and provided each protected area with a set of 1:50 000 maps and a list of the UTM coordinates of each substation. These data were also incorporated into our GIS (Fig. 12.1).

To verify the results of the ranger-conducted surveys, biologists with the Wildlife Research Division (RFD) have begun surveys of four hypothesised populations: the Western Forest Complex, Khao Yai, Taplan/Pong Sida and Khao Ang Rui Nai/Khao Sok. A Global Positioning System is used to record the route of the survey and the locations of tiger and other large-mammal signs. Observation data recorded are the same for both surveys and include the date, place, observer, geographical coordinates in UTM units, the type of observations, and measurements of tiger and leopard tracks and scrapes (Smith *et al.* this volume Chapter 13).

To assess the prey base available to tigers we examined the number of prey species that occurred in a protected area. Prey abundance is a more appropriate measure to assess the carrying capacity of tiger habitat, but obtaining these data across all the protected areas in Thailand was beyond the scope of this initial tiger assessment. The major tiger prey species were among the 16 large mammal species recorded during the field surveys. In each terrestrial reserve, we categorised prey diversity as low, moderate and high. Reserves where only wild boar and

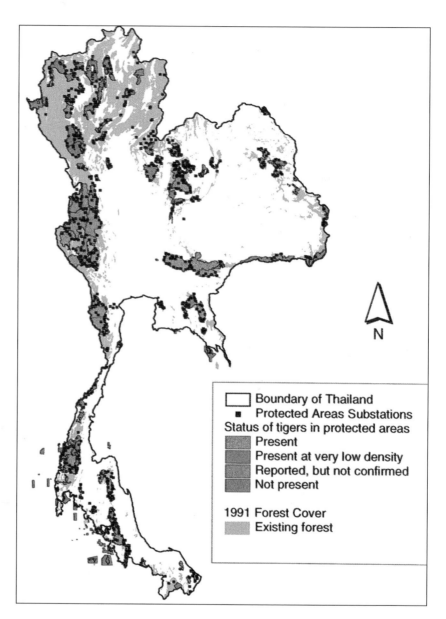

FIGURE 12.1
There are 681 ranger substations distributed throughout the terrestrial national parks and wildlife sanctuaries in Thailand. These are staffed throughout the year, even though there are no roads to some of them, and roads to others are not passable during parts of the year. In a protected area system of approximately 69 000 km² with few roads, the ranger staff at these substations provide the only practical network for monitoring tigers and other biodiversity.

barking deer occurred were classified as having low prey richness. Reserves where wild boar and barking deer plus either sambar or one species of larger bovid in the genus *Bos* or *Bubalus* occurred were classified as having moderate prey diversity. Reserves with at least one large bovid, plus sambar, wild boar and barking deer were considered to have high prey richness.

Results and discussion

Tiger population structure

Manually interpreted 1991 Thematic Mapper satellite data show 26% of Thailand is covered by forest. The protected area network includes 45 wildlife sanctuaries and 65 terrestrial national parks and covers 68 849 km². This represents 52% of the forested land or 13% of Thailand's land area. Based on seven expert group workshops in which 562 protected area staff filled out questionnaires, subsequent field surveys by the ranger staff of the protected areas, and additional surveys by three wildlife biologists of the Division of Wildlife research and J.L.D. Smith, tigers currently occur in 46 reserves (43 356 km²) in Thailand; their status is still unknown in another nine protected areas (Fig. 12.2).

Within each reserve the area actually occupied by resident breeding tigers is also unknown. Surveys in Mae Wong National Park, Umpong Wildlife Sanctuary and Khao Ang Rui Nai Wildlife Sanctuary found no tiger sign in large portions of these reserves. For example, in March of 1996, Budsabong Kanchanasaka and Bishnu Lama, a tiger biologist from Nepal with 20 years experience surveying tigers, surveyed the southern part of Mae Wong National Park. They found tiger sign only in the interior parts of the park, more than 12 km from agricultural communities along the Park's eastern border. A second survey of the same area in February 1997 by Sewai Wanghungsa of the RFD confirmed the earlier survey. In February and April of 1997, Sewai Wanghungsa also surveyed Umpong Wildlife Sanctuary, which was adjacent to Mae Wong on the west. He found a situation similar to that for Mae Wong; tigers occurred only in the interior, eastern portion of the sanctuary near the border with Mae Wong. Similarly four surveys in Khao Ang Rui Nai Wildlife Sanctuary by Sewai and J.L.D. Smith found no tiger or ungulate sign in the western portion of the reserve. The absence of tigers from parts of reserves makes it impossible to estimate tiger numbers based on the size of reserves and a general estimate of tiger density.

Thailand is unique because many of the parks and sanctuaries are adjacent to each other, forming protected area complexes. Based on the spatial pattern of reserves and forest cover we hypothesise that there are 15 tiger populations (Table 12.1). The largest of these is the Western Forest Complex, which includes the Huai Kha Khaeng/Thung Yai World Heritage Site and encompasses 12 parks and wildlife sanctuaries (15849 km²). Five of Thailand's tiger populations extend into neighbouring countries. The Salawin, Western Forest Complex and Kaeng Krachan populations are connected to tiger habitat in Myanmar (Fig. 12.2). The Hala Bala population is contiguous with a large block of habitat in Malaysia and currently there is a project to develop this area as a trans-boundary conservation unit. The Huai Sala population is adjacent to a large forested area in Cambodia where three protected areas are proposed. Although Huai Sala is too small to maintain a viable population of tigers, this area should continue to have tigers as long as it remains connected to the larger proposed protected forests in Cambodia. Inspection of forest-cover maps and remote sensing data suggest that the existing forest matrix in Thailand and Myanmar links the four populations on the western border. The 10 other populations not on the border of Thailand are clearly isolated from each other.

Prey richness and 'potential' tiger density

Because neither relative nor absolute prey abundance has been estimated in the 46 reserves where tigers occur, we assigned an overall habitat quality rating to reserves based on prey richness as a means of roughly estimating tiger carrying capacity. For these 46 reserves where tigers occur, prey richness was classified as low at 14 reserves (8407 km²), moderate at 14 reserves (11 425 km²) and high at 18 reserves (23 524 km²). Because the percentage of habitat within reserves that will support tigers is unknown, we reiterate that a population estimate is simply a very crude estimate. We can predict carrying capacity of existing reserves if poaching and human encroachment were reduced. Based on information from Rabinowitz (1993) and Saksit's camera trapping data from Huai Kha Khaeng (unpublished data) we estimate a tiger density of

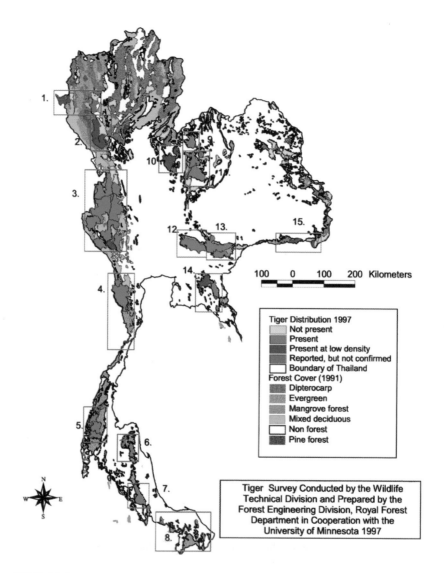

FIGURE 12.2
The distribution of 15 tiger populations based on known occurrences of tigers within the protected area system of Thailand. In the 110 terrestrial protected areas of Thailand, there are 46 protected areas where tigers are reported and nine others where their status is still unknown. The extent of reserves occupied by resident breeding tigers is not known, but parts of many reserves are devoid of tigers due to very low prey density.

one tiger/67 km² for high prey diversity reserves, one tiger/100 km² for moderate prey diversity reserves, and one tiger/200 km² for low prey diversity reserves. This gives a potential tiger population estimate of approximately 500 tigers in Thailand (Table 12.1). Using these densities and size of protected areas, we predict that the largest tiger population occurs in the Western Forest Complex, and suggest that the complex could support approximately 180 breeding adults; in contrast 9 of the

Table 12.1. *The estimated potential breeding population size of 15 hypothesised tiger populations in Thailand*

	Population name	Area (km²) occupied by population				Potential pop. size
		High	Moderate	Low	Total	
1	Salawin[a]		875	1591	2466	17
2	Mae Tuen			2610	2610	13
3	Western Complex[a]	8028	5515	651	14194	178
4	Kaeng Krachan[a]	3404	200	161	3765	54
5	Khlong Saeng	1881		646	2527	31
6	Khao Luang		570	809	1379	10
7	Khao Banthad		1267		1267	13
8	Hala Bala[a]	694	506		1200	15
9	Phu Luang		848	850	1698	13
10	Thung Salang Luang	1262			1262	19
11	Phu Kieo	2526			2526	38
12	Khao Yai	2169			2169	32
13	Thap Lan	3080		480	3560	48
14	Khao Soi Dao		745	1089	1834	13
15	Huai Sala[a]		583	316	899	7
		23044	11109	9203	43356	501

[a] Population probably extends into neighbouring countries.
Based on the amount of high (1 tiger/67 km²), moderate (1 tiger/100km²), and low (1 tiger/200 km²) quality tiger habitat within the protected areas occupied by each tiger population.

15 hypothesised tiger populations probably can support approximately 20 or fewer breeding animals (Table 12.1). The six larger tiger populations account for 76% of the total tiger population estimate. Because there have been no field surveys of forest habitat outside protected areas, the land area supporting the nine smallest populations may be larger than estimated. These forest areas should be included in future field surveys to increase the accuracy of the estimate.

Current survey efforts

In March 1997 detailed surveys of Huai Kha Khaeng and Thung Yai wildlife sanctuaries were initiated to develop a method of establishing the extent of tiger and prey presence within reserves. A total of 10 days surveying resulted in observations recorded at 169 sites. Tiger sign was observed at 41 sites over a total of 155 km of trails and river courses that were surveyed (Fig. 12.3). We plan to continue surveys and to use remote sensing and GIS to develop a survey strategy based on the likelihood of encountering tigers.

Current problems and opportunities

Efforts initiated as part of Saving Thailand's Tigers have identified a number of significant knowledge gaps and conservation problems that need to be addressed. These include: (1) tigers occur in small populations in a fragmented landscape; (2) many of these populations span several protected areas that are managed by separate RFD divisions; the largest populations occur in trans-border areas; (3) management at the level of ecosystems is at its infancy in Thailand; and (4) rangers stationed throughout the protected area system have no biodiversity monitoring responsibilities. Additionally we have discovered that many Thai people are unaware that tigers still occur widely in Thailand.

Our current data on tiger distribution are limited

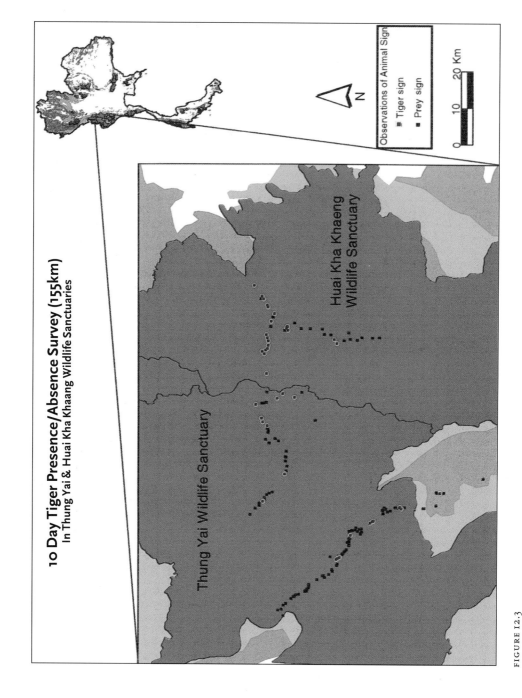

10 Day Tiger Presence/Absence Survey (155km)
In Thung Yai & Huai Kha Khaang Wildlife Sanctuaries

Thung Yai Wildlife Sanctuary

Huai Kha Khaeng
Wildlife Sanctuary

Observations of Animal Sign
■ Tiger sign
■ Prey sign

N

0   10   20 Km

FIGURE 12.3
Methodology to survey for the presence and absence of tigers in protected areas of Thailand is being tested during preliminary surveys of Thung Yai and Huai Kha Khaeng wildlife sanctuaries. Ten days of field surveys by the staff of the Wildlife Research Division have covered approximately half of the area of these two sanctuaries.

Indochinese tiger photographed by 'camera-trap' in Huay Kha Khaeng National Park, Thailand. Such photographs help in establishing the structure of tiger populations as an aid to planning to secure their future.

to the presence and absence of tigers in reserves. With this information we have generated a 'hypothesised' metapopulation structure of 15 populations within a highly fragmented forest landscape. Using data from Rabinowitz (1993) and Saksit (unpublished), we predict that only two forest ecosystems in Thailand can support >50 breeding individuals. This is not too different from India where, using even higher density estimates (Smith 1993; Karanth

1995), we similarly estimate only two populations with >50 breeding individuals. If the remaining smaller populations in Thailand continue to lose habitat or their prey base, the result is likely to be an increase in local population extinctions. Unfortunately, tiger habitat continues to degrade due to subsistence hunting and illegal conversion of forests to agricultural land.

To prevent further reduction in tiger population

sizes, management plans are needed for each tiger population (Smith *et al.* 1992).

Planning at this larger scale is a departure from current practice and implementation of these plans will require cooperation of national park and wildlife sanctuary chiefs that administer two different divisions of the RFD. The first such effort has been formation of the Western Forest Complex Coordinating Committee (WFCCC). It will coordinate the management of the largest tiger population in Thailand. Its mission is to train rangers and establish management activities for the entire complex of 16 reserves. The WFCCC is also the first attempt at ecosystem management in Thailand. Although the concept of ecosystem management is receiving global endorsement as the primary approach to addressing the complexity of biological, social, economic and political dimensions of biodiversity conservation (Grumbine 1994; Boyce & Haney 1997), it is still new in Thailand.

Tigers provide a good example of why management at a large scale is important. Most individual protected areas in Thailand, and throughout the tiger's range, are too small to maintain viable tiger populations in isolation. The ecosystems where tigers occur in large enough numbers to ensure long-term viability typically span 4–12 protected areas in three different government jurisdictions. Furthermore, human activities beyond the boundaries of most tiger populations influence the health of the forest ecosystems in which they live.

The trans-boundary tiger populations along the border with Myanmar provide an opportunity to conserve tigers within the largest natural ecosystem in mainland South and Southeast Asia. Large ecosystem reserves are needed to provide greater protection from rare ecological catastrophes that can potentially extirpate small populations. The importance of establishing and protecting a few megareserves has not been articulated to decision makers in Thailand or elsewhere in Asia. If existing and potential megareserves are going to be preserved, their desirability and purpose must be understood by policy makers and the public. It is important to be proactive because it is usually too late to make these arguments after development

plans have been made that will fragment these areas.

Thailand's protected areas are large and many are grouped into complexes of adjacent reserves where there are few interior roads. To monitor biodiversity and human activities across such inaccessible ecosystems requires the help of rangers stationed across the network of 681 substations (Fig. 12.1). To enlist their help the RFD must redefine their professional responsibilities to include biodiversity surveys and other monitoring activities. Rangers will need additional training, basic field equipment and health insurance (to cover the risks of injury from encounters with poachers) if they are expected to take on these duties. Training was started at a series of large-mammal workshops in 1995, but a formal ongoing training programme to institutionalise increased professionalism among rangers has yet to be established. A shift of professional responsibilities of this magnitude cannot be accomplished without a long-term commitment by the RFD.

Many Thai people are unaware that tigers still occur widely in Thailand. In 1995 we produced a poster showing the distribution of tigers in Thailand and held a national poetry and essay contest on tiger conservation. More recently, the RFD, in collaboration with the Seub Nakhasathien Foundation, organised a series of Tiger Weekends at Huai Kha Khaeng Wildlife Sanctuary for business leaders and educators. These weekends provided an opportunity for RFD staff and the business community to discuss conservation issues and the role of the private sector in contributing to national conservation efforts. These programmes need to continue because business leaders are beginning to politically and financially support conservation programmes.

Another effort to educate the Thai and global public about tigers is a trilingual (Thai, Chinese and English) Tiger Home Page – Saving Thailand's Tigers – which has been established by the RFD in collaboration with Kasetsart University and the University of Minnesota. The first objective is to provide the public, and especially students, with information on the status of tigers in Thailand. The home page is being expanded to include an interactive map viewer that will allow people to query maps to display the distribution of tigers and other large mammals. It will include information about individual tiger populations and explain why these populations should define ecosystem management units that integrate tiger conservation with the overall goal of conserving Thailand's biodiversity. A second objective is to link citizens and non-governmental organisations (NGOs) in Thailand to a global network of conservationists who want to coordinate efforts to reduce the use of tiger products and tiger poaching. As we continue to develop education programmes, we recognise that outreach efforts need to extend beyond tigers to establish in the mind of the public the link from tigers to the health and integrity of Thailand's ecosystems, including sustainable economic development and the well-being of Thai people.

Future prospects for wild tigers can be enhanced by outreach efforts to establish the link between tigers and the health and integrity of ecosystems in the mind of the public.

13

Metapopulation structure of tigers in Nepal

James L. David Smith, Charles McDougal, Sean C. Ahearn,
Anup Joshi and Kathy Conforti

Introduction

Tigers were once widespread across Asia, from the Russian Far East westward to Turkey and southward to southern India and the Indonesian archipelago. As human populations and development increased rapidly after World War II tiger habitat became increasingly restricted and fragmented. Currently there are an estimated 160 tiger conservation units or TCUs (Dinerstein *et al.* 1997). Based on more detailed field surveys in Nepal, Thailand and Vietnam, the number of discrete, isolated tiger populations is probably considerably greater than approx. 160 (Smith *et al.* 1987a, in press; J. L. D. Smith *et al.* this volume Chapter 12). With the world's human population increasing at a rate that will double the number of people in the next 50 years, human competition for the land base that supports tigers and other biodiversity will also increase, causing further loss, degradation and fragmentation of tiger habitat. Fragmentation is the most serious form of habitat loss because very small losses can divide larger tiger populations into much smaller units that have a lower probability of persisting. The issues of fragmentation and habitat degradation cannot be addressed until conservationists and resource managers shift their scale of management from individual parks and sanctuaries to larger scale units encompassing entire tiger populations (Smith *et al.* 1998). This shift in perspective is important because managers currently confine their activities to protected areas and the immediate surroundings. However, fragmentation often occurs at a considerable distance from the boundaries of parks in degraded habitat that may not otherwise appear to have high conservation value.

There are several factors that hinder a landscape approach to tiger and overall biodiversity conservation. First, lack of detailed information as to where tigers and other animals occur makes it difficult to identify where conservation action is needed. Secondly, conservation efforts have focused on establishing and managing protected area systems. Unfortunately, in Nepal and across most of the tiger's range, national parks by themselves are not large enough to support viable tiger populations. Furthermore, other forested and wild areas are often under different administrations and consequently are managed for different objectives. It is important to shift management from protected areas to ecosystem or landscape management (Grumbine 1994; Christensen *et al.* 1996) so that entire tiger populations are treated as a single management unit. This larger scale perspective will help alleviate some of the competition between biodiversity conservation and resource use, and it also identifies where habitat restoration will achieve the greatest positive effect.

This chapter focuses on field assessment techniques needed to map the distribution of tigers, estimate tiger density and monitor the condition of habitat. We discuss the use of some of the basic ecological research methodologies developed during a long-term study of tigers at Royal Chitwan National Park (Seidensticker 1976; McDougal 1977; Sunquist 1981; Mishra 1982; Tamang 1982; Smith *et al.* 1983, 1987a,b, 1989; Smith & McDougal 1991; Smith 1993; Smith *et al.* 1998) and apply these methods to report the distribution and status of the four tiger populations in and adjacent to Nepal. We conclude by identifying the key areas where habitat restoration is needed and propose a system for monitoring tiger habitat that combines remote sensing and ground surveys of prey abundance. These approaches will increase the benefits of current restoration programmes by targeting areas where these efforts can significantly increase population viability.

Study area and methods

Geography of the Nepalese lowlands

Tiger habitat in the 1930s was continuous for over 1800 km along the Siwalik Range, or outer foothills of the Himalaya, from west of Corbet National Park India through the lowlands of Nepal, Sikkim, Bhutan and north-eastern India (Gee 1964). Today this belt of lowland forest still runs the entire length of Nepal, but it has become highly fragmented. Interior river valleys, called *duns*, formed where the outer or Siwalik Range separated from the middle hills or Mahabharat Range. Ample rainfall and the rich alluvial soils of the dun valleys support diverse vegetation characterised as tropical moist deciduous, riverine and upland forest, interspersed with tall grasses (Stainton 1972; Tamang 1982). This mosaic of rich habitat in the duns has resulted in a diverse large mammal fauna, including abundant populations of four species of deer and wild boar that are the primary prey of tigers (McDougal 1977). Historically, some of the highest densities of tigers on the Indian subcontinent occurred in these duns. In 1938–9, for example, 120 tigers were killed in the Chitwan Dun in a single four-month royal hunt (Smythies 1942). The large continuous population of tigers was characterised by high-density foci in the duns and lower densities in the upland sal forest habitat in the Siwalik Hills. With expanding human populations, increased development and road construction, this forest belt became gradually fragmented, resulting in habitat barriers between the major dun valleys and an isolated metapopulation structure (Levins 1969).

Determining tiger presence and absence

To survey the extent of tiger populations remaining in Nepal we used satellite image maps to examine all potential tiger habitat in the lowlands of Nepal (Miller *et al.* 1989). These satellite maps were used to plan a survey strategy that was both cost effective and efficient. Our objective was to determine if tigers were present at each survey site and to identify resident breeding animals. To determine if tigers were actually present, we observed tiger sign or, rarely, tigers themselves. Tiger sign consisted of tiger tracks, scrapes and urine sprays, and natural and domestic tiger kills (Smith *et al.* 1989). Tracks were classified as tiger if they had pad widths ≥7.5 cm and as leopard if pad widths were ≤6.0 cm. Any tracks <6.0 cm co-occurring with tracks of adult-sized tigers were recorded as a cub less than six months old accompanied by its mother. Scrapes wider than 21 cm were classified as tiger and those ≤15 cm as leopard. Urine scent spraying was classified as tiger if it was >75 cm high. Kills of tigers were distinguished from those of leopards by examining the extent of broken bones in a carcass: when eating, leopards do not break as many bones as do tigers, and leopards also remove more meat, leaving a cleaner, more intact carcass.

In Nepal, the quickest way to document tiger presence is to ask local villagers if they have seen tiger sign. To learn if they had recently seen tiger sign, we typically interviewed herdsmen, people who gather minor forest products, women who collect firewood and fodder and forest rangers. If they had, we asked them to show us the sign. This approach often allowed us to find tiger sign quickly. When there was no local information on tiger presence, we searched streams, ridge tops, animal and human trails and forest roads for tiger sign. We usually obtained a local guide to help us search more efficiently. When we found tiger sign we recorded our location with a Geographic Positioning System (GPS) or noted it on an aerial photo or satellite image. We also recorded the type of tiger sign and cover type and quality.

Once information from informants was verified we questioned them about sign observed over the past year. Our goal was to determine if the habitat at the survey site should be classified as breeding or occasionally used, non-breeding habitat. We used two criteria for establishing breeding habitat:

observations by informants or knowledge of tiger cubs or their tracks. When there was no direct evidence of young, we classified an area as breeding habitat if several informants reported that tigers occurred in the area all year round. It is conceivable, but highly unlikely, that an area that did not support breeding tigers was used throughout the year by a series of transient tigers. If we had doubt about classifying habitat as breeding or non-breeding, we revisited the site in a different season and continued to obtain more information (Smith *et al.* 1998).

For survey sites that were classified as breeding habitat we assumed that two animals, a male and female, were resident and that the minimum home range of a breeding pair was an area within a 6-km radius of the survey site within the same quality habitat (Sunquist 1981; Smith *et al.* 1987b). Based on this assumed minimum area, our next survey site was at least 6–12 km from the previous site. Where the habitat appeared to be similar to the previous survey site on satellite photos and by ground inspection, we often travelled more than 12 km between survey sites. If tigers were determined as breeding at both sites we extrapolated that the intervening area was also breeding habitat. The information was organised in a geographic information system that includes visual as well as tabular data (Ahearn *et al.* 1990).

Estimating tiger density

We used and attempted to evaluate three approaches for estimating tiger numbers: a density estimate based on the known home range size of a group of individuals; a density estimate based on capture-recapture using camera traps; and a total count based on recognising the size and characteristics of adjacent resident tigers.

Density estimates were based on the combined contiguous home ranges of resident animals. From 1974 to 1982, radiotelemetry (Smith *et al.* 1983), camera traps and pugmarks were used to monitor a group of resident animals in a 252-km² study area in the north-central part of Chitwan. Tigers were located in a haphazard fashion from the ground and from the air three to five times a month from January 1977 to June 1980 for a total of more

than 14 000 radio locations. They were located intermittently from July 1980 to June 1982. After mid-1982, camera traps and pugmarks were used to continue monitoring a group of resident animals in the western portion of the former study area (152 km²). From 1974 to 1996 McDougal and his staff used radiotelemetry, direct observation and photographs of tigers to confirm the identity of animals whose tracks they observed. The identities of animal tracks were confirmed in an ad hoc fashion; a single camera was used and it was placed in an area for several weeks at a time. This area was patrolled daily so field staff became experts on the tigers that regularly tripped the single camera trap. To provide a more systematic test of the field staff's ability to recognise all the animals within their study area and to distinguish these animals from transients, a formal blind test of their ability was conducted in 1996. A person (K. Conforti) experienced in camera trapping, but unfamiliar with the study area, set up camera traps in the study area. Trail Master cameras were placed in the late afternoon and revisited the following morning between 6 and 8 a.m. The apparatus consisted of a 35 mm camera, an infrared light source that emitted a pulsed beam, and an infrared light sensor connected to a microprocessor that triggered the camera after a certain number of disrupted light pulses. Each morning, after inspecting and measuring the tracks that crossed the light beam, the field staff either identified the individual or indicated that the tracks were not clear enough in the substrate to be identified. Photos from the camera traps were subsequently identified against a stock of photos taken over several years. The photo identity was then compared with the pugmark identity recorded the morning after the photo was taken to determine the accuracy of identifying individuals based on the size and shape of their tracks.

Home ranges of individuals and a combined home range of a group of contiguous animals were calculated by bounding the outer locations with a convex polygon (Mohr 1947a). The polygon was modified to exclude all habitat that was clearly not used by tigers (e.g. cultivated lands outside the park).

Three densities were calculated for tigers living in: (1) riverine forest and grassland, (2) riverine and lowland sal forest habitat, and (3) a mixture of lowland sal forest and open sal woodland and Siwalik Hills habitat. For each habitat complex a combined home range was calculated and this area was divided by the number of resident males and females to obtain a density estimate. Population size was calculated by summing the estimates for each habitat complex within a population.

Prey abundance

To assess the prey base available to tigers we selected representative sites where we measured relative prey abundance, horizontal cover, and type and extent of human use. For relative prey abundance we counted pellet groups and droppings of ungulates and primates within 10-m² circular plots placed at 25-m intervals along 0.625 to 2.5-km transects for a total of 25 or 100 plots per transect (Smith 1984). Leaves were carefully raked from circular plots and we attempted to identify discrete pellet groups. Older groups had fewer pellets than newer groups, but were counted equally. Sometimes groups were spread out in a line when an animal was moving while defecating and these lines were classified as a single group. Pellet groups were assumed to accumulate from October (the end of the wet or monsoon season) until surveys were conducted from February 21 to April 5.

Sources of pellets were classified by size class. The small-prey class was composed of barking deer and chousingha; the medium-sized class included chital and hog deer; and large prey included sambar, swamp deer and blue bull. We also identified three other classes of droppings: wild boar, primates and forest bovids. For habitat patches less than 1 km in length we established transects by choosing a random direction within 30° of the longitudinal axis of the patch; otherwise we chose a random compass

A tiger stealthily approaches to within 10 to 20 metres of its prey, explodes in a gap-closing rush and, with carefully timed manoeuvres, delivers a killing neck bite. Many attempts are unsuccessful.

heading within 30° of north. Transects were conducted by a three-person team.

Classifying tiger habitat quality from digital satellite data

To evaluate the efficacy of satellite data in assessing the status of tiger habitat, we classified a Thematic Mapper (TM) scene of central Nepal (row 151, path 41, March 1988). Our goal was to resolve vegetation classes and habitat quality that supported resident breeding tigers with habitat that did not. An unsupervised cluster analysis classification procedure was used to establish 30 spectral classes that were then assigned to a total of 11 vegetation classes (Lillesand & Kiefer 1994). Classes were assigned vegetation cover types based on the knowledge of J. L. D. Smith who spent four years surveying forest lands through the interpretation of 1:50 000 colour infrared aerial photographs and by conducting ground surveys. The vegetation classes were then grouped into good and poor quality tiger habitat based on discriminant function analysis of the threshold scores needed to support breeding tigers (Smith 1984). Finally, to relate information on tiger populations collected for each of the seven administrative units that cover the study area with habitat quality, the ratio of good to poor habitat for each unit was calculated (Smith *et al.* 1998).

Results

Metapopulation structure

We mapped the distribution of tigers in Nepal based on surveys at 68 sites between 1987 and 1997 (Fig. 13.1). Our initial survey of western Nepal, from Chitwan to the western border of Nepal, was done by a team of five people in 20 days. Subsequent surveys covered eastern Nepal from Chitwan to the eastern border of Nepal (1991, 1994), and reconfirmed the presence of tigers in western Nepal at or near the 1987 survey sites, or searched in entirely new areas in western Nepal (1993, 1994, 1996, 1997). Tigers were found at a total of 43 sites: 15 survey sites were in protected areas and 28 in national forest lands. We do not include data from Chitwan and Bardia

National Parks and Sukhla Phanta Wildlife Reserve because Chitwan was the centre of a long-term study and the other parks have been repeatedly surveyed. At four of the survey sites outside reserves, tigers were observed by McDougal and Smith in 1987, but apparently were extirpated by 1993. In the same five-year time frame, tigers recolonised the Bubai Valley after villagers were resettled out of the valley and it was incorporated into Royal Bardia National Park in 1987 (Fig. 13.2). In 1987 we found no tiger sign in the main part of the Bubai Valley, but tigers began recolonising the valley within one year of the resettlement. By 1993 tigers were found throughout the Bubai (Man Bahadur pers. comm.).

Four viable populations and one remnant population form a metapopulation unit of what was formerly a large contiguous population (K. M. Tamang, pers. comm.) that extended from Corbet National Park, India, eastward across Nepal to Bhutan and Assam. Three tiger populations are named after the protected areas that form the core area of each, i.e. Sukhla Phanta, Bardia and Chitwan. A fourth population occurs just south of the Nepalese border in Dudhwa Tiger Reserve, India. We also investigated reports of a remnant population in the Trijuga Valley east of Chitwan (Fig. 13.2). Tigers were present in 1978 and we believe fewer than five breeding animals occupied the area in 1994. Fragmentation of wild habitat has resulted in four strong barriers to dispersal that we define as intensively cultivated and settled land greater than 1 km in width. These gaps occur at the Baghmati River, the city of Butwal, the Nepalese border area adjacent to Dudhwa National Park, and to the west of the city of Mahendranagar (Fig. 13.2). There are wider, weak dispersal barriers of degraded forest adjacent to these strong barriers. A weak barrier is habitat that tigers may disperse through but is highly degraded with a very low prey base of mostly domestic livestock. In the areas we designate as weak dispersal barriers we have not located tiger sign, but there are occasional reports of tigers in these areas. The 'weak' barrier between Trijuga and Chitwan is 192 km, and between Chitwan and Bardia it is 196 km. Based on interviews with villagers and foresters, and the personal knowledge

Table 13.1. *Eighteen individuals identified in 91 photographs taken over a three-week camera trapping period in March and early April 1996*

| | Known | | | | Unknown |
	Adults	Subadults	Juveniles	Cubs	transients
Males	1		4		1
Females	5	1	2		2
Unsexed				2	
Totals	6	1	6	2	3

of C. McDougal who has spent more than 30 years travelling through the forests of the Terai of Nepal, we estimate that the strong dispersal barriers we identified have isolated these populations for three to five tiger generations (>21 to >35 years). The Dudhwa population was the most recently isolated population; it became fragmented by a Nepalese decision in the mid-1970s to cut a 2-km strip of forest along its southern border to reduce illegal log smuggling. We conducted ground surveys along the northern border of Dudhwa Tiger Reserve to confirm that the cultivated strip of land extended the entire border of the Park, isolating it from national forest land in Nepal.

Population estimates

We used data from 26 radio-collared tigers and 13 animals whose tracks we recognised to determine tiger home range size. Then we used single and combined home ranges to estimate the population density of tigers living in different combinations of habitats. During 20 years of field work in Chitwan we tested our ability to identify tracks by verifying the identity of animal tracks with visual, camera and radio locations. In April 1996 we conducted a more formal test of our ability to recognise individuals from their tracks by identifying tracks at 18 camera traps, 15 of them along forest roads and three along trails. The area of a convex polygon that encompassed the camera traps was 15.8 km². This polygon was within the territories of five resident females with a combined home range of 70 km². The camera trapping effort included 249 trap

nights spread over nine sampling periods between November 1995 and June 1996. The number of trap nights per sampling period varied from 25 to 30. The number of trap nights per site varied from 4 to 21, with a mean of 13.8. The maximum number of cameras set on a single night was 11. A total of 91 photos of identifiable tigers was taken (1 per 2.74 trap nights). Of the 18 different individuals identified, 7 were photographed only once. The others were re-photographed 2–24 times (Table 13.1). At one camera site, 12 of 18 individuals were photographed. On 54 occasions there were distinct tracks of adult tigers, which C. McDougal and a wildlife technician identified. They were correct on 53 occasions. The one misidentification was a transient adult that they identified as a newly resident young female. Based on this level of accuracy in our blind test, and ad hoc camera trapping verification between 1977 and 1996, we feel confident in identifying tigers by their distinctive tracks. Here we report home range sizes of resident tigers from track data as well as radio locations (Table 13.2).

Because all home ranges were scent-marked at home range boundaries and defended, home ranges and territories were considered synonymous (Smith *et al.* 1987a, 1989). From 1976 to 1982 territory size was based on animal locations obtained from radio tracking, visual sightings and pug mark identification; from 1982 to 1996 data consisted primarily of pug mark tracking, photographs and occasional visual observations. Territories in less productive pure sal forest and the rugged Siwalik Hills were the largest. The smallest occurred in prime tiger habitat that was a mixture of alluvial grassland and riverine forest (Table 13.2). Male home ranges varied more in size than female ranges. They ranged from 151 km to 23.9 km and encompassed the territories of one to seven females (Table 13.2).

To estimate density we calculated the average home range size of a contiguous group of adult tigers of both sexes that were living in Chitwan (Smith *et al.* 1998). Our estimates varied from one tiger/14 km (*n*=7) living in riverine forest and grassland to one tiger/20 km (*n*=11) for animals living in a mixture of riverine forest, grassland and

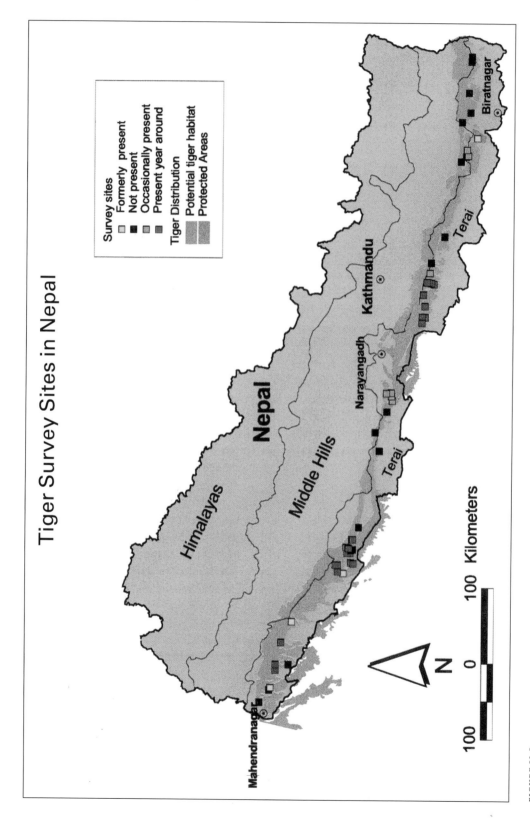

FIGURE 13.1
Tiger survey sites in Nepal.

Metapopulation structure in the central Himalayan foothills

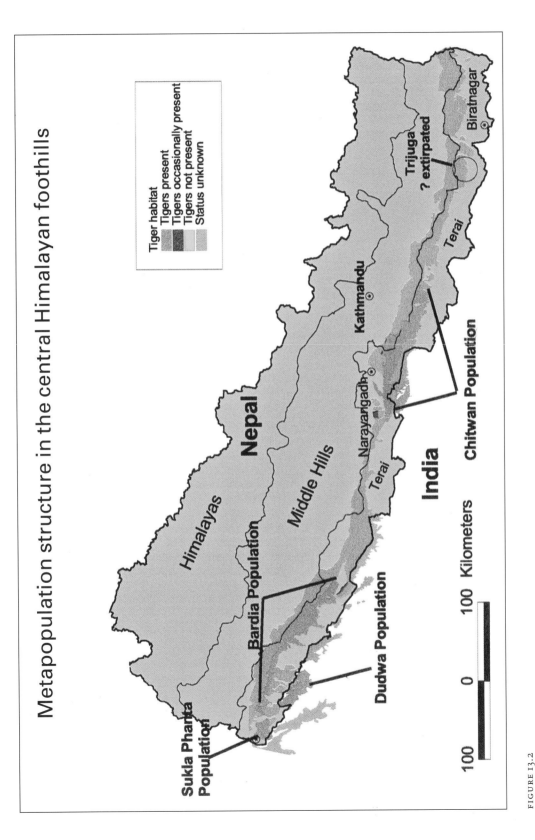

FIGURE 13.2

Metapopulation structure in the central Himalayan foothills.

Table 13.2. *Home range (HR) sizes (in km²) for resident tigers in Chitwan, 1976–96*

Males	Maximum HR	Mean HR	Minimum HR
105	151	105.6	60.2
102[a]	—	77.7	—
SB	—	70.3	70.3
127	—	58.7	—
kB	—	40.9	—
BB1	37.8	35.3	32.7
BB2	73.4	62.4	51.4
DB	92.3	68.7	45
NB	105	95	23.9
Means	78.6	68.3	51.2
Females			
101	51.0	34.7	18.4
103	16.5	16.5	16.5
106	17.7	17.7	17.7
109	19.0	19.0	19.0
115	26.7	19.6	19.2
JP	14.7	14.0	13.3
122	27.2	20.8	18.2
118	17.9	17.9	17.9
PP	35.5	28.4	16.7
BP	20.0	20.0	20.0
JP2	40.0	40.0	40.0
Means	26.0	22.6	19.7

[a] Data for 1 year only.

Table 13.3. *Population estimates based on tiger densities in, and areas of, different quality tiger habitat are compared to total counts made between 1995 and 1997 by McDougal and the Department of National Parks and Wildlife Conservation*

	Area (km²)	Estimate (density × area)	Total count
RCNP			
Alluvium	100	7	
Alluvium/low sal	158	8	
Upland sal/Siwaliks	785	21	
Total	1053	36	40
PWR			
Upland sal/Siwaliks	499	13	8
RBNP			
Alluvium	92	7	
Upland sal/Siwaliks	840	23	
Total	932	30	32
RSPWR			
Alluvium	155	11	14
Upland sal/Siwaliks	165	4	2
Total	320	15	16
Overall totals		94	96

Three density estimates were used: 1 tiger/14 km² for alluvial habitat; 1 tiger/20 km² for mixed riverine and lowland sal forest; and 1 tiger/37 km² for upland sal and hill habitat. Population size from these density estimates were compared to total counts of tigers in Royal Bardia (RBNP) and Royal Chitwan (RCNP) National Parks, Parsa (PWR) and Royal Sukhla (RSPWR) Phanta Wildlife Reserves in 1994, 1995 and 1996 by McDougal and the Department of National Parks and Wildlife Conservation.

low lying sal forest, and one tiger/37 km (n=2) for tigers living in upland sal forest and Siwalik Hills. To estimate population sizes for the protected areas we applied these density estimates to similar habitat in each of the protected areas (Table 13.3). These estimates are compared to total counts made of each of the protected areas by C. McDougal and the Department of National Parks and Wildlife Conservation between 1995 and 1997. We made no density estimates for tigers living outside protected areas; however, from the 43 locations where tigers occurred, we estimated the area of national forest land occupied by tigers and calculate the *potential* carrying capacity of tigers in these forests by applying our sal forest and Siwalik Hills density estimates from Chitwan (Table 13.3).

We classified a Thematic Mapper (TM) scene of central Nepal (row 151, path 41, March 1988) where we had conducted field surveys of tigers and their prey (Fig. 13.3). The 30 spectral classes were grouped into 11 vegetation classes. Seven classes that included open sal forest, tall grass, mixed grasslands, upper bench sal, mixed deciduous forest, Siwalik Hills and dry river courses were categorised as good quality tiger habitat; the remaining four that included degraded deciduous forest, denuded hills, degraded sal forest and agriculture/pasture were classed as poor quality tiger habitat (Fig. 13.4). The ratio of good to poor quality habitat ranges from 0.86 to 0.16 across the seven forest districts that

Table 13.4. *Good quality habitat consisted of open sal forest, dry river courses, grasses, tall grasses, mixed deciduous forest; poor quality habitat included degraded shrubs and grasses, degraded sal forest and Siwalik habitat*

Forest district	Tiger presence	Good habitat (%)
Chitwan	Yes	73
Parsa Reserve	Yes	86
Parsa Buffer	Yes	71
Bara	Yes	54
Badrini	Occasionally	46
Bagmati	No	27
Janikpur	No	16

we classified (Table 13.4). The four areas with the highest ratio of good quality habitat supported breeding tigers; of the other three, one was used only occasionally and two were not used by tigers (Table 13.4). Examination of ratios of good to poor habitat shows that when the ratio drops below 0.54 breeding tigers no longer occur.

Discussion

Metapopulation structure
Cover maps of the extent and quality of habitat, knowledge of dispersal behaviour and field surveys are needed to determine metapopulation structure. The best and usually most current information of forest cover is recent satellite imagery (Miller *et al.* 1989). Combined with field surveys, satellite imagery establishes the size, geographical configuration and degree of connectedness of tiger habitat (Ahearn *et al.* 1990; Davis *et al.* 1990). This information, coupled with data on prey abundance and current land use, allows wildlife managers to roughly predict the present and future probabilities of genetic change between populations and the viability of individual populations. The importance of mapping tiger habitat and human land use activities is to devise landscape scale plans to protect an adequate land base for tigers (e.g. by land use zoning, D.G. Miquelle *et al.* this volume Chapter 19,

and by habitat restoration, E. Dinerstein *et al.* this volume). The restoration of 16.5 km² of wild habitat in the buffer zone of Royal Chitwan National Park resulted in reoccupation of habitat by tigers and their prey. Now that the ecological, social and economic aspects of restoration of wild habitat in the Terai of Nepal are better understood, restoration is a potentially important tool in the prevention of future fragmentation of tiger populations. For example, mapping of tiger metapopulation structure allowed us to identify a previously ignored critical area at the base of an 87-km² habitat peninsula from which tigers have become recently extirpated. A restoration effort at this site could prevent further isolation of a 400+ km² habitat peninsula that extends west of Royal Bardia National Park almost to the Sukhla Phanta and Dudhwa tiger populations.

Even though our efforts to map tiger metapopulation structure are still crude, population sizes are estimated to vary from <20 to <200 breeding animals. These efforts are derived by multiplying the highest known density of tigers that live in excellent tiger habitat in Nepal and southwestern India (Smith 1993; Karanth 1995) by areas of suitable contiguous tiger habitat occupied by individual tiger populations throughout Asia (Dinerstein *et al.* 1997; Smith *et al.* 1998; Smith *et al.* this volume Chapter 12; V. Thapar unpubl. data). Based on these estimates, more than 85% of the tiger populations have <25 breeding animals. Modelling tiger population persistence shows that as populations decline from 50 to 25 breeding animals their ability to withstand genetic, demographic and environmental perturbations decreases in a non-linear fashion (Kenney *et al.* 1995). At approximately 25 breeding animals, models based on data from the Chitwan population show a high probability of persistence for 100 years, but a low probability of persisting for 200 years. When even a short-term strong systematic event (e.g. moderate to high levels of poaching) is added, the probabilities of extinction can increase dramatically (Kenney *et al.* 1995). These authors also showed that when extinction is the final outcome of these events, it usually occurs more than 10 years and often 30–60 years

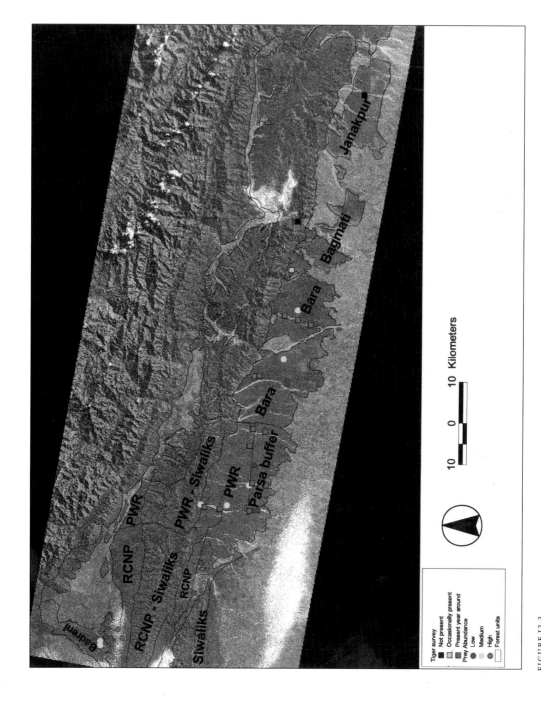

FIGURE 13.3
Thematic Mapper (TM) scene of central Nepal showing tiger and prey survey sites.

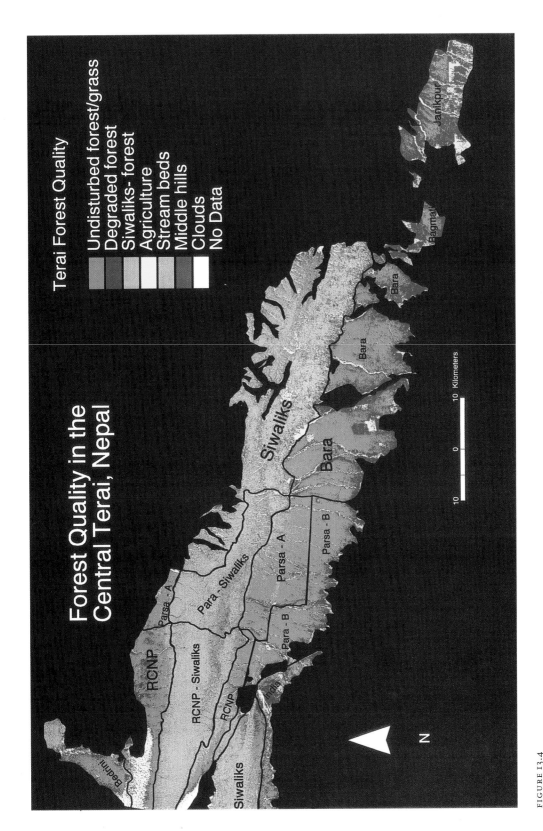

FIGURE 13.4
Automated classification of the Thematic Mapper (TM) scene in Figure 13.3 showing forest cover and quality.

later. In other words, we can stop the cause of a systematic decline and believe we have successfully responded to a crisis, but lose a population to extinction several decades in the future. In setting goals of the minimum land base needed to support tiger populations we need to take into account rare events in the same way that flood control planning considers the probability of water levels of a 1 in 50 and a 1 in 100 years flood. It is essential to have some very large reserves that can withstand the 1 in 100 years catastrophic event.

Few very large areas serve as macro reserves and there is realistic potential for creating more. Most of the existing areas are composed of one or more core-protected areas linked by forest lands and often spanning international boundaries. The importance of establishing such areas as very large ecosystems management units is that such units have a higher probability of withstanding major environmental perturbations, such as large-scale forest fires, hurricanes, disease and poaching. The primary threat to these macro reserves is development. Roads, dams and large-scale plantations often target the periphery. These ecosystems include not only protected areas, but land under many other uses (e.g. agricultural systems in secondary forests in Indonesia). Additionally, more of these large systems can be created by restoring habitat, especially in areas that will prevent or reverse habitat fragmentation. D. G. Miquelle *et al.* (this volume Chapter 19) and E. Dinerstein *et al.* (this volume) discuss two essential ecosystem management tools, land-use zoning and working with local user groups to restore habitat.

Rebuilding depleted prey bases

Securing the land base for tigers is not enough. We must also ensure that this land continues to support tigers. For example, a population survey by C. McDougal in a national forest area east of Bardia found only four tigers in an area that we calculated to have the potential of supporting 12 resident tigers. Habitat degradation and the resultant reduction in the prey base needed to support tigers and direct poaching of tiger prey are the hypothesised explanation for low tiger numbers in this case. K. U. Karanth & B. M. Stith (this volume) have similarly

documented prey depletion. Geographic Information Systems are a tool that provides managers with the ability to assess, manage and monitor the quality of tiger habitat by combining field research with remote sensing and landscape scale information. For example, using Thematic Mapper TM data to classify forest cover and to categorise its quality as tiger habitat, we found there is a small difference in the ratio of good to poor habitat that distinguishes between areas where tigers live and breed year round and areas that are only occasionally used by tigers. This led us to hypothesise that there may be a threshold in the distribution of good to poor habitat which reflects an overall reduction in prey base below which breeding does not occur (Table 13.4). Establishing a variance for these thresholds by examining a number of other regions will be an important objective of any future research. Once these thresholds are established, analysis of a temporal sequence of satellite images will make it possible to track changes in forest condition and thus anticipate changes in an ecosystem's ability to support tigers.

Field skills

The method we outline for mapping metapopulation structure is efficient and objective and requires only modest skill. Field staff do not have to recognise individuals based on their tracks as we have done, nor do they need to classify the sex and

Information from radio-collared tigers assisted Nepal in implementing landscape-scale management for its tiger populations.

age class of animals; they must simply distinguish between tiger and leopard tracks using several measurements. This is not an inherently difficult task given that female tigers weigh twice as much as male leopards and have considerably larger tracks. The pad width of the larger front track of a male leopard is 1.5–2 cm less than the pad width of a female tiger's rear foot. Efficiency in surveying is greatly improved by identifying good informants (e.g. livestock herders, people that gather forest products, or forest and protected area rangers) who know the area and can show the survey team tiger sign they have recently observed. Once a person has shown a survey team tiger sign, he/she becomes a credible informant to question about tiger sign in subsequent years. We surveyed a number of local people at each site to cross check information.

A critical issue we have not completely resolved is the intensity of field surveys. Distance between surveys and the intensity of site survey effort are limited by budget and personnel constraints. We spaced our survey sites at intervals of one to two times the length of a tiger home range (6–15 km). When information at a site is negative, a decision must be made as to when sufficient effort has been made to conclude that tigers are not present. It is difficult to provide a pre-determined criterion of how much effort is appropriate because it will depend on the skills of the survey team and other factors. Stipulating the number of kilometres of trails to search or the number of people to interview does not necessarily add rigour or increase statistical validity to a survey; on the contrary, given that each mapping project will be under finite time and personnel constraints, stipulating survey effort can greatly reduce the overall efficiency of a regional or national survey by reducing the total number of sites that can be surveyed. Because survey effort will vary, it is critical to document survey effort and methodology. This information is referred to as metadata, and includes a description of survey methodology, location in geographical coordinates, kilometres walked, condition of the substrate, days since last rain, as well as locality and number of people interviewed.

Is identifying individuals from their tracks a viable technique?

Data from tracks have been used extensively in our research over the past 15 years (Box 13.1). After we stopped regularly radio-collaring tigers in 1982 we used data from tracks extensively in our research. Our blind test of track identification and identity from photographs in 1996 established the validity of pug mark identification. These autocorrelated location data reveal details of home range use that are difficult to obtain from radiotelemetry (e.g. routes of travel, trees inspected or scent marked and association with other tigers). However, the skill level required makes the technique difficult to replicate elsewhere; it was acquired over many years and in a situation where we had 32 radio-collared animals and camera traps to verify the identity of individuals. In the future tiger researchers will be able to collect even greater amounts of autocorrelated data in a repeatable fashion using GPS collars that combine UHF transmitters and GPS receivers in a collar. This approach will allow researchers to combine information from tracks, radio locations and GPS to overcome the limitations of each type of data.

Conclusion

Beginning in 1986 the government of Nepal tacitly began implementing ecosystem-scale management for the Chitwan tiger population. In that year Parsa Wildlife Reserve was established to the east of Chitwan to expand the total area of tiger habitat under the highest level of protection (Smith 1993). Protected tiger habitat was again increased in 1989 when the government of Bihar, at the urging of the Nepalese government, created Valmiki Tiger Reserve adjacent to Chitwan. These additions expanded the protected area component of the Chitwan tiger population to 75% of the population's entire distribution (2543 km^2) and created a potentially multi-jurisdictional tiger management unit. A similar model has been developed in Russia. This ecosystem approach needs to be extended across the tigers' range.

Box 13.1 You can tell some tigers by their tracks with confidence
Charles McDougal

Since 1980 I have been conducting a long-term tiger monitoring project as a Smithsonian Research Associate, concentrating on a study area of 100 km² and immediately adjacent areas in Nepal's Royal Chitwan National Park. When the study began seven of the tigers being monitored had active radio-collars, and these I tracked as long as the batteries lasted. Meanwhile I and my two wildlife technicians developed a reliable technique to identify adult resident tigers by diagnostic features of their pugmarks. The technique relies on finding and recording the clear impressions of all four of the tiger's feet. This is essential because the distinguishing features that facilitate identification may be found on any of the four feet, though in our experience they are usually found on the forefeet, which are more liable to sustain injuries when fighting with conspecifics or attacking large prey.

It is important to mention that the technique is to a large extent site-specific. In western Chitwan where the study area is located, there is extensive good substrate for the recording of pugmarks along numerous river and stream courses, where it is relatively easy to encounter clearly defined tracks. Initially on an *ad hoc* basis, and during the past two years on a systematic one, I have used camera traps as a cross-check on pugmark identification.

Pugmarks should be recorded on moist, compact sand, or on a thin but ample layer of dust covering a hard surface such as a road. They should not be recorded on loose substrate or in mud. Three measurements are taken of each foot: (1) pad width, (2) total width, and (3) total length. The gaps between the toes and the pads should be measured. Assuming that the pugmark is fully indented in the substrate, and that it is not distorted, the minimum measurements should be taken from a series of tracks.

It is possible to distinguish the tracks of an adult male tiger from those of an adult female by these measurements. The forefoot pads of a male are >9.7 cm (n=42), those of a female are <9.3 cm (n=28); the hindfoot pads of males are >8.5 cm (n=51), those of females are <8.5 cm (n=45); the total width of the hindfoot of a male measures >11.0 cm (n=48), that of a female is <11.0 cm (n=44), unless splayed wider by an injury. Comparing an 18- to 24-month-old juvenile/subadult male with an adult female, the forepad width of the male will be greater, >9.5 cm, as will the hindpad, >8.5 cm, and the total width of the hindfoot will also be greater, >11.0 cm. In addition the toes, especially those on the forefeet, will be larger and more rounded.

To go a step further, is it possible to differentiate between adults of the same sex from their pugmarks? The answer is yes, especially if the animal's feet have been injured, leaving particular patterns of the toes in relation to each other and to the pad. In prime habitat in Chitwan, where density is high, aggressive interactions are not uncommon.

Figure 13.5 shows the distinctive pugmarks of two resident adult females with adjacent home ranges. On the night of 12 February 1996 there was a serious fight between Dhajaha and Assaly. Subsequently it was found that the toes on Assaly's forefeet were very splayed out (Figure 13.5a) , especially on the right forefoot; the left hind was also splayed. Dhajaha (Figure 13.5b) also sustained injuries, leaving toe number one on the right forefoot forward of its previous position and with a non-retractable claw. All but two of the eight females in the study area and adjacent to it have similar peculiar features of their pugmarks facilitating their identification. The exceptions are recently settled females less than four years old whose compact feet show no effects of injury. Nevertheless, with care it is possible to identify these animals as well. I do not claim that every tiger can be identified by their pugmarks. We only claim to identify the resident adults of a limited area, many of which have been monitored for years.

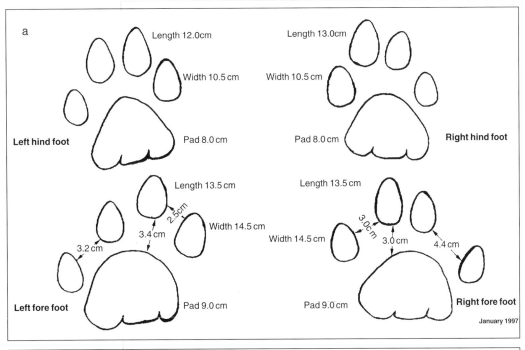

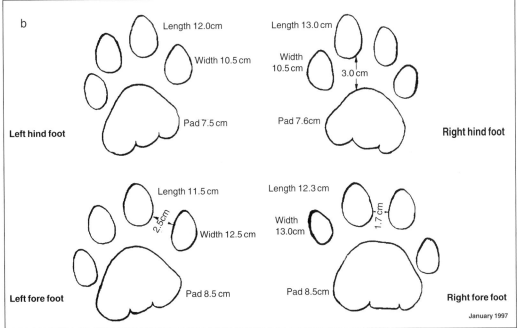

FIGURE 13.5

a. Pugmark measurements of adult female tigress 'Assaly' in Royal Chitwan National Park, January 1997. **b.** Pugmark measurements of adult female tigress 'Dhajaha' in Royal Chitwan National Park, January 1997.

Approaches to tiger conservation

Overview *John Seidensticker, Sarah Christie and Peter Jackson*

> Were it not for reserves we surely wouldn't have many tigers left. But obviously
> protection alone is not enough.
>
> *(George Schaller in Sunquist 1996)*

Reserved land has been an essential tool in tiger conservation, but

> The conservation of species and undamaged habitats is like a three-legged stool.
> Each leg is necessary but not sufficient. The legs of the conservation stool are
> sustainable use of renewable resources, species recovery, and habitat preservation.
> Conservation can progress by focusing on each of these, defining their limits,
> developing improvements, and preventing dysfunction.
>
> *(Humphrey & Stith 1990)*

In the previous section, authors emphasised that long-term success in managing reserved lands with viable tiger populations lies in linking these areas with surrounding landscapes through ecological and genetic mechanisms that promote and preserve natural landscape dynamics. The tiger is in its present precarious state in part because of the growing imbalance between the expanding numbers, needs and wants of people and the limits of the capacity of landscapes to support these wants and needs. Making the economic case for inclusion of tigers in land-use formulations means showing that the benefits gained from including the tiger are as good as, or better than, those to be had by converting the habitat for other purposes. A core hypothesis that emerged from the presentations and deliberations at Tigers 2000 concerned the linkage between the needs of the tiger and the welfare of the people living near it. Human needs are a part of any ecological system and securing a long-term future for wild tigers requires consideration of the needs of tigers and people together. In Part IIIA–C, authors emphasise the need to shift the domain of our conservation theory, doctrine and practice from what goes on inside reserve boundaries to the landscape, the region and the world beyond. Our efforts to secure a future for wild tigers will necessarily include economic, cultural and ecological resources, environmental education and active recruitment

and retention of partners locally, regionally, nationally and internationally, working together to achieve a common purpose – saving the tiger.

Zoos are obvious partners in securing the future of wild tigers. Zoos and related institutions have secured a future for tigers living there (Meacham 1997). And in the last three decades zoos have provided many of the human intellectual resources and the animal populations that led to the formulations of small-population biology. Our expanding understanding of the risks confronting small-sized animal populations and the technologies to assist in overcoming those risks, such as more precise understanding of genetics and the development of assisted reproduction technologies, have been of central concern for conservation biologists.

The World Zoo Conservation Strategy, '...urges the entire global zoo network and all other conservation-orientated networks to integrate and intensify their efforts towards their mutual goals. This great mustering of all available resources will be crucial to giving our Earth's biosphere the best possible chance for survival of its biological wealth of genes, species and ecosystems' (IUDZG–CBSG 1993). Yet, there has been and remains a tension between those who look after tigers living in zoos and those who have invested their time in the stewardship of wild tigers. Ronald Tilson and Sarah Christie conclude that the goals of both tiger conservation paradigms are the same; securing a future for wild tigers. They argue that many of the organisational principles and lessons from efforts to make tiger breeding programmes in zoos goal-driven can be employed by the larger tiger conservation community in learning to work together to save the tiger.

As editors we come from both sides of this issue, and believe that the tension is not trivial and at times is crippling for tiger conservation. It is useful here to explore the historical and cultural contexts of wild and zoo tiger conservationists as a means of separating the people issues from the problem, focusing on interests and inventing options for future partnerships. In our experience, stewards of wild tigers see in tigers a fundamental essence of nature, living in ideal natural places. Much of the moral authority for saving the tiger stems from the fact that tigers and their 'Edenic' habitats represent an external source of non-human values against which human actions can be judged with little ambiguity (Cronon 1996). Tigers are living symbols of power and grace and wild Asia.

Tigers are always value-laden and it is we humans who affix value in the way we perceive them. We do so in many different ways. For those of us who have seen, or who someday want to see, or simply place great value on tigers living out their lives in an ideal nature, tigers living in zoos can be seen as the antithesis of wild tigers. In this view, zoo tigers are living unnatural lives and tigers are a commodity being exploited for financial gain. We contend that there is nothing unnatural in the awe on a child's face the first time s/he really sees a zoo tiger. Wildlife films and beautiful colour photographs of tigers can be informative but simply do not replace living, breathing, roaring tigers. We think of zoo tigers as ambassadors for wild tigers. This is how the vast majority of people will ever witness the wonder of a tiger and, as Tilson and Christie report, the numbers of zoo visitors are staggering. Zoos have embraced the environmental movement to support endangered species conservation, and zoo visitors are an essential tiger conservation constituency. Zoo visitors and the public at large must be a partner in saving the tiger,

because it is the public who supports the legal framework that protects tigers and it is the public which foots much of the bill. Zoos have moved substantially beyond being entertainment centres and are becoming conservation parks with goals focusing on conservation and education; in fact, zoos are playing an enormous role in environmental education (Conway 1997). Zoos see their role as a support, not a substitute, for wild tiger populations. Many of the authors contributing to this volume are supported by, or have received support and training from, zoo-based programmes.

Will zoo populations ever serve as a source of tigers for 'reintroduction' into the wild? Christie and Seidensticker outline the constraints and opportunities for reintroducing tigers. Reintroduction of tigers is not a conservation technology that will be used in the immediate future, but enough has been learned recently from the reintroduction of other large carnivores that it remains a viable option if the ecological, valuational and organisational challenges are met. Certainly zoo tigers are an insurance policy against ever losing the tiger completely, but the goal for zoo-based tiger conservationists and the stewards of wild tigers alike is to provide the support needed to maintain sustainable tiger populations on their home ground. The challenge for all of us, as William Conway (1997) has so eloquently pointed out, is to keep these special places where tigers now live from becoming places where tigers go to die.

The first call to save the tiger led to one of the most famous and extensive wildlife conservation campaigns ever undertaken. With the passage of CITES and better protection for tigers in range states through the 1970s and into the mid-1980s, there was some success in efforts to save the tiger (Seidensticker 1997). But beginning in the late 1980s, tiger parts, especially tiger bone, began turning up in staggering quantities, indicating that a major haemorrhage in tiger numbers was occurring. In 1993, 'The International Symposium on the Tiger' held in New Delhi, India, and 'The Amur Tiger: Problems Concerning Preservation of its Populations', a symposium held in Khabarovsk, Russia, both concluded that the tiger was in crisis because of poaching and trading of its parts to supply the demand from traditional East Asian medicine (TCM).

There has been a focused and sophisticated response to this increase in tiger poaching by a number of NGOs, working with governments to stem the trade in tiger parts and products. Ginette Hemley and Judy Mills trace the history of this new tiger crisis. They can see a beginning of an end in this trade. Significant headway has been made in curtailing the legal trade in tiger bone, but tigers are still being poached and tiger products are still being sold as ingredients in several traditional East Asian medicines. This stubborn residual trade remains to be eliminated. Tackling the demand problem means taking tiger bone out of TCM, and enlisting would-be consumers of tiger bone in tiger conservation efforts is the key concept. TCM is a health care system with ancient roots that hundreds of millions of people depend on; the question is not whether there will be or should be TCM, but how to ensure that TCM will not endanger the survival of the tiger and other endangered species. Important first steps for tiger conservationists are to reject an anti-TCM stance and understand the need for treatment of the pains for which products containing tiger bone have traditionally been used; to help promote the acceptance of substitutes for tiger bone in TCM; and to talk with TCM specialists about the conservation status of tigers and the role TCM trade has played in endangering the tiger's survival. An alternative suggestion has been to 'farm' tigers to supply parts and products. Aside from the animal welfare and other emotive objections to this, there is no way to tell a wild tiger bone from a 'farmed' tiger bone; debate on the possible effects of such a legal market on the illegal one is both heated and inconclusive, but the bottom line is that without a purposive approach to remove the demand for tiger bone, the future for wild tigers is very bleak. Hemley and Mills believe that the next few years are pivotal.

Operation Amba (the name for the tiger used by the Udegai people of the Russian Far East) has added a new dimension to the toolbox of tiger conservation. Steven Galster and Karin Elliot recount how an international effort unfolded to deter the lethal slide of the Amur tiger into extinction in the Russian Far East. A conventional approach to containing the poaching of Amur tigers was unsustainable for financial and managerial reasons; rather, they describe Operation Amba as a strategic defence of the tiger with PsyOps (psychological operations) as a key component. Operation Amba has made the critical difference to the survival of the tiger in Russia. With low densities of humans and roads, the context for a protective scheme for the Amur tiger in the Russian Far East is strikingly different from that in other reaches of the tiger's range, but there are important lessons to be learned from Operation Amba: involve local people, develop skilled personnel and equip and pay them on time, secure political backing, communicate with relevant agencies abroad and maintain good public relations.

The wildlife trader or middleman is the link between the market and a poacher with a dead tiger in hand. Operation Amba focused on identifying and neutralising tiger traders in the Russian Far East. Ashok Kumar and Belinda Wright emphasise the need to focus on removing the city-based wholesaler in tiger parts in controlling tiger poaching in India. There is essentially no demand for tiger parts and products within India; the demand comes from outside. They estimate that a tiger a day is poached in India and this will continue as long as the demand for tiger parts continues and as long as punishment of city-based traders is not imposed, which it rarely is. Kumar and Wright also argue that tiger conservationists must take up and effectively address the concerns of the people who live with tigers on a daily basis.

Tiger conservationists have seemed preoccupied with determining the number of tigers and doing so has been problematic and contentious (Seidensticker 1997). When tiger habitat has been identified it has usually been referenced as the amount in reserves, although it is commonly accepted that many more tigers live outside reserves than in them. There has not been an equal emphasis on determining the amount of habitat remaining for tigers throughout their range in Asia, or what the essential components of that habitat might be. The importance of adequate prey density as a habitat component was emphasised in Part II, and Karanth and Stith showed that with an adequate prey base a moderately sized tiger population could compensate for even relatively high mortality regimes imposed by poaching. In Part IIIC, Wikramanayake and his colleagues present an entirely new way of thinking about tiger conservation options and introduce the idea of tiger conservation units – TCUs. This is an important conceptual change in thinking about the future of wild tigers. They shift tiger conservation planning from a taxonomic to an ecological basis because a conservation strategy must account for the behavioural, demographic and ecological variation present among populations across their range. The underlying principle is that tigers are adapted in all these ways to the ecological conditions where they live. They propose that we think in terms of tigers living within five bioregions, to achieve fundamental goals in conservation biology; maintenance of representation and also conservation of the range of communities in which tigers occur. They identify major habitat types within each bioregion, and introduce a process to identify both remaining blocks of habitat where tigers may still be able to live and the essential linkages within these habitat blocks or TCUs. Then they prioritise these blocks through a process that orders risks and opportunities, and suggest that the highest probability of securing a future for wild tigers can be achieved by focusing conservation efforts on the top 50 of the 160-odd blocks of potential habitat scattered among the bioregions and habitat types. This is the first time we have had maps of Asia that accurately reflect where the tiger may survive in the next century (Figs. 18.2–4).

The habitat of the Amur tiger was not included in the deliberations of Wikramanayake and his associates because it was already under study by Dale Miquelle and his colleagues. The process of including the tiger in long-term land-use plans is further advanced in the Russian Far East than anywhere else in Asia. The inaccessibility of the Sikhote-Alin Mountains where most of the remaining Amur tigers live has been a primary factor in their protection, but these mountains are now becoming more accessible as road density increases to support the extraction of timber and

other resources. Using the ecological criteria they reported on in Part II for sustaining the Amur tiger population, Miquelle and associates define and implement a planning process and propose a habitat protection plan for the Amur tiger. Reserves would be linked by multiple-use zones, providing connectivity and habitat for 70 reproducing females overall. With the ecological criteria fully developed, they move on to examine the political criteria necessary for sustaining the Amur tiger in the future and the essential factors that enable people and tigers to share forest resources. In very explicit terms they lay out exactly what they believe will constitute a habitat that will sustain the Amur tiger into the future; a process which has not so far been replicated anywhere else in the tiger's range. This is a process that will maintain the largest TCU in the world.

When the first call to save the tiger was heard in the late 1960s, India, like other countries, responded with a top-down, command and control programme in which the tiger was treated as a public good. New protective legislation was passed. Habitat for tigers was provided through specially designated reserves that have grown from the initial 8 to 23 at the time of writing. In these reserves every effort has been made to separate tigers from people, usually by restricting commercial harvest of forest products in these reserves and by relocating inhabitants. The underlying assumption was that these reserves would serve ' . . . to increase the tiger population to optimum levels through improvement of the biotope . . . this situation will provide a breeding nucleus [of tigers] from which surplus animals can migrate to surrounding forests' (Task Force, Indian Board for Wild Life 1972). The limitations of a heavy-handed, top-down approach, especially with dissipating central power in India, have become apparent as Valmik Thapar describes. Because of human population growth in and around tiger reserves, people and their livestock have been spilling into reserves rather than tigers spilling out as was originally envisioned. Tigers are very sensitive to dispersal barriers (Smith 1993), and, because of the small size of active tiger reserves, tigers living in them are at considerable risk from the potential impacts (demographic, genetic, environmental and catastrophic) of stochastic processes. Valmik Thapar reports that after 25 years the much-vaunted tiger reserve system initiated in India at the inception of Project Tiger is in tatters. There are some notable exceptions, such as Nagarahole, but many reserves (e.g. Panna) are rapidly becoming more isolated and degraded through massive extraction of resources to support growing populations of people, and their livestock, living in and around them. While the tiger remains a public good and national treasure in India, the people who live near tigers on a daily basis are having an increasing loud and negative voice in the tiger's future.

This contribution is a cry to arms for tigers in India. Although Valmik Thapar does not talk about this in his chapter, he watched the increasing threats to tiger habitats driven by the needs of local people as he studied tigers in the Ranthambhore National Park during the 1980s. In his book *Tigers, The Secret Life* (1989) he railed against a government that seemed not to care or at least did not give the tiger any priority status. The final chapter of his book is a prescription on how to better the lives of those who live near tigers. He did not just write about it, he then did something about it. With his teacher, Fateh Singh Rathore, and others he set up the Ranthambhore Foundation which is dedicated to bringing both the public and private sectors together to better the lives of those who live near Ranthambhore. And he established 'TigerLink', a tiger support network that strives to bring together the many people throughout India and the world that believe there should be a place for wild tigers. Herein lies hope for a future for wild tigers when such efforts are duplicated and adapted to local circumstances throughout the tiger's range.

Often in tiger conservation, planning and actions have focused on a single threat, such as poaching, to the exclusion of the larger problems tigers face in the shifting social and political Asian landscape. The lands where tigers live contain resources – timber, minerals, hydropower – needed to fuel rapidly expanding Asian and world economies and the extraction of resources requires an improved transportation network, including the widening and straightening of roads. Bittu Sahgal has prepared a map (Fig. 20.2) that highlights – wildlife 'hotspots' is his word – the locations of planned or ongoing major projects threatening tiger reserves and other tiger habitats. The power of his map is that in a single page you can grasp how all-invasive and overwhelming these threats are to the future of the tiger in India. Largely financed through multi-national and international structures, these direct pressures are being brought to bear on tiger lands in India in response to the needs and wants of people who have never heard a tiger's roar in the still of the night or seen a tiger track along a dusty forest road. These are not the local people but those of us who live in India's and the world's urban and suburban centres. Most of these people may be personally neutral or even supportive of sustainable tiger populations but are unknowing participants in the tiger's continued demise. What is so frustrating for the stewards of wild tigers is that these root challenges to securing a future for wild tigers are largely beyond the control of local people and in many instances even beyond control at national or international levels. Securing a future for the tiger on its home ground requires broad understanding of how international forces are threatening the tiger's future. No legal mandate can completely exclude economic considerations from the conservation process. Making the case to preserve the tiger and tiger habitat specifically and biodiversity generally means showing that the benefits to be had from preservation are as good or better than those to be had by converting the habitat for other purposes. This is where partnerships and linkages with local people specifically and the private sector in general will make all the difference. The lessons from Tigers 2000 were that the place to begin is in securing local partnerships based on sustainable land-uses and recovery because the long-term interests of both tigers and local people are usually at risk from these extractive pressures.

Our final two chapters address the issues of how to make it worthwhile for people to live in proximity to tigers and how to increase their tolerance of tigers. These are examples of large- and small-scale investment in tiger conservation to the benefit of the local people. We believe these two chapters represent a breakthrough that will begin a shift in the save-the-tiger paradigm, a new roadmap to take tigers and their significant habitats into the next century.

Kathy MacKinnon and her associates outline how the Global Environmental Facility (GEF) is supporting projects that attempt to improve protected-area management and ecodevelopment for communities living within and adjacent to protected areas. At their core these projects seek to take the pressure off reserves and to rebuild natural capital around them. Strategies include planning for micro-credit, alternative livelihoods, conservation agreements, special programmes for joint forest management, education, conservation-awareness building and monitoring; all designed to promote better living conditions and public support for conservation at the local level. The GEF is spending tens of millions of dollars in this effort around reserves in India, Indonesia, China and Laos.

Do the GEF-sponsored programmes adequately address the needs of tigers as well as those of people? This contentious issue was expressed and discussed at Tigers 2000 because the needs of the tiger are not explicitly recognised or addressed in the GEF programme. Our advancing understanding of the ecological criteria that must be met to sustain tiger populations as outlined throughout these chapters will move this discussion from a theoretical plane to pragmatic on-the-ground assessment, consultation and action in each case. Ullas Karanth has demonstrated how this understanding can be furthered with an ongoing, statistically valid system to monitor the density of undulate prey and tigers, based on his work using trained volunteers in India's Nagarahole National Park.

Eric Dinerstein and his associates report on a small-scale project that promotes local guardianship, including that of tigers, on degraded lands adjacent to the Royal Chitwan National Park in Nepal. They recognised that the main threat to biodiversity in Chitwan is the poverty of the surrounding villages, which hold 290 000 people and have a population growth rate of over 2% a year. Lacking suitable alternatives for the fulfilment of

their daily subsistence needs, many local residents believed that their livelihoods were threatened by the creation of the park in 1973. As is the case with many tiger reserve areas with a weak economy and a lack of supplies from outside, people in Chitwan are dependent on the park for fodder and firewood. This project focuses on rebuilding natural capital by developing a sustainable supply of fodder and firewood, and increasing habitat for wildlife, in the park's degraded buffer zone. The local community was responsible for resource management and protection in the Park's buffer area. By providing economic incentives and developing self-sufficiency, villagers have created alternative sources of the forest resources previously available only in the Park. They have also regenerated 16.5 km² of prime habitat for the tiger and the Asian greater one-horned rhinoceros. This wildlife resource has attracted ecotourists, and local communities are making money for community improvement, such as schools and health clinics, by charging fees to local hotels that use the regeneration area for wildlife viewing. This also provides funds for gas digesters that encourage stall-feeding of cattle with improved milk yields; this facilitates dung collection and reduces the size of the free-ranging cattle population around the reserve. Poaching in this area of the Chitwan Valley has significantly decreased. These authors provide principles for adapting this programme to other areas in tigerland. Indeed, this Chitwan site welcomes local groups from other TCUs who want to learn more about approaches that link tiger conservation with local development. As neighbours watch the communities involved with this programme prosper, the project is expanding regeneration activities from the site level to the landscape scale. Dinerstein (pers. comm.) calculates that when the full potential of this programme is reached there will be enough regenerated habitat to double both tiger and rhino populations in the Chitwan region. Chitwan is recognised as a primary success story in tiger conservation; the lessons from this valley provide one roadmap for securing a future for wild tigers.

Linking *in situ* and *ex situ* tiger conservation

Amur tigers romping in a
five-hectare habitat at the
Minnesota Zoological
Garden. Modern zoos
design tiger habitats with
complexity and space to
stimulate exploration and
other natural behaviours.
This is where most people
will ever see a breathing,
roaring tiger.

14

Effective tiger conservation requires cooperation: zoos as a support for wild tigers

Ronald Tilson and Sarah Christie

Introduction

Today there is an unsettling polarisation between international conservation organisations and their programmes for wild tigers and the international zoo community and their programmes for zoo tigers. This is ironic because both share an identical mission – continued viability of wild tigers in their natural habitat. This polarisation is expressed as a communication failure with significant repercussions. We attempt to articulate the prevailing philosophy of our respective zoo tiger programmes in North America and Europe in order to dispel the clichés that have been historically used and are currently invoked to characterise and dismiss contributions from the zoo community. We also report on how similar programmes in tiger range countries are developing and how they might contribute to *in situ* conservation efforts. The world's zoos are learning the value of co-operation in the management of zoo programmes, and to a lesser extent in the generation of support for *in situ* tiger conservation, and with this background we hope to illustrate how the traditionally isolated fields of *in situ* conservation (wildlife agencies and field programmes) and *ex situ* conservation (zoos and zoo programmes) can also work together in the species' best interests. In today's world, close co-operation among individuals in these two fields will lead to more effective conservation action for tigers. If we can learn to work together, perhaps we can learn how to help save the tiger together.

The tiger's distribution in Asia ranges from Indonesian tropical rainforests, to seasonally dry evergreen forests of India, east to Vietnam and north to Russian temperate forests. This vast range across differing climates, habitats and cultures marks the tiger as enormously adaptable (Schaller 1967), as should our efforts be in attempting to preserve it. The story of the tiger and its plight is as complex as the many languages of the peoples who live near the tiger's forests. The one common bond throughout Asia is that the tiger is both revered and feared as a symbol of great power and strength by those who live near it. Now we are in danger of losing this living symbol of Asian wilderness.

It is generally agreed that wild tiger numbers are declining nearly everywhere. A considerable proportion of wild tigers live outside of the boundaries of the protected areas designed to secure them (Dinerstein *et al.* 1997; E. D. Wikramanayake *et al.* this volume). Tigers are unlikely to survive the imminent loss or deterioration of their forests. Those who poach tigers and the animals on which they feed pose another threat. All five remaining tiger subspecies are threatened with extinction; the Bali, Caspian and Javan tigers were lost between 1937 and 1980 (Seidensticker 1987; Nowell & Jackson 1996) and the South China tiger is precariously close (Tilson *et al.* 1997), an extinction rate of one subspecies every 15 years.

The problem for tigers is as simple as it is brutal. Resolution of these problems is much more complex (V. Thapar this volume). Across all of Asia, wild tigers have had their vast forests and grasslands destroyed for timber or conversion to agriculture. What are left are small islands of forest surrounded by an ever-increasing and relatively poor human population trying to make ends meet. They in turn place even more pressure on wild tigers and their habitat. Firewood is collected, domestic livestock graze in the forests, and common tiger prey – wild pigs and deer – are shot and snared for food by poachers. Worse yet, tiger bones and other body parts used in traditional Chinese medicine now command premium prices on the international black market (G. Hemley & J. Mills this volume), and poachers now poison water holes or set steel wire snares to kill tigers. It is a cruel death (*Animal Traffic: 31 Tigers.* A series of films by R. Orders & A. Bondy).

For the most part, responsible forestry and wildlife departments are too understaffed and underbudgeted to be effective against this onslaught. The realities of ever-decreasing tiger numbers, increasing human utilisation of remaining habitat and continuing demand for tiger body parts fuel the decimation of wild populations. Simply put, tigers are disappearing in the wild (Jackson & Kemf 1994; Jackson 1996). If we continue to maintain the status quo, then we run the risk of losing all wild tigers (Seidensticker 1997).

We would like to thank the organisers of the Tigers 2000 symposium (the Zoological Society of London and Esso UK) for funding one of us (R.T.) to attend the conference, and our respective zoos, the Minnesota Zoo and London Zoo, for their continued support in developing tiger conservation programmes in North America and Europe respectively, and in Asia together. The Sumatran Tiger Project and the Tiger Information Centre are supported by the Save the Tiger Fund, a special project of the National Fish and Wildlife Foundation in partnership with Exxon Corporation.

The potential of zoos

In the past, large and small cats were nearly always kept in scandalous conditions in zoos, often locked alone in concrete and steel cages. It was little different from being in jail. Row upon row of cat cages still remain at many zoos, a distasteful reminder of an era when animals were collected and displayed as living trophies to be gawked at by a seemingly uncaring public. The food fed to the animals was atrocious as well and, because it lacked essential minerals, vitamins and amino acids, there was little chance, if any, of successful reproduction. Social and psychological requirements of the species were also poorly met, thus further suppressing reproduction. But during those times this was thought to be unimportant because replacement animals could readily be collected from the wild, and animal dealers were doing brisk business (Tilson 1991).

Times have changed, but the image has not. Of the more than 90 000 mammals currently known to be held in the world's zoos via their registration in the International Species Inventory System (ISIS), 84% are captive-born (ISIS 1997). Nevertheless, today's zoo, no matter how modern it tries to be, is still burdened with the misfortunes of its past. Zoos are continually characterised with the following mis-statements:

▸ 'Zoos have nothing to contribute to conservation; they are only interested in collecting animals from the wild' (quotes too numerous to list). In truth, virtually all wild-caught tigers entering zoos nowadays do so without money changing hands and at the request of forestry authorities; these are orphaned cubs, animals confiscated from illegal holders, or tigers that kill livestock or people around reserves – often young males attempting to disperse.

▸ 'Zoos funnel conservation efforts and resources away from field programmes' – if they spent even a fraction of their money protecting wild animals there would be no need for zoos (e.g. Varner & Monroe 1991; Balmford *et al.* 1995 1996; for counterpoint see Hutchins & Wemmer 1991; Tilson & Sriyanto in press).

▸ 'Zoos justify their captive programmes by saying they are going to reintroduce tigers to the wild, but, even if this were possible, it would be extremely complicated, very expensive and face strong local human opposition' (P. Jackson pers. comm.). The 'genetic lifeboat' concept is only one of the

Javan tigers in the Berlin Zoo in the early 1900s. In Victorian zoo design, tigers and other animals were displayed as interesting objects and no emphasis was placed on stimulating natural behaviours.

conservation functions of zoo tiger populations; in any case, recent work with other big cats (Belden & McCown 1995) has shown that reintroduction is indeed an option (see S. Christie & J. Seidensticker this volume, Box 14.1).

Over the past 200 years a large number of zoos have been established in all parts of the world, and a great diversity has arisen among these institutions. Of particular importance is that conservation is moving ever closer to centre-stage as the central theme of all zoos (Wiese & Hutchins 1994). Zoos are destined to evolve into conservation centres or fail, because otherwise they will lose the trust of their primary constituency, the public visitor. Many leading zoos have already taken the first steps along this path; many other zoos still need to be stimulated, guided and helped to follow.

To assist in this revolutionary process, the World Zoo Conservation Strategy (IUDZG/CBSG 1993)

was drafted. It emphasises three major initiatives by which the zoo community can achieve its goals. It is not by coincidence that the first of these goals outlined in this document, referred to here as the Strategy, is the conservation of endangered species populations and their natural ecosystems.

The global zoo network

The Strategy is directed primarily at the more than one thousand zoos that are organised in zoo associations. Collectively the zoos of the world are visited annually by at least 600 million people, which constitutes approximately 10% of the current world population. In Asia alone, where the tiger lives, zoos are visited by 308 million people annually (IUDZG/CBSG 1993). This sheer mass visitation to zoos makes them excellent institutions to increase public awareness of the irreplaceable values of nature.

Education is therefore an essential conservation task of zoos.

It is estimated that the thousand-plus federated zoos of the world collectively house some one million animals, predominantly higher vertebrates, and practically every one of these zoos has one or more tigers in its collection. There are more or less 1250 studbook-registered tigers (Müller 1995) in these zoos, with many more tigers numbering in the thousands kept in private ownership and circuses. It is important to note that the Strategy takes the position that the commercial trade in animals taken from the wild must cease as soon as possible as a source for acquisition of zoo animals. In addition, zoo populations should be managed to support the survival of species in the wild in accordance with *The IUCN Policy Statement on Captive Breeding* (1987) and the Convention on Biological Diversity (Glowka *et al.* 1994). The Conservation Breeding Specialist Group (CBSG) of the Species Survival Commission of the IUCN forms a link between the World Conservation Union and zoo-based *ex situ* species conservation. It enhances liaisons between zoo programmes and the *in situ* conservation activities of the IUCN, the SSC's many specialist groups, other international conservation organisations and governmental agencies. One component of the plan is to link captive tiger programmes in each tiger range country via a global network while there is still time.

This global network – the Tiger Global Conservation Strategy (GCS) – was initiated by CBSG in 1992 and now includes the regional zoo associations of America, Canada, Europe, Russia, India, Southeast Asia, Australasia, Japan and China. The purpose of this group is to integrate tiger conservation activities among the world's zoos by providing a communication network and a strategic framework for the most efficient use of resources for the species. Programme goals include international captive programme coordination, integration of conservation management strategies for captive and wild tiger populations, and generation of financial and technical support from the world zoo community for *in situ* tiger conservation efforts. The most significant point is that the world's zoos have launched a comprehensive management strategy for captive tigers in support of wild tigers (Tilson *et al.* 1993). It is a process that needs considerable and frequent repair, but it is a process that is in place, and it is built on the premise that the only way we are going to succeed is by working together cooperatively, towards common goals set by consensus decisions.

For wild tigers to survive it will be crucial that the majority of the world's tiger authorities reach consensus on how to move beyond defining the problems affecting tigers to resolving these problems. This process was initiated at the Tigers 2000 symposium (Seidensticker 1997). Previous reports have focused on various components of defining and resolving these problems. Some of these documents include A Global Tiger Conservation Plan (Seal *et al.* 1987a), Indonesian Sumatran Tiger Conservation Strategy (Ministry of Forestry 1994), Saving the Tiger: A Conservation Strategy (Norchi & Bolze 1995), A Habitat Protection Plan for Amur Tiger Conservation (Miquelle *et al.* 1995), Wild Cats: Status Survey and Conservation Action Plan (Nowell & Jackson 1996), and A Framework for Identifying High Priority Areas and Actions for the Conservation of Tigers in the Wild (Dinerstein *et al.* 1997). Our task is now to reach consensus and congruency so that specific tiger *in situ* conservation strategies can be developed for each tiger range country and bioregion.

Most wild tiger populations throughout Asia are too small and fragmented to ensure their survival in the future (Seidensticker 1997; E. D. Wikramanayake *et al.* this volume, Chapter 18). Small populations are at great risk of extinction from disease, catastrophes and effects of inbreeding (Gilpin & Soulé 1986). To counter the trend towards ever smaller wild populations, wildlife managers must be prepared to link isolated populations through land-bridges (Dinerstein *et al.* 1997) or transfer genetic material among populations through translocations or reintroductions (see later), or by using emerging but as yet unperfected techniques in assisted reproduction – embryo transfer, artificial insemination and gamete freezing – which may provide necessary options when they are fully developed (Wildt *et al.* 1995). All of these processes are constrained with a combination of inherent issues – behavioural incompatibility, disease, costs and feasibility – as well as the scepticism of those who believe it cannot be done. Regardless of what can or cannot be done, conservation of wild tigers requires that we envision innovative strategies for managing fragmented wild populations, because not all wild tiger populations are of sufficient size to be left on their own. In the case of the South China tiger, most of the conventional options seem to have already been exhausted (Tilson *et al.* 1997).

Back to nature: zoo animals for reintroduction

The World Zoo Conservation Strategy emphasises – in accordance with the IUCN Guidelines for Reintroductions (IUCN 1998) – that reintro-

ductions, when properly applied, can bring great benefits to natural biological systems. The Strategy reported that reintroductions (and restocking projects) have been undertaken with more than 120 species. Although it is too early to assess the outcome of all of the projects, 15 projects have established self-sufficient populations. In a more specific review of carnivore reintroductions, Reading & Clark (1996) list 55 attempts of which 29 were judged successful by the practitioners (no fixed definition of success was used). Tiger conservationists should review these case studies for possible future direction while there is still time and sufficient number of tigers.

From our perspective the primary mandate of captive tiger programmes is to reinforce, rather than replace, wild populations. The Tiger GCS implicitly states that the protection and continuance of *in situ* populations is the highest priority (Tilson *et al.* 1993). All too often the general public expects

A South China tiger in the Guangzhou Zoo. The Chinese Zoo Association is developing a long-term management plan for the 50 South China tigers living in zoos. This exceeds the number remaining in the wild.

individuals of endangered species at their local zoo to be released back into the wild. For tigers, this is not the case today for biological, political and practical reasons. Rather, the captive population needs to be perceived as a reservoir of genetic material representing the species, not just individuals. They represent a 'genetic insurance policy' against extinction in the wild. The potential biological role of these individuals is to help re-establish populations that have vanished from their natural range or to revitalise wild populations that have genetic or demographic problems (Wiese *et al.* 1994). Until then, our primary concern must be to continue to maintain, or even increase, what is left of the tiger's natural habitat.

The Strategy specifies that reintroduction, restocking or translocation projects will always be carried out in agreement with IUCN guidelines and within regulations of the conservation authorities of the recipient countries, as well as in co-operation with IUCN/SSC Specialist Groups for Reintroduction, Veterinary, Conservation Breeding, and others relevant to the situation. Taken together, these documents list a daunting set of preconditions and necessary precautions, precluding any haphazard approach, and zoos must operate within these constraints.

The frozen zoo: assisted reproduction and genome banking

The use of techniques for assisted reproduction also has the potential to enhance the management of *ex situ* populations, and can aid in the retention of maximum genetic variability through time (Wildt *et al.* 1995). Cryo-preservation of gametes and embryos, or Genome Resource Banks (GRB), may also serve as genetic reservoirs, and can act as a third component of species conservation, with *in situ* and *ex situ* populations as the primary and secondary components respectively. These technologies are not viewed by zoos as a resolution to the extinction process, but as additional alternatives for the long-term conservation of species when natural processes fail. We emphasise here that these

techniques are not yet tested, and thus are not yet feasible.

We agree with the Strategy's caution against overestimating the conservation value of cryo-preserved genetic material. Living populations particularly are essential to preserve non-genetic learned behaviour patterns, for these may be crucial to species survival. Thus, deep frozen genetic material may be useless without living populations. In contrast, because GRBs also contain bio-materials, subspecies differentiation – and possibly even genetic characterisation of local wild tiger populations – by molecular DNA analysis may someday provide information crucial to the tiger's survival (J. Wentzel *et al.* this volume).

Sharing knowledge: crucial to conservation

The Strategy calls on the global zoo community to continue adding its special knowledge to nature conservation, and whenever possible to increase the amount of this contribution. Veterinary knowledge compiled by zoos is already invaluable in the field; without it, safe anaesthesia of wild tigers for radio-collaring studies would not be possible. Zoo knowledge of the biology of small populations will become more and more relevant to conservation of wild species as habitats are further reduced in area and the geographic ranges of species are increasingly fragmented. Tigers are particularly illustrative of this dilemma. Information on what is happening with tigers in each range state should be distributed as widely as possible throughout the conservation community.

Accurate information is vital to making good decisions. In the case of tigers, situations change rapidly to crisis proportions – e.g. tiger poaching in India (A. Kumar & B. Wright this volume), and in Russia (S.R. Galster & K.V. Eliot this volume), and information needs to be widely and quickly circulated to formulate a coherent conservation response. In part, and in co-operation with as many organisations as possible, this is underway at the Tiger Information Centre (http://www.5tigers.org), where a centralised database of field research,

Box 14.1 Is re-introduction of captive-bred tigers a feasible option for the future?
Sarah Christie and John Seidensticker

Most remaining populations of wild tigers are small, and much of their habitat is fragmented (Dinerstein *et al.* 1997). In such small populations, inbreeding and the consequent reduction in heterozygosity may expose the effects of deleterious alleles, thus generating the 'extinction vortex' – a positive feedback loop between the size of the population and the average fitness of its members (Caughley 1994). No mammalian species has yet been shown to be unaffected by inbreeding (Lacy 1997), and supplementation of the gene pool in small populations of wild tigers may therefore be desirable in the future. It may also be necessary to replace tigers in areas from which they have disappeared altogether as a result of the extinction vortex or other factors. Long-term, holistic tiger conservation strategies should include options capable of addressing these aspects of the problem.

In any future scenario where supplementation or reintroduction is deemed desirable, three options are available: translocation, release of captive-born animals, and genetic supplementation through artificial reproductive techniques. The last of these could in theory, when fully developed, be used to replace either of the first two options in cases of supplementation; but it is beyond the scope of this short discussion, which considers the relative suitability of wild and zoo-bred tigers for release. Reintroduction is now accepted as a conservation tool, provided that it is carried out in accordance with the IUCN Guidelines for Reintroductions (IUCN 1998). Preconditions such as control of the causes of decline, availability of sufficient suitable habitat and food, a commitment to long-term post-release studies and many more must first be met. In the case of the tiger such conditions are as yet far from fulfilled, and it is argued here only that tiger reintroduction is feasible in the right circumstances, not that it should be attempted in the near future. For reintroduction of large carnivores such as the tiger, the possibility of adverse reactions from local people must be given special attention; accusations that conservationists put wildlife before people can do immense damage.

Genetically, there may be little advantage in choosing either wild or captive tigers for such a project. In behavioural terms, it is often argued that zoo-bred animals would be incapable of learning how to survive in the wild, and that translocated wild tigers would be more suitable.

Recent work has shown that captive-born large carnivores *can* learn to adapt. Basic hunting behaviour is essentially instinctive, as is readily demonstrated by the ability of domestic cats to kill without instruction from their mothers (and see below). Given a pre-release training period using appropriate species, there is little doubt that zoo-born tigers could learn to hunt, though it is acknowledged that the finer points – such as appropriate resource use inside the home range – may usefully be learned from a wild mother.

Captive-born large carnivores have been used, for example, in red wolf (Phillips 1995), European lynx (Smielowski 1996) and European wolf (Badridze 1992) reintroduction projects; and in a feasibility study for reintroduction of the Florida panther (Belden & McCown 1996). Between 1993 and 1995, 19 mountain lions of varied origin were released in northern Florida. Ten of these were wild-caught and released within three months of capture, three were wild-caught and held in captivity for some years prior to release, and the remaining six were captive bred. The cats caught deer within a week of release and successfully hunted throughout their time in the wild (J. Lukas, quoted in *Cat News* (Anon.1996c)). Seven of the nine captive-bred or captive-held cats settled within two months as opposed to only three of the ten wild-caught animals. The captive-bred animals settled closer to the release site, and all of them settled in a social system while only two of the wild-caught specimens did so. They also showed significantly lower mortality and higher breeding success – three of the six captive origin cats were known to have bred at the time of writing (R. Belden pers. comm.) while there was only inconclusive evidence of breeding for three of the ten wild-caught animals. The disadvantage of captive-born mountain lions was, predictably, that they were significantly more likely to be involved in lion-human conflicts.

Based on these observations, captive-born large cats might in fact be better candidates for reintroduction than those from the wild – provided it were possible to overcome the problem of familiarity with humans, and the concomitant problem of negative perceptions on the part of local people. One way to achieve this with tigers would be to use not zoo-bred tigers themselves but their offspring, born in large natural enclosures deep in suitable habitat. Pairs of zoo tigers would be allowed to breed in such enclosures, with live prey of appropriate species supplied and sight and sound of humans kept to an absolute minimum. The cubs would reach dispersal age relatively unaccustomed to humans and knowing what and how to kill. At this stage, the parents would be removed and the juveniles radiocollared and given veterinary check-ups prior to release. In order to ensure that released tigers completely avoided human habitation, negative reinforcement might be advisable, and two options are touched on below. Released tigers would be closely monitored and any that did cause problems with humans would be immediately removed.

Aversive conditioning of large carnivores has been attempted through both conditioned taste aversion and electric shock. The two processes are very different and much remains to be done to discover how effectively each can contribute to reducing carnivore/human conflicts, but both merit further investigation. Conditioned taste aversion, the inhibition of food consumption by association of illness with taste, is a well-documented behavioural phenomenon (Riley & Baril 1976) but is sensitive to variations in methodology, which have tended to confound field testing (Forthman *et al.* 1985). Forthman *et al.* list several factors which which must be controlled, of which the most important is that the illness-inducing substance (lithium chloride) should be undetectable by taste or odour. The objective is an aversion to the food item (here, domestic livestock), and if the additive can be detected the animal will simply avoid that substance in future while continuing to consume the untreated item. Intensity of illness is also important but the desirable effects of high dosages are reduced if they also induce vomiting; delivery in cellulose-acetate-coated capsules, which dissolve only on reaching the duodenumem, can eliminate both taste and vomiting.

Peripheral stimuli such as noise and electric shock inhibit not consumption but approach behaviour (Forthman *et al.* 1985). Remote-control shock collars are used in police dog training, and collars triggered by crossing underground wires confine domestic dogs to their owners' property in the US. Few applications of this technique to wild carnivores are on record, but one wolf reintroduction project in Georgia, Europe, reported some success (Badridze 1992). Shocks were administered to the wolves at the sight or scent of humans or domestic livestock and Badridze reports that avoidance behaviour began to appear on the fifth or sixth day and was established after 20 days.

Badridze's approach also had the virtue of involving local people in the project. There was considerable hostility to begin with, and two project wolves were shot in their enclosures. By using local people as volunteers in the avoidance training, the researchers demonstrated that wolves were not in fact the ravening beasts of popular myth; attitudes were completely changed and some volunteers continued to assist in monitoring the released wolves for several years (Badridze 1994). Though tigers have killed humans while wolf damage is limited to livestock, the key point is that villagers' perception of wolves as being more vicious and dangerous that they are was successfully altered, which provides welcome encouragement for large predator conservation in general.

A tiger reintroduction would, like all reintroduction projects, be difficult, time-consuming and expensive. The evidence argues, however, that it is feasible in the right circumstances. Reintroduction of tigers is not a tool that will be used in the foreseeable future, but the problems of securing a future for the tiger are vast and we must be prepared to use every tool available to us if we are to succeed in the long term. The necessary resources for this– well-managed populations of pure-bred tigers in zoos – must be developed and maintained into the next century in case of need, along the way playing a vital role in constituency-building for tigers around the world.

Box 14.2 The 5Tigers Website: a tiger information centre
Ronald Tilson, Janet Tilson and Anne-Marie Alden

The Tiger Information Centre was established at the Minnesota Zoo and went on-line on 28 September 1995, the same day the Exxon Corporation and the National Fish and Wildlife Foundation publicly launched the Save the Tiger Fund. The Information Centre currently includes a site on the World Wide Web (http://www.5tigers.org) and an information telephone line (1-800-5TIGERS) for North America. It was created to provide the public, scientific and conservation communities with an international forum for exchanging information relevant to the preservation of wild tigers across Asia and in zoos world-wide. Multiple levels of information, ranging from general to scientific, are continually updated to provide the most current information about tigers available. By the time of the London Tigers 2000 symposium, traffic on the website was averaging 120 000 hits per week.

The 5Tigers Website currently includes information on tiger distribution, status, natural history and threats; tiger conservation organisations, programmes and field projects; zoo tiger information, organisations, a husbandry manual, genome resource banks and regional tiger management reports; scientific publications, conferences and workshops; book and video reviews, bibliography, pictorial essays and newspaper abstracts of tiger-related articles; information and games for children, question and answer e-mail capability, and a gallery of tiger art by kids. We redesigned the Web site for a cleaner look, faster downloading times, easier navigability and a new dimension for the educational content via lesson plans and interactive learning modules.

Although there is not yet one centralised database for tigers, the growing power of the World Wide Web is making one possible. With this goal in mind, we present the articles on tigers from all the IUCN/SSC Cat Specialist Group's newsletters *Cat News* (courtesy of Peter Jackson) and TigerLink (courtesy of Valmik Thapar) and have posted materials from the World Wide Fund for Nature's TRAFFIC on the trade in tiger parts in Asia. Conservation organisations that do not have the time or money to publicise their efforts can send us materials to be posted on the Web site. Other sources of information include conservation journals, newspapers and occasional articles on field research, and there are links to many other sites concerned with tiger conservation.

Accurate information is vital to making good decisions. Events in the tiger range countries of Asia change rapidly, and this information needs to be widely circulated and easily accessed. This can be achieved in part by developing and maintaining a centralised database of field research, public

The Tiger Information Center

The Tiger Information Center is dedicated to providing information to help preserve the remaining five subspecies of tigers. To learn more about tigers, just click on the topic below. If you have more questions, email us.

- What's New
- Tiger News
- All About Tigers
- Tiger Adventures
- Tigers in Trouble
- Teachers' Resources
- Conservation
- Make A Donation
- Cubs 'n Kids
- Who Are We?
- Research Center
- Search Our Site

News Flash: "Habitats: Realm of the Tiger," a multi-media education kit produced by the National Geographic Society and sponsored by Exxon, is now being offered.

FIGURE 14.1
Tiger Information Centre Home Page (http://www.5tigers.org).

policy and anecdotal reports of trends affecting tigers throughout Asia. The general public needs current information in order to understand the problems that tigers are facing. This is also true for professional field workers in tiger range countries, and globally based tiger conservation organisations.

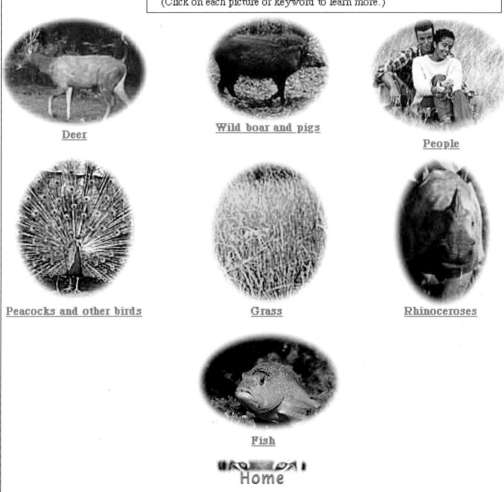

What do tigers eat?

Which of these things do you think tigers typically eat?
(Click on each picture or keyword to learn more.)

Deer

Wild boar and pigs

People

Peacocks and other birds

Grass

Rhinoceroses

Fish

Home

FIGURE 14.2
The children's sections on this website include games and quizzes as well as interactive educational materials. Teachers' guides are provided and the pages are classified by school grade. The page illustrated can be found at http://www.5tigers.org/handbook/b2.htm.

In summary, the 5Tigers website was created with no ownership in mind. It belongs to everyone who cares about tigers and is accessible to anyone in the world, at any time, and at no cost. We invite your participation and welcome your suggestions.

public policy, and reports on trends affecting tigers throughout Asia is being assembled (R. Tilson *et al.* this volume, Box 14.2).

Zoo tiger exhibits have huge potential to stimulate interest and concern for tigers in their visitors. The larger zoos in tiger range states have annual visitor numbers well in excess of one million, and these are of course the places where public concern is most urgently needed. Many zoos in the West already have tiger graphics oriented towards conservation, and further progress in this direction is highly desirable. In addition, North American, European and Australasian zoos are well placed to provide assistance for range country zoos in this area.

Support for *in situ* conservation

A further contribution is the potential for zoos to foster and support direct links with *in situ* conservation programmes. One example is the conservation partnership of the Zoological Society of London/London Zoo with the Sumatran Tiger Project/Minnesota Zoo's connection to protect wild tigers and their habitat in Sumatra. The *in situ* efforts include a long-term study of tiger biology (N. Franklin *et al.* this volume), rapid evaluation of tigers, tiger prey and tiger habitat across Sumatra (in progress), education and conservation efforts to reduce tiger-human conflicts (P. Nyhus *et al.* this volume, Box 10.2), deployment of a tiger rescue team to deal with 'problem' tigers (the first such tiger was brought in in 1997), production of a Geographic Information System (GIS) map of Sumatra's protected area system (in progress), and empowering existing programmes to protect remaining tiger habitats (Tilson *et al.* 1997).

There are compelling reasons why it makes sense for zoos to embrace *in situ* conservation. Zoos are competent business centres, so the costs of administration do not absorb hard-to-find conservation funds; zoos can therefore sell their programmes to donors by declaring that all of the funds will go directly into 'grass-roots' conservation activities. Zoos are also ideally placed to attract

funding from corporate sponsors, as they can publicise the resulting conservation work to their visitors in their graphics, and to an even wider audience via the media, as well as through their membership magazines. No other kind of conservation organisation has this capability.

A new dimension to zoo support for *in situ* conservation was demonstrated by the UK Federation of Zoos' 1996 'Tiger Week' initiative, in which 35 member zoos joined forces to raise funds for tiger projects in the field. The co-operative, country-wide nature of this event led to over 300 mentions in the local and national media, including national TV, and very effectively publicised both the plight of the tiger and the role of zoos in its conservation. Over £75 000 was raised for conservation measures in Russia, Indonesia and India (Christie 1997). The conservation potential of this kind of co-operative effort between zoos is huge, and the success of Tiger Week should not have gone unnoticed by other regional zoo associations.

Towards a new integration

Zoos are urban-based institutions that can uniquely integrate their three major conservation tasks of environmental education, research, and species and habitat conservation. Combined with the enormous public visitation and interest in zoos and the ever more intensive co-operation within the worldwide zoo network, this integrated approach has an enormous but untapped potential to contribute to conservation. This potential should be used to the very best advantage. When this is realised, the zoo community will be a powerful ally to conservation agencies, authorities and networks: in building conservation partnerships; in raising public awareness; in generating funding for *in situ* work; and in the provision of a behaviourally, physically and genetically healthy back-up population as an insurance against catastrophe in the wild. The World Zoo Conservation Strategy strongly emphasises that co-operation, coordination and interaction in all conservation efforts are the only means for success. The time has passed for diverse conservation groups to insist dogmatically upon their individual viewpoints. The Strategy therefore urges the entire global zoo network and all other conservation-oriented networks to integrate and intensify their efforts towards their mutual goals. We seek here to extend this idea to an interdisciplinary level that hopefully would provide links between the *in situ* and *ex situ* programmes. To reiterate, if we can learn to work together perhaps we can save the tiger together.

The trade in tiger parts and what to do about it

An undercover agent with a tiger skeleton, posing as a buyer. She is working with organisations tracking trade in animal parts.

15

The beginning of the end of tigers in trade?

Ginette Hemley and Judy A. Mills

Introduction

Conservationists have become increasingly concerned in recent years about the extensive trade in
endangered and threatened species to destinations
in East Asia, where wildlife parts are used in traditional medicines. Key among targeted species is
the tiger, which perhaps more than any other animal
epitomises one of the most daunting challenges in
international conservation today; that of stemming
a trade rooted in thousands of years of tradition,
now driven by economic forces unprecedented in
world history. At the heart of this issue are two very
divergent, and sometimes competing, interests
– wildlife conservation and traditional Chinese
medicine (TCM). As advocates of these interests
begin to grapple with a common problem – the rapid
decline of the world's wild tigers – it is becoming
clear that unprecedented co-operation and mutual
understanding are essential to secure the tiger's
future.

While experts generally agree that loss of habitat
and prey are the most important long-term threats to
the tiger's survival (Seidensticker 1997), poaching
for trade of tiger body parts, particularly bones for
use in traditional medicines, is now considered by
many to be the most immediate danger to the
world's remaining wild tiger populations (Mills &
Jackson 1994; Jackson & Kemf 1996; Hemley &
Bolze 1997). Because of its illegal nature, the trade
is difficult to quantify. However, import and export
statistics and anecdotal evidence compiled in the
late 1980s and early 1990s suggest a significant
increase in seizures of tiger bone and other products
during this period, as well as a growing commerce,
particularly in East Asia, of manufactured tiger medicines (Mills & Jackson 1994; Mulliken & Haywood
1994). Although the Convention on International
Trade in Endangered Species of Wild Fauna and
Flora (CITES) decided in 1975 to ban international
commercial trade in all tigers and their parts except
the Amur tiger, this protection did little to stop the
illicit commerce. International trade continued at
significant rates even after CITES banned trade in
all tigers and their parts in 1987.

What sustained trade in tiger parts despite an
international ban? Why were laws so slow to stop
this trade? Answers to these questions are critical to
the development of a comprehensive tiger conservation strategy. While tiger conservation efforts
have until now focused largely on protecting tigers
and their habitat, they must also take into account
the dynamics of black-market commerce and the
socio-economic issues inherent in human demands
for medicine.

The authors wish to thank Karen Baragona of
WWF-US, Samuel Lee and Sean Lam of TRAFFIC
East Asia, Andrea Gaski of TRAFFIC USA, and
Teresa Mulliken, Crawford Allen, and Steven Broad
of TRAFFIC International for their invaluable
assistance in completing the research for this paper.

A brief history of the tiger trade

In reviewing the tiger's status over the last half of the
twentieth century, conservationists have pointed to
two major 'crisis' periods. The first occurred during
the 1960s and 1970s, when the combination of
habitat loss and over-hunting of tigers – for sport,
for commercial trade in tiger skins, and because of
tiger conflicts with people and livestock – led to
significant tiger population declines, especially in
parts of India, the Caspian region, Indonesia (Java),
and The People's Republic of China (Norchi & Bolze
1995).

A second crisis came to light in the late 1980s
and early 1990s, when several political and

economic developments combined to trigger a rash of poaching that severely impacted many tiger populations. Reports of increased illegal killing of tigers, first in India, then in Russia, became more frequent. Investigations by the Indian and Russian governments and non-governmental organisations revealed a steady stream of incidents of illegal trade in tiger parts, most of which were confiscations of tiger bones destined for the medicinal trade (Mills & Jackson 1994; Galster *et al.* 1996; A. Kumar pers. comm.).

The recent tiger trade crisis

While some observers argue that the international conservation community simply became aware of a problem that had been ongoing for years, others believe the acceleration of poaching was caused by the depletion of tiger bone stockpiles accumulated in China during the 1950s and 1960s, when the tiger was officially classified as an agricultural pest and exterminated throughout its remaining range. When these stocks ran out in the late 1980s, China's TCM industry began looking for new sources (Jackson & Kemf 1996), shifting market pressures to areas where tigers remained relatively numerous, particularly India. At roughly the same time, the collapse of the Soviet Union led to a breakdown in the tiger-protection infrastructure in the Russian Far East, greater economic instability across Russia, and a relaxing of trade controls along the Sino-Russian border (Mills & Jackson 1994). These simultaneous developments brought about substantial increases in both opportunities and incentives to poach tigers to supply the commercial demand for bones.

The international conservation community was caught by surprise when, in 1993, the gravity of the tiger poaching situation first received international exposure following the seizure of nearly 500 kg of tiger bone in India (Mills and Jackson 1994; press release, Associated Press, 1 September 1993). Around the same time, Russia's tiger population underwent its most serious poaching assault in years. Reports indicate that dozens of Amur tigers

may have been killed during the winter of 1993 (Mills & Jackson 1994; Galster *et al.* 1996). Evidence of sales of tiger parts and products also began coming in from Vietnam, Cambodia and North Korea (Mills & Jackson 1994; Martin & Phipps 1996; S. LaBudde pers. comm.; S. Kang pers. comm.; S. Nash pers. comm.).

In early 1993, reports by non-governmental organisations such as the IUCN/SSC Cat Specialist Group, TRAFFIC and the Species Survival Network prompted the CITES Standing Committee to request a formal review of the tiger trade problem. Growing concerns led to a call by the Standing Committee in September 1993 for specific action in China, Hong Kong, the Republic of Korea and Taiwan, where trade in tiger bone and its derivatives appeared to be continuing (CITES Secretariat 1993). The following 18 months brought unprecedented political action in the major Asian consumer markets, which led to crackdowns on the illegal trade. After CITES directives and a threat of trade sanctions by the USA, China, South Korea and Taiwan all banned trade in tigers and their derivatives. Hong Kong and Singapore also enacted strict trade control measures.

Together these actions marked the most concerted international effort ever undertaken to stop illegal trade in tigers and their parts and medicinal derivatives. As importantly, they raised political and public awareness of the gravity of the tiger's precarious situation and the threat posed by poaching and commercial trade. This new-found awareness, combined with new anti-poaching initiatives undertaken in the Russian Far East, is thought to have slowed illegal trade (Global Survival Network 1997). Where tiger bones, skulls and tiger bone medicines were openly displayed in China, Hong Kong, South Korea and Taiwan in the late 1980s, they had largely disappeared by 1995.

Illegal trade in tigers and their parts and derivatives did not stop altogether, however. While the open sale of tiger-based medicines in several East Asian countries was reduced significantly, a residual trade continues in some key consuming markets such as China (Mills 1997), and tigers continue to be poached. With such low numbers and

Asian folk myth ascribes aphrodisiac powers to soup made from these tiger penises, which will sell at about $350 a bowl, but there is far more profit in tiger bones and it is the bone trade which drives tiger poaching.

highly fragmented populations, tigers probably cannot survive the pressure of even this residual trade for long.

Understanding the root of the problem: human health care

Tiger trade is more complex than the basic forces of supply and demand. The chief complication is that tiger parts and derivatives provide medicine for treating painful illnesses such as rheumatism. While these medicines are traditional, they are nonetheless respected by more than one-fifth of the world's population. They have been used to provide health care, which is regarded as an inalienable right by most humans, for at least a thousand years (Bensky & Gamble 1993) and perhaps as long as 4000 years (Gaski & Johnson 1994).

Over the centuries, nearly every part of the tiger has had a prescribed benefit, either in TCM or in Asian folk remedies. In parts of India, for example, tiger fat has been considered a remedy for leprosy, while in Lao PDR, tiger claws were used as a sedative (Mills & Jackson 1994). TCM once prescribed tiger blood as a general strengthening tonic, while the tiger's tail was recommended for skin diseases, the eyeballs to treat convulsions and cataracts, and the whiskers to alleviate pain associated with toothaches (Mills & Jackson 1994). The tenets of TCM say the tiger's humerus, the upper front leg bone, is the most efficacious bone (Bensky & Gamble 1993), though all parts of the skeleton have been used in recent years. Referred to as 'Os Tigris' in TCM, tiger bone is prescribed today as a muscle strengthener and treatment for rheumatism (Bensky & Gamble 1993; Mills & Jackson 1994; see Box 15.1). It is utilised in individually prepared remedies dispensed by TCM specialists or manufactured as 'patented' medicines which can be bought without prescription (Gaski & Johnson 1994). Commerce in patented medicines put millions of units of tiger derivatives into international trade in the three years leading up to China's ban on the use and sale of tiger bone in 1993 (Mulliken & Haywood 1994).

For those interested in stopping the commercial demand for tiger bone as a medicine, it is important to distinguish between the legitimate practice of TCM – which is a respected, systematic form of health care – and folk medicines, which are self-prescribed on the basis of folklore and hearsay. Modern TCM texts ascribe medicinal value only to tiger bones. As mentioned previously, the use of tiger bone in medicine is now prohibited in the former major consuming markets of China, Hong Kong, South Korea and Taiwan.

The volume of tiger trade since 1970

The tiger trade, like most trade in wildlife contraband, is extremely difficult to measure. CITES has banned commercial international trade in the parts and derivatives of most tigers since 1975, and by 1994 this ban had been officially implemented in most countries, leaving only secretive black market trade. This progression towards a complete global trade prohibition also progressively reduced the

Table 15.1. *South Korean imports of tiger bone by country, 1970 to October 1993*

Exporting country	Mass (kg)	% of total
Indonesia	3994	44.5
China	2415	26.9
Thailand	607	6.8
Malaysia	493	5.5
India	248	2.8
Singapore	195	2.2
Taiwan	150	1.7
Hong Kong	120	1.3
Others	729	8.3
Total	8951	

Source: South Korean Customs statistics.

amount of reliable data on trade in tiger bone and its derivatives. Therefore, what trade data were available before the international ban was widely implemented may grossly underestimate the actual levels of commerce in the waning years of legal trade.

Mills and Jackson (1994) were the first to attempt to compile a comprehensive picture of the trade. Their report, based on government trade and confiscation records, market surveys and anecdotal reports, showed that key East Asian consumer markets for tiger bone products imported a minimum of 10 tonnes (10 000 kg) of tiger bone between 1970 and 1993. Principal consumer countries were China, Japan, Singapore, South Korea, Taiwan and the USA, which were supplied mainly by Hong Kong, Indonesia, China, Singapore and Thailand. Converting these weight figures to precise numbers of tigers killed is impossible, but one analysis estimates that they represent 500–1000 tigers (Jackson & Kemf 1996).

Few countries have maintained official records of their tiger bone trade, with one noteworthy exception. Until acceding to CITES and officially banning trade in tiger parts in 1993, South Korea maintained customs records of tiger bone imports that provide an interesting snapshot of the trade. As Table 15.1 indicates, from 1970 to 1993 the vast majority of South Korean imports came from Indonesia and

China (Mills & Jackson 1994). South Korea's imports from Indonesia coincide with data charting the steady decline of tiger populations on Java and Sumatra, although commercial trade was not the only pressure on those populations at the time.

China was the second largest supplier of tiger bone to the South Korean market from 1970 until the trade was banned in 1993 (Mills & Jackson 1994). Furthermore, in a brief six-month period from South Korea's announcement of CITES accession until the Convention entered into force there in October 1993, traders amassed 1.5 tonnes of tiger bone from China in an apparent attempt to stockpile bone before trade restrictions took effect (Mills & Jackson 1994).

Trade data also reveal an extensive international trade in manufactured medicines containing or purporting to contain tiger bones. From 1990 to 1992, CITES reports indicate that China exported at least 27 *million* units (=packages) of tiger-based medicines to at least 26 countries world-wide (Mulliken & Haywood 1994). The largest importers recorded were Hong Kong, Japan, Singapore, Taiwan and the USA.

While these trade statistics reveal useful information about supply and demand for tiger products before legal international trade virtually stopped, India and Russia are notably absent from the list of major suppliers. Available records do not explain whether these two key tiger range states became major international suppliers of tiger bone after most national bans took effect, or whether trade with these countries was simply conducted outside recorded channels.

Political pressure prompts progress

Much has happened in the international arena since the alarm bells first sounded in 1993. As noted above, pressure from CITES and the USA catalysed important legislative and enforcement changes in several key East Asian consumer countries. Following the CITES Standing Committee's September 1993 call for action in the major tiger consumer countries, the USA threatened in

September 1993 to impose trade sanctions on China and Taiwan. After CITES missions to East Asia in late 1993 and early 1994 revealed modest progress in China but comparatively few improvements in Taiwan, the USA imposed wildlife trade sanctions on the latter in April 1994. This marked the first time that the USA had ever invoked sanctions under its Pelly Amendment to the Fishermen's Protection Act of 1967, which provides for the President to prohibit importation into the USA of wildlife products from an offending country if advised that the nationals of that country are engaging in trade or removal which diminishes the effectiveness of any international programme for endangered or threatened species. In response, Taiwan instituted a series of trade control measures, including strengthening its laws regulating trade in tiger bone and rhino horn, establishing a new wildlife enforcement unit, undertaking widespread investigations of illegal trade in tiger bone and rhino horn, and launching a national effort to heighten public awareness of the conservation impacts of trade in tiger bone and rhino horn. In June 1995, the USA recognised Taiwan's progress in curtailing trade in these products and lifted the sanctions.

One of the most significant actions taken during this same period was China's May 1993 announcement of a ban on the manufacture and sale of both tiger and rhino parts and medicinal derivatives, and removal of these items from the national list of approved medicinal ingredients (Mills & Jackson 1994; Mainka 1997). At the same time, China undertook a series of enforcement measures nationwide. TRAFFIC surveys of China's market in 1994, 1995 and 1996 of hundreds of medicine shops across China showed that while tiger products were still being sold, they were not widely available and, in many cases, seemed to be products left over from before the ban (Anon. 1994; Mills 1997). Investigators also found a surprisingly high awareness of the ban among shop owners and pharmacists, though many were still willing to sell banned products.

Elsewhere in Asia in 1994, Hong Kong made it illegal to trade in medicine containing or even purporting to contain tiger bone, Vietnam joined

Box 15.1   Tiger bone: a millennium of treating pain
Ginette Hemley and Judy A. Mills

Traditional Chinese medicine (TCM) categorises natural plant and animal ingredients, or 'herbs', by their tastes and aromas, the latter of which correspond to the four energies; cold, cool, hot and warm. The five tastes of sour, bitter, sweet, sharp and salty relate to the five ancient elements of wood, fire, earth, metal and water as well as to the body's principal organs and glands – liver, heart, spleen-pancreas, lungs and kidney-bladder. Some herbs have more than one taste, but the primary taste determines its medicinal use.

The use of tiger bone in TCM goes back at least a thousand years. Known by the pharmaceutical name 'Os Tigris', tiger bone is classified in TCM as a herb that dispels 'wind dampness' from the muscles, sinews, joints and bones. It is considered to have a 'warm effect' which eases 'cold' conditions such as body pain. The cold condition for which tiger bone is most often prescribed is rheumatism, but it is also used to treat muscle weakness, stiffness or paralysis of muscles and bones, and lower back pain. Tiger bone is considered to have acrid, sweet and warm properties, and enters the body through the liver and kidney.

Baked tiger bone and rhino skin ready for sale in Shen Zhen, China.

In TCM, herbs are almost always combined for a specified enhanced or modified effect. Tiger bone, the major known ingredients of which are calcium and protein, is commonly mixed with medicinal plants, particularly the roots of *Achyranthis bidentatae* (Ox's knee; an amaranth species) and *Angelicae sinensis* (Chinese angelica). It is generally taken in medicinal wines, as a powder, in plasters, or in pills, and is normally toasted in oil or vinegar before use.

From a modern biomedical perspective, TCM herbs that are used to dispel wind dampness have analgesic, anti-inflammatory, antipyretic and circulation-promoting properties. Pharmacological and clinical research has indicated that certain tiger bone preparations have an anti-inflammatory effect against experimentally induced arthritis in animals, and an analgesic effect in rats.

Substitutes under investigation include the bones of dog, leopard, bear, pig, goat, sheep, cow and mole rat.

A plaster on sale in a London TCM shop in 1993 listing tiger bone as an ingredient. Such plasters can be found in TCM shops the world over.

Tiger bones and medicines made from them were openly displayed in South Korean shops in 1993. By 1997, South Korea had prosecuted its largest manufacture of tiger-bone medicines for illegal trade and burned the company's remaining stockpiles.

CITES, and Singapore banned the sale and public display of tiger medicinal products. The collective actions taken in East Asia since 1993 to stop trade in tiger bone are unprecedented in history. They have not, however, been enough to completely stop the commerce.

New developments in the tiger bone trade

More recently, there has been evidence of a growing trade of fake tiger bone in China. This trend may reflect the fact that authentic tiger bone is both illegal and scarce. Alternatively, it may indicate that prices have gone up so much since the ban that real tiger bone is simply off-limits for the street trade.

China shares borders with seven tiger range countries – Bhutan, India, Laos, Myanmar, Nepal, Russia and Vietnam. Most border crossings are not well policed, and low per capita incomes on all sides make poaching, smuggling and black market trading attractive options. At the same time, China's central government has limited capacity to initiate and sustain law enforcement initiatives necessary to stop illegal trade (Mainka 1997).

Tiger bone trade in South Korea has virtually stopped since the ban took effect there in 1993. This success is due largely to the fact that only one pharmaceutical manufacturer used most of the tiger bone entering South Korea. Since the ban, top executives of this company have been prosecuted for illegal trade in tiger bone, and the company itself is not sure it can survive the loss of its flagship medicine, which contained tiger bone, and the high costs of defending its executives (S. Kang pers. comm.). Meanwhile, the Korean government has burned several hundred kilograms of tiger bone seized from this company to publicly underscore its commitment to stopping the trade.

In January 1995, Hong Kong increased penalties for trade in anything purported to contain tiger parts to more than $US 650 000 per violation. Enforcement officials report significant seizures and prosecutions, while the judiciary has handed down fines large enough to drive some violators towards bankruptcy. While Japan has yet to similarly ban domestic trade, the USA has done so and Australia's parliament was considering legislative changes at the time of publication. Still, Hong Kong and US law enforcement authorities will require increased funding and staff if they are to utilise their model laws to full potential.

More attention is beginning to focus on countries once considered secondary markets for medicines containing tiger bone. Ironically, these countries include some outside East Asia which were among the most vocal critics of China and Taiwan for their trade in tiger bone. Among them is the USA, which once imposed trade sanctions on Taiwan for similar inaction with regard to stopping tiger bone trade.

Japan stands out as the only key consuming market in East Asia (other than perhaps North Korea) which has not banned all trade in tiger bone derivatives (Mainka 1997). Although Japan prohibits import of tiger bone and tiger bone derivatives under CITES legislation, its legal domestic trade in tiger bone derivatives may provide incentive for traders to smuggle in new supplies to satisfy market demand. While there is no evidence of such smuggling, the risk cannot be denied.

There is also increasing concern over how many so-called tiger bone products actually contain tiger bone. While this has always been an issue for law enforcement cases, it is increasingly important in understanding where tiger bone is actually being used in the post-ban period. Unfortunately, current laboratory testing methods cannot detect minute amounts of tiger bone mixed with other ingredients to make manufactured medicines (Gaski & Johnson 1994).

Meanwhile, TRAFFIC surveys in Australia, New Zealand, the United Kingdom, parts of Europe, Canada, the USA, China and Japan document continuing availability of products labelled as containing tiger bone (Callister & Blythewood 1995; Chalifour 1996; A. Gaski pers. comm.). Most of these products carry labels indicating they were manufactured in China.

In a 1994 resolution, CITES Parties suggested trade prohibitions on all products *purporting* to contain tiger. However, few consumer countries implemented this measure, and the resolution was strengthened at the 1997 Tenth Meeting of the Conference of the Parties in Harare, Zimbabwe (CITES Secretariat 1997). The new resolution calls for measurable improvement in tiger trade controls before the Eleventh Meeting of the Conference of the Parties, through, among other things, strengthened national legislation, the establishment of regional enforcement networks, the development of a forensic protocol for identifying tiger bone derivatives, and the removal of tiger products from official pharmacopoeia and promotion of appropriate alternatives. The CITES Standing Committee has been tasked with assessing, on a country-by-country basis, additional legislative and enforcement measures that may be necessary to stop the remaining illegal trade.

A new trade dynamic emerges

There is no doubt that tiger bones and products labelled as containing tiger bone are more difficult to find in the main Asian markets today than they were in the early 1990s. New laws, better law enforcement and widespread public awareness efforts have decreased trade significantly. However, these successes do not mean that illegal trade does not exist. It does. Tigers are still being poached. Tiger products are still being sold. What remains unknown is the exact nature of the trade in its new form. For example, if most tiger 'bones' seen for sale are fakes and if most medicines claiming to contain tiger bone do not, then where are the bones from newly poached tigers going? Who is buying them? Are they secretly being made into medicines or are they being stockpiled in case legal trade should ever resume or, worse yet, in case the tiger should become extinct and make tiger bones a priceless commodity? Trade monitoring and law enforcement efforts will have to be revised, become more sophisticated, delve more deeply into black markets and, most importantly, receive more substantial funding if the answers to these crucial questions are to be answered so that residual trade does not continue to threaten the survival of tigers in the wild. Nowhere will these measures need stronger support, both financially and in terms of political will, than in China – which seems to remain the key source of medicines containing tiger bone – and in

India and parts of Indochina – which are arguably the most important supplying areas for today's tiger bone trade.

In the case of North Korea, simple baseline information about trade in tiger bone and its derivatives is needed. However, that country's political instability and 'closed door' policy to foreigners will make meaningful trade monitoring difficult if not impossible.

The bottom line: tackling demand

World history and sociological studies both have shown that laws cannot and do not change behaviour completely. This is especially true when laws aim to change what are perceived as inalienable rights, such as choice of religion. Choice of health care is nearly as sacrosanct as religious freedom. Therefore, it is logical that laws banning the sale and use of tiger bone in medicines would not stop this trade immediately or completely; and that the mostly Asian consumers of these medicines would not stop using them just because conservationists in non-Asian countries admonished them to do so. Because of these realities, *enlisting* would-be consumers of tiger bone in tiger conservation efforts is as important as strong laws and law enforcement in stopping commercial trade in wild tigers and their bones.

This can be accomplished in several ways. The first is for tiger advocates to communicate with the TCM industry and key user groups in a language which they can understand and in which they can communicate their concerns about tiger conservation's impact on them and their interests. Specifically, this means that conservationists will need to be able to speak and listen in the languages of East Asia.

It is equally important that all those affecting or hoping to affect the tiger's survival set aside cultural prejudices and myths. Where TCM advocates suspect tiger conservationists of being anti-TCM, media reports often make it appear that conservationists consider tiger bone an aphrodisiac and TCM advocates are naturally offended. These stereo-types and myths prevent communication and co-operation and prolong the threat that TCM poses to the tiger's survival.

In recent years, TCM specialists have begun asking conservationists for more information – in languages they understand – about the need for tiger conservation and the ban on trade in tiger parts and derivatives. In addition, they have asked for more information on viable substitutes for tiger bone researched and endorsed by other TCM specialists, as well as financial support for further research into such substitutes. Many say that they would welcome and utilise a substitute that is natural and proven effective by the research standards demanded by the TCM establishment.

At the same time, it is important that attitudes among TCM specialists and users towards the tiger and other endangered medicinal species be measured by scientific means. Sociological research of this sort will document the demographics of key TCM consumers and opinion makers so that awareness, communications and law enforcement activities can be more strategically and effectively designed and targeted. In addition, repeated demographic and attitudinal research will help map changes in the views and behaviour of potential consumers of tiger bone and other endangered species' parts for medicinal purposes.

A number of initiatives have been undertaken in the last two years to enlist the support from TCM users for tiger conservation. Conservationists and wildlife law enforcement agencies have begun communicating with TCM advocates and users in Chinese, Korean and Japanese. Meetings and workshops have TCM and wildlife advocates talking face-to-face about tiger conservation measures on a more regular basis.

The 1995 International Symposium on Traditional Chinese Medicine and Wildlife Conservation, sponsored by TRAFFIC, WWF and the Hong Kong government, brought together for the first time conservationists, government officials and TCM specialists from nine countries to chart what common ground might be found for enlisting the TCM industry in wildlife conservation efforts. The symposium's conclusions were, as they relate to tigers, both worrying and encouraging.

The lack of understanding among TCM specialists about the status of tigers and the role TCM trade has played in endangering the tiger's survival was sobering for many of the non-Asian participants (Mills 1996; Hemley & Bolze 1997). At the same time, the symposium highlighted a lack of understanding on the part of conservationists regarding the importance of TCM as a legitimate and respected health care system (Mills 1996; Hemley & Bolze 1997). Many non-Asians tend to dismiss TCM as mere superstition. Western clinical research, however, has documented the efficacy of a number of medicinal products (Benksy & Gamble 1993; Gaski & Johnson 1994) and, perhaps more importantly, TCM does not regard Western research as the standard bearer for medicinal efficacy.

As the head of China's CITES Scientific Authority told an audience at the IUCN World Conservation Congress in Montreal in 1996, TCM has been around for centuries and it is likely to endure for centuries to come (S. Wang pers. comm.). Therefore, the question for conservationists is not whether there will or should be TCM, but how to ensure that TCM will not endanger the survival of species such as the tiger.

It is encouraging to note that the Tenth Meeting of the Conference of the Parties to CITES in 1997 saw TCM specialists sitting on the delegations of China and South Korea.

The way of the future: substitutes for tiger bone

The 1995 symposium in Hong Kong underscored one firm piece of common ground shared by conservationists and TCM advocates; that viable substitutes for tiger bone seem to be a solution endorsed by both sides. The only caveat for conservationists is that substitutes should not endanger other species of wild animals or plants. The only caveat for TCM specialists is that substitutes must be TCM rather than 'Western' or synthesised and that they receive endorsement from TCM researchers using TCM research methodologies.

Prompted by a recommendation from the TCM community itself, TRAFFIC and the Chinese Medicinal Material Research Centre of The Chinese University of Hong Kong co-hosted in December 1997 the First International Symposium on Endangered Species Used in Traditional East Asian Medicine: Substitutes for Tiger Bone and Musk. The more than 100 delegates from 16 countries and territories reached a consensus on the need for further research to find effective substitutes for tiger bone. Research presented during the symposium showed that substitutes being considered for tiger bone are the bones of domestic species including dog, goat, pig, sheep and cow, and wild species including leopard, bear and the mole rat (*Myospalax*, or zokor). A wine made from the bones of the mole rat, called sailong in Chinese, has already been introduced into the market in China. However, the challenge, according to researchers, is to find substitutes that are not simply similar but rather identical in effect. Other presenters stressed perhaps even a greater challenge: gaining acceptance of substitutes from both TCM practitioners and consumers. Successful substitutes would have to be effective, low-cost, and without side-effects. At the same time they must not endanger other wild animal or plant species.

In summary, substitutes for tiger bone may significantly relieve residual demand for tiger bone as a medicine. However, the search for and transition to acceptable substitutes will be slow at best and only one aspect of the multi-faceted approach required to reduce demand to the point where it will no longer threaten the small number of remaining wild tigers.

Tiger farming: panacea or Pandora's box?

Where conservationists tend to overlook the complications of finding substitutes for tiger bone, TCM advocates tend to oversimplify the issue of tiger farming. Many, including some CITES authorities in China (J. H. Qing pers. comm.), believe that tiger farming is a solution to saving tigers and putting tiger bones back on the medicinal market. This view of tiger farming as a panacea was underscored by some TCM specialists at the 1995 Hong Kong symposium.

It is important for conservationists therefore to help TCM advocates understand the conservation concerns inherent in tiger farming. Chief among these is the fear that any legal consumption of tiger bone will stimulate demand for bone and sustain the market for tigers poached from the wild. The possibility of bone from poached tigers being passed off as farmed and 'laundered' through such facilities is another inherent danger that seems to be overlooked by TCM advocates hoping to dismiss conservation concerns. There also seems to be a general lack of appreciation of the value of ensuring that wild tigers remain in the wild, which requires the simultaneous conservation of habitat and prey species.

In addition, there are humane issues among the international community that advocates of tiger farming will inevitably be forced by global public opinion to address.

The value of general public awareness

New research by TRAFFIC offices in East Asia and the USA document the fact that the end-use consumers of medicines containing tiger bone usually do not know the ingredients of the medicines prescribed for them (S. Lee pers. comm.; A. Gaski pers. comm.). Few ask for tiger bone. It seems that, in most cases, doctors and pharmacists make the decision as to whether or not a patient will take a medicine containing tiger bone. Hence, initial efforts to improve public awareness of the impact of TCM on tiger conservation should focus on these decision makers.

There is also value in making the general public aware of the connection between tiger bone use and tiger conservation. Ideally, this would lead to end-use consumers requesting medicines that do not contain tiger bone.

Conclusions

Looking to the year 2000, the next two years will be a pivotal time for tiger conservation. Regarding the trade in tiger bone, it is important that new research and communications initiatives build on what has been done and what progress has been achieved, rather than addressing the problem as it existed when it first came to international attention in 1993. The time has come to move past the notion that TCM advocates are the enemy and to view them instead as partners in tiger conservation without whose support the tiger may perish.

The trade that has proved so deadly to tigers can only be stopped with a combination of the following:

1   Passing stronger laws which prohibit not just trade in tiger parts and derivatives but also any items claiming to be or contain tiger parts.
2   Increasing penalties for illegal trade in tiger parts and derivatives to amounts high enough to deter further trade.
3   Launching law enforcement operations which infiltrate the inner workings of the black market for tiger bone, particularly the network of middlemen.
4   Increasing communication about tiger conservation concerns to the TCM industry and its opinion makers in Chinese (Mandarin and Cantonese), Korean and Japanese.
5   Promoting, by way of funding and communication, research and development of substitutes for tiger bone which are endorsed by key specialists in the TCM industry and in key government public health agencies.
6   Increasing international discussion in all relevant languages about the conservation concerns associated with farming tigers for their bones.
7   Conducting scientifically valid sociological surveys in all consumer markets for tiger bone to establish baseline demographics of

potential users, with periodic follow-up surveys to track demographic and attitudinal changes.

8   Monitoring of key consumer markets to assess availability of tiger bone and its derivatives.

9   Launching appropriate outreach activities in supplying countries to inform poachers of reduction in demand for tiger bone in consuming markets.

10 Increasing anti-poaching programmes in all range countries so that supply does not persist despite decreasing demand.

11 Increasing the number and frequency of training programmes for law enforcement officials in all supplying and consuming markets, especially at international border crossings.

12 Establishing reward programmes for law enforcement officials who take action against trade in tiger bone, especially in China and other countries with poorly paid personnel.

16

Roaring back: anti-poaching strategies for the Russian Far East and the comeback of the Amur tiger

Steven Russell Galster and Karin Vaud Eliot

Introduction

'Imminent extinction.' That is how Russian and international wildlife experts described the situation facing the Amur (or Siberian) tiger by 1993. As with other subspecies of tiger, the Amur tiger's critical situation was relayed to the international community through press reports and via the Convention on International Trade in Endangered Species of Flora and Fauna (CITES). In four consecutive meetings in 1993 and 1994, CITES gave high priority to the plight of the tiger. While some hope was held out for the future of the relatively numerous Indian tiger, little was expressed for the remaining 200–300 Amur tigers being pursued by commercial poachers in Russia's wild Far East.

Thanks to a handful of government and nongovernmental organisation (NGO) representatives, who believed otherwise, by 1997 the situation had changed dramatically. Instead of losing 50–60 Amur tigers a year to commercial poachers, experts were happy to report a rising population of Amur tigers and a lower level of commercial tiger poaching in the Russian Far East. This feat was achieved during Russia's post-perestroika transition – a period marked by widespread organised criminal activity – in Primorski Krai, a territory in close proximity to the world's largest tiger-consuming markets. Lessons from this positive experience must be applied to future protective efforts in the Russian Far East if a healthy and sustainable population of Amur tigers is to be maintained well past the turn of the century. Remember the Indian tiger, which was also brought back from a slide towards

extinction two decades ago. India's Project Tiger was initially successful in the 1970s, but then deteriorated during the late 1980s and early 1990s, demonstrating that no endangered wild animal is ever 'saved'. Vigilance is the key to continued survival.

The authors of this chapter represent the Global Survival Network (GSN), an environmental security NGO that works closely with wildlife biologists and enforcement personnel worldwide to develop effective species recovery programmes. GSN co-sponsors anti-poaching operations in the Russian Far East, where we approach tiger conservation from a strategic, 'environmental security' point of view. Therefore, we will examine the objective of the Russian programme, code-named 'Operation Amba', the impediments to this objective (i.e. the threat), the strategy designed to overcome the threat and the tactical plan that the rangers of Operation Amba continue to follow. This analysis will cover four distinct stages of anti-poaching operations, from 1994 to 1997.

Background to the tiger's slide

Wildlife protection is too often analysed in a vacuum. It is necessary to examine the threat to a species within its local political, economic and social context before devising effective ways to overcome it. Russia and its Amur tigers are no exception to this rule. The fragmentation of the Soviet Union caused a breakdown in law and order across Russia and the former Soviet republics. There was an outbreak of crime, which continues today and is likely to continue until the turn of the century despite attempts to control it.

Russian law enforcement efforts have focused on high-level organised crime groups, also known as 'Mafia'. Between 1992 and 1996, drug dealing, plutonium and weapons smuggling, money laundering and the war in Chechnya absorbed the resources and time of Russian authorities in charge of the government's war on crime. Environmental crimes – commercial poaching, illegal timber cutting, toxic dumping, etc. – were not genuinely treated as national security issues and therefore were left in the hands of local authorities. The only central (i.e. Moscow-based) authorities who continued focusing on environmental security issues were in the Ministry of Environmental Protection and Natural Resources, renamed as the State Committee for Environmental Protection, but referred to in this chapter as the Ministry of Environment. (The exception was Alexy Yablokov who served as President Yeltsin's Ecological Security Advisor. Yablokov found it difficult to push an environmental agenda in the Kremlin and eventually went back to his position as Director of the Russian Centre for Environmental Policy.)

During the post-perestroika transition, the Ministry of Environment experienced severe budget cuts, forcing it to leave its local branches across Russia (called 'Ecology Committees') to fend increasingly for themselves. This was particularly true the farther away the Ecology Committee was from the Kremlin. Primorski Krai, home to an estimated 85% of Russia's Amur tiger population, is seven time zones away from Moscow. Among many unfortunate consequences of these budget cuts were layoffs and salary reductions for wildlife rangers in places like Primorski Krai. Those rangers who braved the budget cuts and remained working inside or outside nature reserves (zapovedniki) were left with few resources to fight against well-equipped commercial poachers. Local hunters – hungry for meat – ravaged the populations of deer and wild pig. Opportunistic hunters and self-made middlemen – hungry for money – targeted wildlife whose parts yield high profits on any scale: bears for gallbladders, musk deer for musk glands, wild

ginseng for its roots, and tigers for their pelts, organs and bones.

By the winter of 1993, officials from Russia's Ministry of Environment and the Primorski Krai Ecology Committee estimated that 60 tigers were being poached each year, and that at this rate the tiger could reach such low numbers that extinction was possible by the year 2000. With decades of wildlife conservation experience behind them, the Ministry and local Committee representatives developed a plan to reduce poaching, but this plan required money. Since it was unlikely that any funds could be obtained from the Kremlin, foreign assistance was required.

International cooperation to address an international threat

Thanks to international publicity, by 1993 the plight of all tiger subspecies was well known to the world. The US government and CITES had highlighted the need to protect tigers by threatening trade sanctions against states that continued to trade in tigers. The time was clearly ripe for Russia and the international community to join forces to save the endangered Amur tiger. By December 1993, several foreign NGOs and Russian authorities in charge of protecting the nation's endangered species of flora and fauna met and agreed on a joint two-fold objective:

1   Stabilise the tiger population by the year 2000. 'Stable' was defined as a measurably increasing population. (Several officials in the Ministry of Environment, including Dr Valentin Ilyashenko, Vsevold Stepanitsky and Vladimir Shetinin, thought the tiger could rebound before 2000 if an effective mobile anti-poaching programme were put into place by 1994. American scientists from the Hornocker Wildlife Institute thought the same, and made their case to the US government for anti-poaching assistance by 1993.)

2   Secure sound habitat for a stable population of tigers well into the next century. Sound

habitat was defined as consisting of good tree cover, food (healthy prey base), water and 'connectivity' (eco-corridors).

Interviews with Russian tiger experts revealed a uniform view of the nature of the threat to the tiger:

1 Commercial poaching for the underground market in skins, bones and organs outside of Russia.
2 Poaching of the tiger's prey base (wild pig and deer) for local human consumption.
3 Habitat loss, caused mainly by legal and illegal logging throughout the taiga forest.

Interviews with enforcement officers revealed specific details about this threat assessment:

1 Commercial poaching and trading:
   a. Tiger poaching was being conducted by two sets of hunters: organised poaching gangs and opportunistic poachers. In either case, the poacher would sell the tiger parts (in some cases full bodies) to middlemen operating out of several cities, mainly Vladivostok, Khabarovsk, Ussuriysk, Nakhodka and Plastun.
   b. The middlemen buying and selling tigers were usually Russian, Russian-Korean, or Chinese. Most tiger parts, it appeared, were being smuggled to The Peoples Republic of China, South Korea and Japan.
   c. Tiger parts smuggled to China were usually taken across the border by road or train through only a few channels, namely Pogranichniye and Poltovka, and possibly by air between Khabarovsk and Harbin, China. Tiger parts smuggled to South Korea and Japan were transported by boat from the ports of Plastun, Vladivostok and Nakhodka, or by air from Khabarovsk and Vladivostok.
   d. Commercial tiger poachers and smugglers were often connected to organised criminal groups that would lend firearms, vehicles, or a 'roof' (protection) to poachers. Sometimes the

Mafia group would simply buy the tiger parts from the poacher. Some poachers were unconnected to Mafia groups and resorted to open advertisements of bones and skins for sale from tigers they had killed.
   e. Tiger kills were highest in the winter, when poachers found it easiest to track the animal through the snow.
2 Reduction of the tiger's prey base:
   a. Both legal and illegal hunting were responsible for the decline of the deer and wild pig that serve as the tiger's main diet. The hunter was usually familiar with the taiga and the lack of wildlife law enforcement in and around it.
   b. The reduction of the tiger's prey base caused the tiger to roam further and wider for food. This meant that the tiger – and its killer – could be found almost anywhere.
3 Habitat loss:
   a. Logging meant an increased number of roads into formerly pristine and inaccessible tracts of forest. This in turn facilitated poaching of tigers and other wildlife. Defending the taiga forest became even more difficult. Poachers could now use logging roads to penetrate deep into the taiga within minutes. Some timber company employees became poachers. Once having killed a tiger, the poacher could make a quick exit by the same road, and stood little chance of being stopped by a law enforcement officer.
   b. Actual deforestation affected the tigers' roaming patterns. Where logging occurred, for example, wild pig would migrate in search of ground food produced by trees of the taiga (especially Korean pine cones), and this movement lured some tigers deeper inside the forest in search of them, or farther outside the forest in search of domestic animals.
   c. Overall, deforestation was fracturing key habitat for the tiger. The animal's access

to tree cover for safety, to ungulates for food, and to water in and near the forest was being slowly eliminated.

The plan to stabilise the Amur tiger: Operation Amba

It is important to revisit the goal established by the Russians and their foreign supporters, just as it is critical to frequently revisit the goal of any strategic plan if it is to lead to success. If the law enforcer in the battle does not do this, the battle will eventually be lost to the law breaker. In the case of the tiger, the law breakers (poacher and illegal trader) maintain a focused approach to their objective because that objective is short-term, i.e. to make money quickly and avoid getting caught. Their strategy to reach this goal is equally simple: kill a tiger wherever protection is absent and sell it to a buyer as quickly as possible.

The more difficult goal of the law enforcer has been to stabilise the tiger population by the year 2000. Deciding on the strategy and tactics to achieve this goal was even more difficult. This was where foreign support entered the equation.

Some Russian authorities recommended an ambitious and expensive counter-poaching plan, which we shall refer to as 'plan #1'. They sought to determine where the poachers were striking and to deploy newly hired and trained rangers to these 'problem areas' using newly purchased vehicles, boats, and snowmobiles. From a strategic standpoint, we claimed, this was equivalent to 'containing' an insurgency, which was only possible when the insurgents (in this case poaching gangs) operated in distinct locations. But the tiger and its stalkers were scattered throughout Primorski Krai, inside and outside nature reserves, and around towns and villages. This was a war without borders. Plan #1 would only work if the taiga forest was saturated with reliable, well-trained and well-equipped rangers. That would require much more money than was available, would create massive managerial problems (i.e. supervision of numerous rangers), and was almost certainly unsustainable

from an economic point of view. The tiger's range was just too vast to monitor all at once.

Russian plan #2 recommended a mobile anti-poaching operation. Under this plan, several small teams (four to five men apiece) would rove the taiga in order to: (1) maintain a periodic presence in areas no longer protected by rangers; and (2) follow up on reports of poaching and trading gangs. This appeared to be closer to a workable plan, but it still had holes. Its main problem was that with only several ranger teams roving the huge taiga, the poacher could strike wherever the teams were absent. The poachers and traders would eventually win in this equation. Anti-poaching patrols would end up chasing the poacher after the poacher had already killed the tiger.

Restricted by limited resources, we chose to work with and amend plan #2. The government, in exchange for financial support of this plan, agreed to establish a new, specialised 'Tiger Department' within the Ministry of Environment's branch in Primorski Krai. We code-named the special task of this department 'Operation Amba' ('Amba' being the indigenous Udegai name for the tiger). Money for 15 rangers, a commander and a small administrative staff would be channelled by foreign sponsors into Department Tiger to implement Operation Amba. Amba's strategy was to eventually put the poachers and traders – who at the time were on a very profitable offensive – on the defensive, and to raise the stakes and costs of killing and trading tigers. But how could this be done with only 15 rangers in an area roughly the size of England and Wales combined? In one word: 'PsyOps'.

Stage one (1994): getting started – the strategic defensive

'PsyOps' is a strategic term short for 'psychological operations'. PsyOps have been a key component of counter-insurgency or insurgency operations throughout history. In this case, we sought to make Amba appear to be bigger than it was, and to possess the power of operating invisibly. We wanted the poachers to think that an Amba ranger could walk or drive up behind them at any moment, no matter where they were; or that an Amba informant was positioned somewhere along the trail to the tiger. We wanted the tiger trader to think that Amba informants were everywhere, and that the chance of getting caught and going to jail had suddenly increased dramatically. PsyOps work only when they mirror an expanded version of reality. We sought to create an expanded image of Amba operations by doing the following:

1   Using the element of surprise: Amba patrols were quietly deployed to three distinct areas of the taiga at once. Rangers would randomly check vehicles coming out of the forest as well as hunters walking inside the forest. This activity would be concentrated in three days. Then without a word, rangers would depart for a new area, maybe to return the next week, sometimes the next month.

2   Developing an intelligence network: During these patrols, Amba rangers would inform local citizens and authorities about their mission, inviting them to provide information on poachers and traders. A surprising number of people stepped forward. Over time, a network of informants was formed.

3   Using the press: As the Russians say, 'bad news has wings' and Amba was bad news for poachers and traders. Besides spreading word of Amba by mouth, press coverage was used to maximise publicity of Amba's presence.

The first four months of Operation Amba – January to April, 1994 – were learning ones. The winter of '93/94 turned out only slightly better for tigers than the previous two winters. Amba rangers found themselves turning up at the crime scene after the crime had been committed. Poached tigers and signs of poached tigers were found throughout the Krai. Commander Vladimir Shetinin reckoned that the number of dead tigers discovered by rangers represented between 10–20% of the true amount killed. Between January and April, Amba discovered five tigers shot by poachers.

But several things that Amba achieved during this period paid off later. Local citizens were suddenly aware of Amba's new presence. Poaching would no longer be so easy. Furthermore, infor-

mation was compiled from new Amba agents about the structure of the underground wildlife trade in the Russian Far East. This information allowed Amba to design several undercover investigations targeting major tiger traders. These initial investigations were 'passive' – in other words we did not yet feel prepared to act on an illegal incident on the spot. We watched and learned.

By the Spring of 1994, through joint undercover investigations, GSN and Amba personnel had discovered several major poaching and smuggling operations in Primorski and Khabarovski Krais. Three are worth mentioning in that they were typical of others discovered later. The first involved professional hunters killing tigers, storing them whole and selling them to Chinese citizens in the city of Khabarovsk. The Chinese traders were purchasing the tigers at $5000 cash per frozen carcass. Sometimes only bones were sold to the Chinese at a rate of about $300/kg. The Chinese traders would transport the tiger carcasses or parts back to China by air and by vehicle (also reportedly by train). The tigers were sold to traditional Chinese medicine companies in Heilongjiang and Jilin Provinces in northeast China. When investigators visited these firms in China, they learned that the tiger's bones were mixed with bones from other animals and ground into wines, pills and plasters and sold locally and to companies in Singapore and South Korea.

The second illegal operation involved poachers employed by a governmental department in Khabarovski Krai. These corrupt officials sold dead tigers to members of a Russian-Korean community in the city of Khabarovsk. Running low on tigers in Khabarovski Krai, the poacher would link up with colleagues in Primorski Krai where he would shoot the tiger. The poacher would immediately skin the tiger and maintain the rest of the body in a frozen state (easily done in winter). The poacher would then contact the Russian-Koreans who would arrange to sell the dead tiger to South Korean or Chinese merchants in the city of Khabarovsk. During our investigation, we met the poacher and the traders, and viewed one of their dead tigers in the back of a government truck. The Russian-Koreans could speak Russian and Korean, and a smattering

Operation Amba (tiger) went on the strategic offensive to curtail tiger poaching in the Russian Far East by using small anti-poaching teams to roam the taiga and build a network among local residents for support and information.

of Chinese, and were therefore well situated to act as middlemen. They sold skins for $3000–$4000 on average and the price of bones fluctuated wildly.

The third illicit operation we discovered involved a Russian government employee working with an employee of a joint Russian-Korean logging operation. The logging employee was a driver for the logging company. He had full access to the taiga by the roads his company created and the vehicle he drove. He would carry a rifle with him during his work and poach deer and wild pig, and tigers on the side. Once the tiger carcass was in hand, the poacher used two contacts to sell the dead tigers. One was a customs officer working in the port of Vladivostok who could facilitate a sale and secret export of the tiger to any country by ship. The other contact was an employee of the shipping port in Plastun, which is situated along the eastern seaboard (Sea of Japan). The second contact would sell tiger skins and bones to Russian sailors and Korean workers heading by ship to South Korea or Japan. The tiger parts were easily concealed on board the ships amidst commercial containers and tons of Russian logs.

Amba received no support whatsoever from other enforcement agencies to do follow-up work on these investigations. But during these investigations, Amba learned three important lessons:

1   The line between the poacher and the market was becoming increasingly direct. Some Russian traders had supplanted foreigners as middlemen.
2   Amba could not yet rely on the police and courts to back it up in the field. Amba needed its own roof, preferably from within the government.
3   Networking paid off. By talking with local citizens and authorities, Amba rangers were building up a network of support and information. Not only through the press, but also through these direct contacts, Amba was becoming known throughout major parts of Primorski Krai. Amba was growing bigger and would soon be able to operate invisibly.

Stage two (1994/95): going on the strategic offensive

By late August 1994, a survey among Amba agents and tiger experts (conducted by GSN personnel) revealed that the volume of tiger bone trading had decreased. It was not clear at the time how much of this was attributable to anti-poaching and how much was due to the time of year. Amba undercover agents claimed that Amba's presence was slowing down trading activity by raising the risk of getting caught, which raised the financial costs of trading tigers.

Amba's new vehicles and uniforms projected a professional image of rangers who operated more like police. While in fact many Amba rangers deserved this reputation, as a whole Department Tiger was still very much in the learning stages. Could it have been, we asked ourselves, that tiger traders exhausted their supply of bones and were waiting until the winter of 94/95 to stock up again? Since only time would tell, it seemed wisest to prepare for a new onslaught of poaching by snowfall.

While Amba prepared for winter, GSN took aim at the second part of our objective:; securing tiger habitat for the future. Through sophisticated radiotelemetry technology, Hornocker Wildlife Institute personnel and their Russian scientific colleagues were amassing critical data on tiger roaming patterns and habitat (D. G. Miquelle *et al.* this volume chapters 6 and 19). The Hornocker team was compiling scientifically credible data that would empower environmental officials to halt logging and certain levels of hunting within prime tiger habitat.

But the Kremlin was fighting a war with Chechnya and, at the time, afforded very little attention to environmental degradation in the Far East. The Russian Duma (parliament) took up issues such as commercial poaching of endangered species very slowly, if at all. GSN searched for and found a channel to the Prime Minister, whose advisors offered to draft a national decree on saving the Amur tiger. By the following year this draft decree would become a reality.

On the anti-poaching front, Amba rangers set out to find allies within local enforcement agencies. Their investigations showed that poaching gangs and traders detected by Amba were too well connected to the Mafia for Amba to fight alone. Thus, rangers looked to develop better relations with local police and FSB (formerly KGB) officers. Good relationships developed between two Amba teams (based in Ussuriysk and Luchegorsk) and local enforcement authorities. New investigations were set up in these two areas. With police and FSB support, the investigations could be more 'active' (e.g. once the perpetrator was detected performing an illegal act, Amba and the police could move in for the arrest).

Amba suspected that tiger parts were being smuggled to China by road from Ussuriysk to a border checkpoint called Poltovka. Before Amba could solicit information from the Chinese community in Ussuriysk about tiger smuggling, they received a call from a potential informant. Alias 'Mr. Chang' offered information about wildlife smugglers in exchange for an agreement by Amba to arrest a particular smuggler. Although Amba rangers and the FSB concluded that Chang wanted to use them to undermine a competitor, they verbally agreed with Chang's condition. Chang subsequently informed Amba about the movements of a truck illegally carrying sea cucumbers and tiger parts. The information at first turned up nothing. But after four days of road checks the truck was found. Besides three metric tonnes of sea cucumbers, a tiger skin was found. Customs officers working the border point of Poltovka, where the contraband was impounded, refused to co-operate with FSB and Amba, insisting on forwarding the evidence to the Department of Transportation. Although no one was prosecuted for the incident, the smugglers were surprised and disappointed to have lost perhaps $50 000 worth of contraband. Better interagency co-operation would have led to a successful prosecution.

In another investigation, in the Bikin Valley, GSN and Amba investigators turned up evidence of tiger poaching and trading. In July 1995, undercover investigators negotiated the purchase of a tiger skin, a tiger skeleton and bear gallbladders from a local citizen. The tiger parts came from a one-year-old tiger that had been shot the previous winter. Investigators filmed the tiger parts in his possession and the conversation in which he revealed how he had come to obtain them. The seller was asking $11 000 for the skin and bones together. The film was later shown to local police who searched the dealer's house and found the tiger parts, together with narcotics and dynamite, which he was using to blast bears out of their dens. The dealer went to jail. It was decided that each Amba team should have its own video camera, and later these were supplied.

When local enforcement agencies do not help out with wildlife investigations, the work can be dangerously lonely for Amba. In May 1995 an Amba investigation was compromised by apathetic or corrupt officials, nearly causing one ranger to lose his life. Four Amba rangers discovered a tiger-smuggling channel between the city of Arseniev and Vladivostok, Primorski Krai's capital and major sea port. They learned that hunters near the city of Arseniev (north of Vladivostok) were paid to kill tigers by a Mafia group, which then smuggled the remains to Vladivostok. Further examination by Amba revealed that this channel was also used to trade drugs and arms. Amba rangers took the information they had gathered to police and prosecutors in Vladivostok District. The police told Amba that they already knew about this channel, and that they should 'leave it alone' because the police 'were on top of it'. Several days later, the lead Amba ranger who discovered this smuggling route was attacked by a group of young men outside his home and badly beaten. An investigation by OMON (Police Special Forces) confirmed that the beating incident was linked to the discovery of the smugglers.

Stage three (1995/96): Amba gains a roof

On August 7, 1995, Russian Prime Minister Victor S. Chernomyrdin issued National Decree number 795 'On Saving the Amur Tiger and Other Endangered Fauna and Flora of the Russian Far East'. This high-level political support for tiger protection sent a serious message to Russian law enforcement agencies and courts. The Russian courts had previously shown very little interest in tiger poaching, agreeing to hear only two tiger poaching cases between 1992 and 1995, and no cases relating to trading tiger parts during the same period. In 1995–96 alone, seven people were indicted for these crimes. More importantly, tiger kills were still down.

Taking advantage of their increased legal support, Amba investigators turned up the heat on wildlife traders in Ussuriysk, which informants consistently pointed to as the hub of tiger smuggling activity. Over the course of the year, information was gathered on a disconnected ring of tiger dealers around Ussuriysk who were sitting on a total of 15 tiger skins and approximately six sets of bones. Based on a tip from an Amba informant (an ex-hunter), investigators used fresh American hundred-dollar bills to lure the traders into position for an arrest. It worked twice, and failed a third time. The two traders caught, however, were persuaded to give details on other tiger traders and poachers before they were indicted. Some of the bizarre details included a middleman from Azerbaijan who was trading in tiger parts and leopard skins between Ussuriysk, Central Asia and China.

While the Ussuriysk Amba team focused on underground traders, the Iman Amba team made significant progress deep inside the forest where it was focusing on poachers. In July, after an eight-day stake-out near Novopokrovka, Iman caught a poaching gang involved in all aspects of wildlife smuggling, ranging from bears to salmon. Iman rangers videotaped the entire bust and the tape was used in court. A heavy fine was issued by the judge and the case made headlines in local newspapers.

Recognising that interagency coordination on wildlife law enforcement was still weak, and that Amba's ability to win cases in court against poachers and traders needed to improve, Amba officials invited wildlife law enforcement trainers from the US Department of the Interior to address Amba rangers and customs authorities. The US trainers provided training and technical assistance in the areas of CITES implementation, species identification, anti-smuggling and anti-poaching. A CITES manual was provided in Russian as well. Shortly after this training took place, the CITES Secretariat provided a short course in CITES regulations to Russian environmental officials, including Amba's Deputy Commander.

Stage four (1997): turning the corner

Interagency co-operation and coordination improved in 1997. On April 11 1997, officers working for Russian customs, police and Amba co-operated on an arrest of a tiger dealer who had attempted to smuggle a tiger skin out of Russia through Vladivostok's International Airport.

In fact, Amba investigations have improved over-

all in 1997. When Amba began operations, some of the tiger dealers' internal smuggling routes had been discovered, but the actual destinations for the skins and bones of poached tigers were rarely confirmed; co-operation from police and customs officials working the border points was lacking. In the first three months of 1997, however, investigators discovered a major tiger skin trading route from Khabarovsk to Japan by way of Vladivostok. They also discovered a tiger skin channel by sea from Vladivostok to South Korea. Furthermore, a tiger skin and set of bones were discovered on the ship 'Kapago' in the port of Nakhodka; Nakhodka customs and local police offered to help with the investigation. Investigators also discovered other endangered fauna being smuggled out of Primorski Krai. Amur leopard skins, snow leopard skins, deer musk glands and bear gallbladder are some of the highly priced derivatives of endangered fauna recently detected by Amba investigators. For several years now, Amba investigations have also routinely turned up wild ginseng being smuggled to China.

Smoother operations by Amba are also attributable to improved coordination of outside financial aid. GSN and the World Wildlife Fund (WWF) now coordinate all assistance to anti-poaching patrols. Lack of such coordination can slow anti-poaching efforts. One slow or inaccurate money transfer can ground one anti-poaching team for a month while they wait for money to buy gasoline. Unfortunately, coordination of aid was not so good during the first two years of operations. In the first quarter of 1997, however, Amba was on the road almost constantly, stopping and/or inspecting 1700 vehicles, 1900 hunters and potential poachers, and 76 hunting shacks.

Amba public relations has also steadily improved. In the first quarterly field report of 1997, Amba Commander Vladimir I Shetinin states that, 'video equipment was used to make more detailed assessment of crimes. This film, and information obtained during patrol, is used in the mass media – journalists have used our materials in four articles, two radio broadcasts, and three television shows.'

Shetinin's latest report added a sober note about the situation surrounding the Amur tiger. After detailing the involvement of local officials in tiger trading, he concluded that the Tiger Decree of 1995 was still not being adhered to by all Russian agencies. 'This inaction will make it very difficult to ensure biological diversity in Primorski Krai. This is especially true given the tremendous economic pressure to overuse natural resources at a time when neither legislation nor governmental funds can ensure the necessary protection from environmental degradation, especially of protected nature areas (national parks).'

Such an honest assessment of the situation is what caused most foreign conservationists to dismiss the possibility of saving the Amur tiger. However, on 8 July, 1997, the Office of the Prime Minister informed the Ministry of Environment that the Russian Duma had approved governmental funding of the Tiger Decree of 1995. And the most recent Amba investigations (summer of 1997) confirm that illegal tiger exports are low enough to allow Amba to eventually reduce them further. As Amba and its supporters move forward, it is critical to review the lessons learned thus far, key components of which may be applicable to similar situations elsewhere.

Box 16.1 Lessons for the future: key components of Operation Amba
Steven Russell Galster

In less than four years, for less than $750 000, the Amur tiger has been brought back from the brink of extinction. Looking back at what went right and what went wrong with the development and execution of this species recovery programme, a number of clear lessons emerge. The Russian Far East differs from most other parts of the tiger's range in that human population density is low, and this factor certainly contributed to success; however, many of the lessons learned from Amba may still be applicable to other species recovery efforts.

Local involvement Local support is a key issue. To succeed, the core ideas of a species recovery programme must be locally designed and the programme must enjoy local community support and participation. Amba achieved this in two ways; first, through the media. Just after Amba was formed, the local population came to know about it through media stories announcing the arrival of the new 'tiger protectors'. And second, Amba teams visited communities throughout Primorski Krai announcing their own arrival and explaining what their mission was. In most cases, at least one person in each community would step forward to offer the rangers help with information. Good relations between anti-poaching teams and local law enforcement agencies are also vital; Amba team members do not have the authority to make arrests and it is only by working closely with the local police that offenders can be brought to court.

A drug smuggler apprehended by Amba's Chief Inspector Dukov at a roadblock. Amba teams work with police to catch wildlife smugglers and other criminals.

Skills and equipment Rangers must be able to operate effectively whether catching poachers, collecting evidence on illegal traders, or uncovering smuggling routes. The radios and jeeps Amba needs for the field are bought locally with funds from non-Russian wildlife organisations, including the Global Survival Network's Siberian Tiger Support Coalition and WWF Germany. Good communications and vehicles allow the rangers to efficiently coordinate their anti-poaching operations – when Operation Amba began its work, Amba teams could not communicate unless they were within 15 km of one another. More recently, better radios and faster vehicles have enabled teams operating vast distances apart to converge on poaching gangs in at least six incidents. Training in investigative techniques and wildlife law is also necessary and GSN provided live undercover training, involving posing as tiger skin dealers and utilising covert audio and visual recording, for several Amba rangers. The US Department of the Interior also provided Amba with on-site training in species identification and in anti-poaching and anti-smuggling techniques used in the USA and elsewhere.

Political backing   High-level political support is essential. The issue of Russian National Decree number 795 'On Saving the Amur Tiger and Other Endangered Flora and Fauna of the Russian Far East' in August 1995 made a gradual, but very positive difference to Amba. The apathetic attitude of the Russian courts and law enforcement agencies towards wildlife issues in general gradually changed, resulting in a rise in prosecutions of wildlife criminals. Judges, police and FSB (formerly KGB) agents are theoretically obligated to assist Amba in executing the decree. Although in practice Amba receives less assistance than it should, there is a higher level of law enforcement co-operation, evidenced by an increasing number of confiscations, arrests and prosecutions in the cities of Luchegorsk, Ussuriysk, Vladivostok and Arseniev.

International links   Good communication with relevant agencies abroad is important for two reasons. Financial assistance comes from overseas and must be coordinated and provided in a timely manner – one delayed money transfer can ground a team for weeks. Unfortunately, this happened more than once to the Amba rangers, who had fulfilled their working obligations but were forced to operate without salaries – sometimes for more than 10 weeks straight – due to financial mistakes made by foreign sponsors. Secondly, smuggling of tiger parts by definition involves international borders, so liaison with parallel agencies in neighbouring countries adds considerably to the overall chances of success.

Public relations   Any species recovery programme must have effective public relations, both locally to generate local support and internationally to assist in raising funds. Initially for Amba, the local non-governmental wildlife media agency Zov Taigi did an excellent job of publicising the tigers' plight through their conservation journal, posters and educational materials and several TV and radio shows. More recently, Amba has liaised directly with local and international media outlets, providing them with information, commentary and video material.

Building support for tiger conservation: produced on a shoestring in a basement office, the newsletter *Zov Taigi*, edited by Vasili Solkin and Larissa Kabilik, reports on Amba's work.

Poachers such as these work at the dangerous end of the illegal trade in India, risking capture and prosecution for relatively small rewards. The wealthy traders to whom they sell make far more profit yet it was not until November 1998 that the first conviction against traders was secured in the Indian courts.

17

Combating tiger poaching and illegal wildlife trade in India

Ashok Kumar and Belinda Wright

Introduction

It is a common mistake to confuse wildlife poachers with wildlife traders. For instance, when a seizure of tiger skins or bones takes place in a city in India, the headlines next day scream 'Poacher arrested'. The press is not alone in making such a mistake – at least in the Indian subcontinent, many enforcement agencies and wildlife conservation organisations follow suit. This may not seem terribly important at first. Unfortunately, the confusion is carried through to development of enforcement strategies and allocation of resources, and this has serious implications. It is necessary to distinguish between the role of the poacher and the trader, and to develop different strategies to combat each.

By definition, a poacher is a person who actually kills wild tigers by snaring, electrocution, poisoning of tiger kills with easily available pesticides or, though rarely now, shooting. Poaching is described in some detail in a Wildlife Protection Society of India publication 'India's Tiger Poaching Crisis' (Wright 1997; Box 17.1). The poacher is generally a villager who lives near a forest, travels on foot and knows his forest and its wildlife intimately. The days of city-based desperadoes driving into the jungle to shoot tigers are virtually gone, though the occasional case is still reported. Additionally, there are nomadic tribes whose hereditary profession is to travel from jungle to jungle to poach wildlife. There is considerable specialisation in this. For instance, poachers of tigers and leopards will rarely kill elephants for ivory, though they may target lesser cats. Bird trapping is also a specialised skill. Knowing these finer nuances helps in the development of appropriate enforcement strategies.

Villagers commonly put poison in the carcasses of livestock killed by tigers or leopards. Their basic purpose was to rid the area of the predators, but of late they have discovered that tiger skins and bones are saleable. In addition, there are representatives of city-based traders, who actually travel to villages around the forests, leaving behind poison and a promise to pick up the goods next time around.

Wildlife traders are city-based, Mafia-type gangsters, such as those in drug running cartels. They are the money-men at the heart of the trade. As in trade in most commodities, the trade in wild flora and fauna is controlled neither by the primary harvester nor by the retailer, but by the wholesaler.

Anti-poaching strategies

Control over poaching requires a two-pronged approach. The first strategy should address local villagers and forest dwellers. Tiger Reserves in India are invariably surrounded by village communities who were denied access to the forest for their fuel-wood and fodder requirements when the areas were given the highest level of protection as National Parks. Many have taken to illicit cutting of trees in the forest, and some to killing tigers for profit.

To earn the trust of these communities, a series of measures broadly termed 'eco-development' will have to be implemented to relieve pressure on the reserves. These include development of timber and firewood plots, thatch grass and grazing land and improved cattle breeds which can be stall-fed, as well as income-generation programmes (e.g. basket-weaving, appliqué work, etc.) to replace wages lost due to restricted timber operations. Wildlife tourism is another activity that should provide income for people living near Tiger Reserves; it

Box 17.1   Tiger poaching statistics in India
Belinda Wright

The Wildlife Protection Society of India (WPSI) has maintained a database of reported seizures of tiger parts and discovery of dead tigers since January 1994. The central and state governments do not collate information on tiger poaching cases and most of the figures recorded are the result of reports received by WPSI from enforcement authorities throughout India, operations carried out by WPSI and other sources. WPSI also has records of a large number of tigers that were 'found dead' in various protected areas in India. Without verification of poaching evidence these deaths have not been included in this report.

Up to July 1997, 67 tigers had been found dead, most of them poisoned, if not by professional poachers, often by farmers. Agricultural pesticides, which are freely available to boost food production, are put into the carcasses of livestock killed by tigers. This can lead to the death of a tigress and her cubs. Traders are also known to provide farmers with poisons and return to pick up the carcass of any tiger killed. Furthermore, poachers may deliberately release a cow in a tiger area, and poison the carcass when it is killed in order to get the tiger parts. Poisons may also be put in ponds where tigers are known to drink. This can lead to the death of other wild animals.

Seizures by the police and wildlife authorities included 118 skins and 29 skeletons or skulls, over 363.6 kg of bones, over 606 claws, and 16 teeth, mostly the prominent canines, which are turned into charms and pendants. In addition there were nine seizures of miscellaneous items, such as a clavicle bone, a relict which 'floats' in the flesh unattached to the skeleton and has long been a favourite charm, a mounted trophy, a tiger paw and whiskers.

It is not possible to draw any clear conclusions from the statistics, except that they indicate a high level of poaching and illegal trade. As with most illegal activities, e.g. the drug or art trades, dead tigers and seizures clearly represent only a fraction of the deaths and of items being traded clandestinely. The annual totals fell towards the end of the period, but WPSI attributes this to traders becoming more wary following a series of successful 'sting' operations in which poachers and traders were located by informers and consequently arrested by enforcement authorities.

The central Indian State of Madhya Pradesh stands out as the site of most tiger deaths and seizures. However, the State, with over 900 tigers reported by a 1993 census, has nearly twice as many tigers as any other State and is an obvious target for the illicit trade.

The data in Table 17.1 have been summarised from those in the main database, a sample section of which is shown in detail in Table 17.2.

Table 17.1. *WPSI tiger poaching data for India, 1994 – August 1997*

State	Tigers killed	Tiger parts seized					
		Skins	Skeletons and skulls	Bones (kg)	Claws	Teeth	Misc.
A. Totals for each state							
Andhra Pradesh	13	5					
Arunachal Pradesh	1	1	1				
Assam	3	1	1				
Bihar	5	6					1
Delhi		10	1				2
Jammu & Kashmir		3					
Karnataka	7	2	1				1
Kerala	3	1					
Madhya Pradesh	8	45	21	>318	318	9	4
Maharashtra	7	12		?			1
Manipur	1						
Mizoram	1						
Orissa	3	2					
Punjab							1
Rajasthan	2	2	1		4		
Tamil Nadu	2						
Uttar Pradesh	6	23	4	>84			
West Bengal	6	10			295	7	
Totals	68	123	30	>402	617	16	10
B. Totals for each year							
1994	26	20	10	>262	145	9	3
1995	20	49	14	>85.6	443	7	3
1996	12	32	1	26	29		1
1997 to August	10	22	5	28.5			3
Totals	68	123	30	>402	617	16	10

Table 17.2. *Tiger poaching data in India, January–March 1995*

Date	Place	State	Tiger skins	Tiger bones/parts	Remarks	Source
January						
early	Nr Dudhwa TR	U.P.	1 tiger	—	Poisoned	BASTF
early	Khawasa, S. Seoni	M.P.	1 skin	? kg – 'bones of 2 tigers'	Seized with chital & python skins; 13 people arrested. Fresh skin.	M.P. Tiger Cell
early	Behri village, Barghat, nr Sagar	M.P.	1 tiger	+	Poisoned, found with sloth bear carcass.	M.P. Forest Dev. Corp.
late	Pench TR	M.P.	1 tiger	+	Found dead (electrocuted).	M.P. Forest Dept., Seoni
February						
early	Khajuria village, nr Dudhwa TR	U.P.	2 skins	9.6 kg	Seized from trader; fresh skins.	U.P. Police/BASTF
2	Kodiya, Dist. Seoni	M.P.	—	10 claws	Ramnath + 2 others arrested.	M.P. Tiger Cell, Bahrai Police
4	Nr Chhindwara	M.P.	—	1 skeleton (subadult)	Found buried nr Dudgan village.	M.P. Forest Dept., Seoni
5	Aria, Seoni Dist.	M.P.	—	? kg bones	Sunderlal + 2 others arrested.	M.P. Tiger Cell, Kurai Police
6–7	Bandra village	Maharashtra	—	? kg bones	Tiger killed recently in M.P.	M.P. Forest Dept., Seoni
7	Karmajhiri, District Seoni	M.P.	—	? kg bones	Nirpat + 5 others arrested.	M.P. Tiger Cell, Kurai Police St.

March

Date	Location	State	Quantity	Additional	Details	Source
3	Baihar, nr Kanha TR, Dist. Balaghat	M.P.	2 skins	3 skeletons	4 poachers/traders arrested at bus stand – Chuttan & co; tigers killed recently.	M.P. Tiger Cell, Baihar Police
3	Ukwa, Dist. Balaghat	M.P.	—	? kg bones	2 poachers/traders arrested – Chhagan & Prahalad.	M.P. Tiger Cell, Roopjhar Police Station
3	Malajkhand, Dist. Balaghat	M.P.	—	3 claws	3 poachers/traders arrested – Anil, Khemlal & Jhadulal.	M.P. Tiger Cell, Malajkhand Police Station
3	Comp.P-259, Rorighat forest beat, Satpura N.P.	M.P.	1 dead tigress	Plus flesh/intestines of 2 other tigers	**1 tigress + 2 female cubs poisoned – skins & bones recovered various locations (see entries – 23 Apr + 26 May). 6 poachers/traders arrested.	Director, Satpura N.P., M.P. Forest Dept.
7	Morcha, Bargi Phata, Kisli n Bichia, Mandai & Kacharni villages – Dist. Balaghat & Mandla	M.P.	3 skins (incl. 2 cubs)	35 kg bones (approx. 5 skeletons, incl. 5 skulls & 2 cubs) 60 claws	Huge operation – more than 23 poachers/traders arrested. 6 leopard skins & 2 leopard skeletons also seized; possible duplication of figures.	M.P. Tiger Cell
8	Namdapha TR	Arunachal Pr.	1 tiger	—	Poaching case.	Field Investigator
8	Dharavasi, Dist. Balaghat	M.P.	1 skin	—	3 poachers/traders arrested, incl. Mhd Rafiq; fresh skin.	M.P. Tiger Cell, Lalbarra Police Station
15	Puri	Orissa	1 skin	—	Skin of white cub seized, origin unknown.	Orissa Forest Dept.
16	Baresnar, Compt 3, nr Palamau TR	Bihar	1 tiger	+	Shot in 'hakua shikar', 4 people arrested.	Bihar Forest Dept/ Nat.Con.Soc.
29	Bandipur TR	Karnataka	1 tiger	—	Poaching case.	Field Investigator

could also involve them in the management of the park. There is a risk, though, of over-commercialisation around reserves, and the drawing in of more people, which can very easily become counterproductive.

To ensure that development programmes for local communities do not disturb wildlife, attempts should be made to stabilise or divert conflicts. Electric fences and crop protection ditches, along with effective compensation schemes for cattle kills and crop damage, offer plausible ways to achieve this. Tigers or other predators that are identified as dedicated livestock killers may have to be removed. An information-gathering network and reward system should be established around each tiger reserve to expose greed-driven professional poachers.

However, these methods are unlikely to be sufficient to deal with hardened criminals, who are often based in towns and cities some distance from the tiger reserves. Tackling these will necessitate a much higher standard of patrolling, communications, information gathering and reward schemes – a tight enforcement strategy, in short. Monsoons are the best opportunity for professional poachers to strike. During this season, when many Tiger Reserves become flooded, communication systems break down and patrolling becomes very difficult. Nevertheless, in these periods, patrolling by day and by night, on foot and on elephant-back, is essential. Special commando units that can traverse the forests for several days at a time need to be trained.

Before any of this can be achieved, there is an urgent need to boost the morale of wildlife guards by implementing a series of welfare measures. These are well documented in a 1994 report of a committee for 'Prevention of Illegal Trade in Wildlife and Wildlife Products' (Government of India 1994). The committee recommended detailed and far-reaching measures that covered four critical aspects: (1) enlisting local people for protection of wildlife; (2) an enhanced enforcement strategy; (3) motivation and welfare for field staff; and (4) measures to prevent illegal import and export of wild animals and their products.

These measures will largely address problems in the 23 official Project Tiger Reserves. Unfortunately, the censuses indicate that two-thirds of India's tigers are outside these reserves, and, although there are tigers in many other reserves, very little protection is in place in most of them, while some tigers are not even in protected areas. Nevertheless, the last census identified many places supporting tigers which function as essential biological links with tigers in the reserves. These areas need to be developed as special tiger conservation zones and assigned mobile strike forces for patrolling and vigilance. However, policing of these areas, by itself, will not be sufficient to ensure a future for the tiger there. Local people, as around the official tiger reserves, can be recruited to contribute significantly towards intelligence gathering and habitat protection, if they are involved in forest management and accorded a share of the benefits. The village panchayat (council) is an effective set-up for achieving this. It provides the ideal framework for community participation and decision making. Through a system of incentives and awards to vigilant people, it will be possible to check tiger poaching outside tiger reserves.

Anti-trade strategies

While all Tiger Reserves have some kind of a system in place to control poaching, excellent in some reserves and poor elsewhere, there is virtually no enforcement mechanism against the city-based traders. It is not surprising that traders in tiger parts have not taken wildlife laws seriously, and continue to operate with impunity. Data gathered on poaching and the wildlife trade show that infringements of wildlife laws are widespread. The volume of the trade and smuggling are much larger than was earlier believed. The logic is clear; why should a traditional trade have died out, even if it was made illegal, when punishment for infringement has hardly ever been imposed?

The trade in tiger bones destined for China was considered small until the middle of 1993. Then, at the end of August, nearly 400 kg of bones was seized in Delhi, with an associated promise to

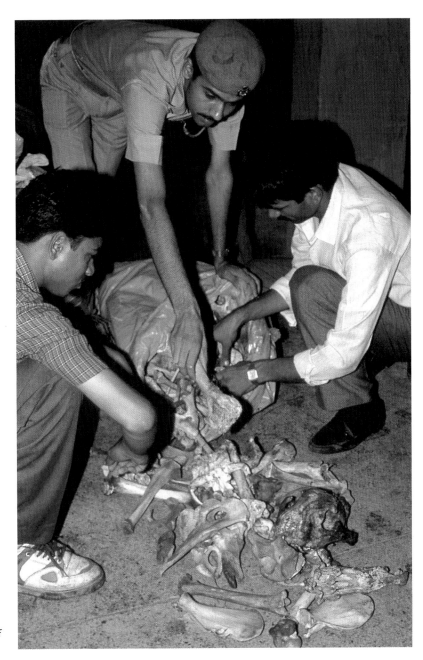

This seizure of tiger bones in New Delhi, India, in 1993 helped to bring the scale of the tiger poaching problem to the attention of the international media.

supply a further 1000 kg over two months – equivalent to the products of nearly 90 adult tigers. The conclusion is clear; thorough investigative techniques are needed to penetrate and put a stop to the trade. Its magnitude and what species are involved need, urgently, to be assessed.

Can wildlife laws be enforced in India? The answer is a cautious 'Yes', and here is what is required. First, demand for tiger products needs to be brought under control by persuading people in user countries that continued consumption may lead to the extinction of the species involved. They

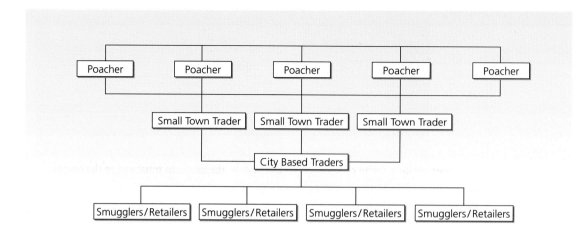

FIGURE 17.1
Flow chart of the links in illegal trade of wildlife and wildlife products.

must find alternatives in their own interest and forswear tiger products. Education campaigns must be directed to these people through their compatriots, in their own languages, and in relation to their own cultures, if they are to be effective. Secondly, people who break the law should be brought to justice and punished. Research by the Wildlife Protection Society of India (WPSI) has revealed that, in the last ten years, not a single person accused of a tiger-related offence has, until very recently, been convicted anywhere in India. And this despite the fact that hundreds of poachers and traders have been caught red-handed. The legal cell of the Society has taken up this challenge and WPSI lawyers are pursuing 22 cases through the courts to support prosecution of wildlife crimes.

Clearly, the legal system must be more responsive to the need of the hour. There are signs that this is slowly happening in India. The judiciary (and the media) are now alive to threats to the environment in general, and to the survival of the tiger in particular. The Wildlife (Protection) Act of India gives every individual the opportunity to institute a court case, giving non-governmental organisations (NGOs) and individuals the chance to play a critical role in invoking the law. The higher courts of India

can play a constructive role by directing lower courts to take up wildlife cases on a time-bound basis or by ensuring that wildlife cases go to designated courts. Here again, NGOs and individuals can take court action whenever they feel that justice is being delayed, and thus denied.

Lastly, we have to revamp enforcement strategies. Enforcement authorities have so far been concentrating on either the poacher or the retailer. To solve the problem, wildlife enforcement agencies must direct their efforts towards snapping the key link in the chain of this trade: the city-based trader (Fig 17.1). If you arrest a tribal man with a tiger skin you have caught only one person – another from the family takes his place. The same is true of the retailer. But when you catch a wholesaler, you snap a link in the chain and, for a while at least, the chain is broken. The number of these wholesalers and financiers, the masterminds of the wildlife trade, is not very large. It is these links that need to be eliminated. As a result of the Delhi seizure of tiger bones in August 1993, the wildlife trade from Ladakh to Tibet virtually shut down. Such cross-border trade is entirely dependent on barter. After the supply of tiger bones and bear bile was stopped for a while, *shahtoosh* (the wool of the Tibetan antelope) no longer came in from Tibet by this route. And in Delhi, two gangs of wildlife wholesalers were neutralised.

With the removal of one wholesaler, we believe you achieve the equivalent of catching something like 15–20 forest-level primary poachers or retailers. Furthermore, as the wholesalers know each other, news spreads that the heat is on. Up to now the maximum pressure on them has been around Delhi and Calcutta. That leaves major trading centres like Bombay, Madras, Patna and Kanpur largely untouched. Good leadership, and training and equipment of law enforcement personnel to undertake surveillance and intelligence gathering will be key ingredients for successful field operations against wildlife wholesalers. But after that, there is once again the need to invoke the judiciary. Wildlife law enforcement does not consist merely of dashing rangers trading bullets with poachers, or undercover James Bond types busting wildlife traders. Ultimately, the problem must end in the courts.

A tiger wanders through a ruin in Ranthambhore National Park, a small island in a sea of people. Wild tigers can only be sustained in this small tiger conservation unit (TCU) or any TCU with leadership, and the cooperation of local residents working in partnership with vigilant local tiger stewards.

People, tiger habitat availability, and linkages for the tiger's future

With tiger habitat dwindling throughout their range, all areas containing tigers are worthy of conservation effort and support.

18

Where can tigers live in the future? A framework for identifying high-priority areas for the conservation of tigers in the wild

Eric D. Wikramanayake, Eric Dinerstein, John G. Robinson, K. Ullas Karanth, Alan Rabinowitz, David Olson, Thomas Mathew, Prashant Hedao, Melissa Connor, Ginette Hemley and Dorene Bolze

Introduction

Threats to tigers in the wild have intensified over the past five years due to rampant poaching of tigers and their prey, and intensive logging and encroachment into the little remaining natural habitat available over most of the tiger range (Nowell & Jackson 1996). Unless we act now to curb these threats through sound, well-designed conservation measures, the long-term prospects for survival of tigers in the wild are grim.

Because tiger habitat is dwindling throughout the range, all areas containing tigers are worthy of conservation effort and support. Unfortunately, funds and numbers of trained personnel remain inadequate to protect all tiger populations and habitats. Thus, we need to identify remaining tiger populations and habitats with the highest probabilities of long-term persistence to ensure that the limited resources are allocated to these areas on a priority basis. We must identify the most cost-effective projects and the most appropriate places, timing and levels of effort required to allocate conservation resources, and base these decisions on sound principles of conservation biology and landscape ecology. Here, we present a framework based on the ecology and behaviour of tigers that helps to identify and prioritise tiger conservation areas across Asia.

Priority-setting framework for tiger conservation

Approach to the framework

Our approach is ecology-based and represents a major departure from the taxonomy-based approach, which uses putative subspecies to maintain representation (Seal et al. 1987a; Seal 1991). The ecology-based approach ensures that, in addition to the inherent genetic variation, the ecological, behavioural and demographic variation among tiger populations is also represented in a portfolio of prioritised conservation units. Thus, we maintain representation by conserving examples of tiger populations living in distinct habitat types across the extant range.

We also depart from other approaches that set priorities largely on the perceived inviability of a population that falls below a certain critical threshold size (Lande & Barrowclough 1987). Experiences from Nepal, India and the Russian Far East show that tiger populations severely reduced by poaching can recover if given adequate protection, and if food and water are available (Mishra et al. 1987; Panwar 1987; State Committee of the Russian Federation for the Protection of the Environment 1996). Furthermore, much of the life-history data required to conduct population viability analyses (PVAs) for tigers are lacking, and gathering the required data for all tiger populations is prohibitive in terms of time, manpower and funds (Nowell & Jackson 1996). We do not automatically assume that tiger populations that fall below a certain number are no longer viable or even that there is a clearly established number that serves as a defining indicator of the long-term viability of tiger populations across their range. We recognise that the probability of long-term persistence is determined in part by factors intrinsic to the population (e.g. population size, reproductive success, demographic structure), but also in part by factors extrinsic to the population itself (e.g. short- and long-term trends in habitat fragmentation, human impact, local development and commitment to conservation), and have thus adopted a convenient and biologically sound alternative to the PVA. We assess the population status of tigers as a demographic trend over the previous decade, rather than attempting to determine viability from absolute numbers. Because tigers are difficult to census accurately, these general trends are likely to be as precise as numbers estimated from censuses. Moreover, these trends represent population trajectories, which are more significant to conservation than numbers.

Underlying principles of the framework

Several basic principles of ecology and of conservation biology of tigers underlie our approach to this priority-setting initiative. We consider that tigers are ecologically, behaviourally, demographically and genetically adapted to the ecological conditions and bio-geographic realm in which they live. For instance, tigers living in the Sundarbans mangrove habitat are adapted to hunting on tidal islands in a brackish delta. Tigers native to the six-metre-tall grasslands in the subtropical alluvial habitats of the Gangetic plains live at much higher densities and hunt different prey than do tigers in the boreal forests of the Russian Far East. Preserving these adaptations, or tiger phenomena, is as important as conserving populations of putative subspecies.

We also consider the root causes of the decline of tigers across their range to be habitat fragmentation and degradation, and intensive poaching of tigers and their prey. These threats affect the integrity of the habitat, impoverish the biological communities in which tigers live, and reduce tiger populations.

Most existing protected areas in Asia are small (Dinerstein & Wikramanayake 1993). Since core protected areas are essential for tiger survival, landscape-level planning in large habitat blocks should entail restoration and creation of protected core areas, buffer zones and dispersal corridors in a large mosaic of other land-use options (Karanth 1991; Karanth & Sunquist 1995; Nowell & Jackson 1996). Thus, tiger conservation plans should extend over larger landscape units, and we incorporate landscape features as an integral part of our assessment of tiger habitats.

Reclamation or restoration of large tracts of tiger habitat in many areas of its previous range is unlikely in the near future; thus, the immediate responses necessary to ward off local extinctions are to protect remaining habitats and to reduce poaching pressure both on tigers and on their prey. These goals can be achieved by a combination of protection measures, managing protected areas and extending conservation measures beyond protected areas. The latter will include financing and supporting anti-poaching information networks, adequate staffing and training needs for managing protected areas and reserved forests, and development and implementation of Integrated Conservation and Development Programs (ICDPs) in and around protected areas.

Framework structure

Our framework is hierarchical (see Fig. 18.1). This structure ensures representation of all extant tiger populations throughout their range, the full spectrum of habitat types found across tropical Asia and the Russian Far East, as well as the distinct species assemblages, predator-prey dynamics and ecological processes associated with these diverse ecosystems.

To adequately capture the variation of tigers throughout their range, we first divided the extant tiger range into five bioregions – Indian subcontinent, Indochina, Southeast Asia, South China

Table 18.1. *The geographic boundaries of the five bioregions*

1 *The Indian subcontinent*: from South India to the Indian and Nepalese Himalaya and foothills, and east to the Burmese transition zone.

2 *Indochina*: from the Burmese transition zone into southern China and south to the Isthmus of Kra.

3 *Southeast Asia*: from south of the Isthmus of Kra to the southern tip of Sumatra.

4 *South China*: low-lying areas where tigers still persist.

5 *Russian Far East*: Manchurian China, North Korea, northward into Khabarovski Krai.

The demarcations of the Indian subcontinent, Indochina, and Southeast Asia bioregion boundaries are modified from MacKinnon & MacKinnon 1986.

Identifying, protecting and linking prey-rich, disturbance-free areas, where tigresses can rear their young, is essential to sustaining wild tigers in human-dominated landscapes.

and Russian Far East (Fig. 18.1, Table 18.1). This first layer of stratification ensures that tiger populations from different bioregions will not be combined and compared, allowing us to capture broad patterns of genetic variation of tiger populations across their bio-geographic range.

We then identified a total of eight tiger habitat types in the five bioregions:

1 Alluvial grassland / moist deciduous forests.
2 Boreal taiga.
3 Mangroves.
4 Subtropical and temperate upland forests.
5 Temperate mixed conifer and broadleaf forests.
6 Tropical dry forests.
7 Tropical moist deciduous forests.
8 Tropical moist evergreen forests.

These tiger habitat types reflect habitats where tigers are considered to play different ecological roles, and represent a second layer that allows us to capture the ecological, behavioural and genetic differences of tiger populations within and among bioregions.

Then, within each bioregion, we identified Tiger Conservation Units (TCUs). A TCU is defined as, '*a block or a cluster of blocks of existing habitats that contain, or have the potential to contain, interacting populations of tigers.*'

A TCU can consist of several adjacent blocks of habitat among which tigers can disperse. We considered adjacent habitat blocks to comprise a TCU if they are: (1) linked by degraded scrub forests, tall crops (e.g. sugar-cane), plantations with dense growth or canopy cover (e.g. coffee), or river or stream courses since tigers can disperse across these altered habitats or along river courses; (2) separated by less than 5 km, since field studies in India suggest that 5 km is about the threshold of cleared/open land that tigers may cross. Thus, habitat blocks separated by more than 5 km of cleared or open land were delineated as distinct TCUs.

A TCU need not be restricted to nor contain protected areas, but instead includes the entire landscape of natural habitats over which tigers may disperse and become established.

Table 18.2. *Classification of Tiger Conservation Units (TCUs)*

Level I TCU: A TCU offering the highest probability of persistence of tiger populations over the long term. They are essential for a global tiger conservation strategy. Level I TCUs share the following attributes: large blocks of habitat suitable for tigers and prey with adequate core areas; low to moderate poaching pressure on tigers and prey species either as a result of remoteness or vigilant protection. (45–70 points)

Level II TCU: A TCU offering medium probability of persistence of tiger populations over the long term. They contribute to a bioregional tiger conservation strategy. These TCUs share the following attributes: moderate- to large-sized blocks of habitat suitable for tigers with adequate core areas; moderate to high poaching pressure on tigers and prey species but with potential for implementing effective anti-poaching measures in the near future. (32–44 points)

Level III TCU: A TCU offering low probability of persistence of tiger populations over the long term due to its small size, isolation from other habitat blocks containing tigers, and fragmentation within its respective major habitat type. With intensive management and protection, Level III TCUs can harbour small populations of tigers. They are most important to national tiger conservation strategies. Level III TCUs share the following attributes: small blocks of habitat suitable for tigers with small or no core area; high poaching pressure on tigers and prey species that endangers conservation efforts (<32 points)

TCUs requiring immediate surveys: Any TCU that potentially contains extensive blocks of appropriate tiger habitat with or without protected core areas, but data on habitat quality, poaching pressure, or population status for the most important habitats within the TCU are lacking.

TCUs were delineated using vegetation maps provided by the Asian Bureau for Conservation and the World Conservation Monitoring Centre. These data represent the most up-to-date digital maps of remaining habitat across the range of the tiger in the Indo-Malayan realm. We also overlaid anthropogenic features, such as roads, railroads and settlements, from the Digital Chart of the World (a global Geographic Information System database), to assess the spatial relationship between habitat blocks and potential dispersal barriers. If habitat blocks were completely separated by these anthropogenic features, the blocks were placed in different TCUs. To calculate the total intact area of a TCU, we subtracted the polygons of altered habitat within TCUs from the total area using the package ARC/INFO.

Next we scored and ranked the TCUs based on three variables that we consider to represent the root causes of the decline of tigers. The variables are:

> *Habitat integrity* – which includes the size, degree of degradation, fragmentation and connectivity of tiger habitat blocks (Appendix 4A).

> *Poaching pressure* – which indexes the intensity of illegal hunting and potential for its control (Appendix 4B).

> *Population status* – which indexes tiger abundance and recent population trends within each tiger habitat type (Appendix 4C).

We suggest that these three variables are powerful predictors of the long-term persistence of tiger populations.

We solicited regional and local expertise to assess and score TCUs, and corroborated these evaluations with published and unpublished reports reviewed during this study.

The index scores reflect the reversibility of the primary threats to tiger conservation. We considered loss of habitat integrity to be the most difficult to reverse, and thus weighted this variable twice as high as poaching pressure, which can be turned around more easily. Poaching pressure, in its turn, was weighted twice as high as population status, reflecting the observation that tiger populations can rebound quickly if they and their habitat and prey are protected over sufficiently large areas. Habitat integrity, poaching pressure and population

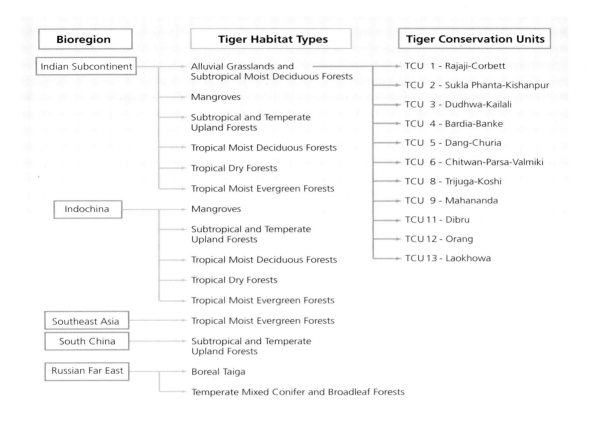

FIGURE 18.1
Framework structure showing hierarchical relationships between bioregions, tiger habitat types, and tiger conservation units (TCUs). The TCUs are shown only for the alluvial grasslands and subtropical moist deciduous forest tiger habitat type.

status were, therefore, weighted according to a 4:2:1 ratio.

The sum of the three variable scores was used to rank and prioritise TCUs in four categories (Table 18.2). Because the TCUs were nested within tiger habitat types and bioregions (Fig. 18.1), the scores and category levels for TCUs were compared only within tiger habitat types of the same bioregion. For instance, we did not compare TCU scores from the Indian subcontinent bioregion with TCUs from the Southeast Asia bioregion, nor did we, within the Indian subcontinent bioregion, compare TCUs from the alluvial grasslands tiger habitat type with TCUs from the tropical dry forest tiger habitat type. This approach ensures better representation of the biological differences of tiger populations, tiger-related predator-prey dynamics, and regional patterns of biodiversity across the tiger's range.

If information on the status of habitat integrity of large TCUs that have the potential to be Level I or Level II priority units was insufficient, they were flagged for immediate ground surveys and/or information verification. Index scores for unknown poaching pressure or population status were, however, assigned a value just above the median for the respective indices to ensure that a TCU will still be retained in the portfolio if it qualifies as a Level I or II based on the habitat integrity score, reflecting the relative reversibility of the threats.

TCUs were assigned to the dominant tiger habitat type represented in the TCU. If a TCU included two or more tiger habitat types, the tiger habitat type that represented the major part of the TCU was the primary tiger habitat type and the next largest tiger habitat type the secondary tiger habitat type.

We also created a decision rule to require at least three TCUs in each tiger habitat type for each bioregion to be the minimum number necessary to maintain representation. If fewer than three TCUs were categorised as Level I, II, or Immediate Surveys for a particular tiger habitat type, Level III TCUs with the highest scores in that tiger habitat type were elevated to Level II.

We named Level I, II, III and Immediate Survey TCUs after significant protected areas (e.g. 113 Virachay-Xe Piane-Yok Don). If the TCU contains one or more protected areas, it was named according to the largest/'most significant' protected area. Similarly, TCUs that straddled one or more international boundaries were listed first under the country that contained the largest or best-quality block of habitat within the TCU. If there were no protected areas within the TCUs, they were given a name to signify regional distribution (e.g. 43 Orissa Dry Forests).

If a habitat block included extensive roads, railroads, settlements, etc. (as indicated by the Digital Chart of the World overlays), we considered it to be degraded, and scored it in the appropriate category (i.e. 5a, 6a of Appendix 4A).

The shape of the TCU was also considered in the scoring process. Long, narrow-shaped TCUs that would not offer adequate protection of core habitats were considered to represent suboptimal habitat (see Dinerstein *et al.* 1997 for details).

Scope of framework

This priority-setting framework is meant to be a coarse-grained assessment that ranks and high-

lights tiger conservation areas in order to bring to the attention of international donors where investments in tiger conservation activities might yield the biggest dividends; bring to the attention of conservation biologists and policy makers the important areas for tiger conservation on a regional scale and highlight general conservation activities; and allow conservationists to develop tiger conservation strategies with regional and global perspectives.

This assessment is *not* meant to provide detailed conservation activities or plans at local or site-specific levels; or supplant detailed conservation planning at the national level already being conducted in some range states.

However, we urge that any local conservation management plans and efforts should be based on landscape-ecology principles, to the extent possible, and developed within a regional or global conservation framework. By directing limited, valuable resources more carefully and scientifically in time, space and sequence, the most important tiger populations will be protected and allowed to flourish throughout their range.

The next logical steps from this assessment are to focus on the priority areas that we have highlighted at a regional scale and scale down to enable on-the-ground conservation at a TCU level. The information presented here can be better refined to increase accuracy at these smaller scales, (e.g. 1:50 000 or less), which would be more suitable for developing conservation management plans.

To make the process as transparent as possible for planners, the complete databases are published in Dinerstein *et al.* (1997).

Results

We identified 25 TCUs as Level I (16% of all TCUs), 21 as Level II (13%) and 97 as Level III (61%) across the Indian subcontinent, Indochina, and Southeast Asia (Figs. 18.2–18.4). With the exception of two smaller blocks of habitat on the Russian-Chinese border, the Russian Far East remains a single Level I TCU. We did not have detailed information about

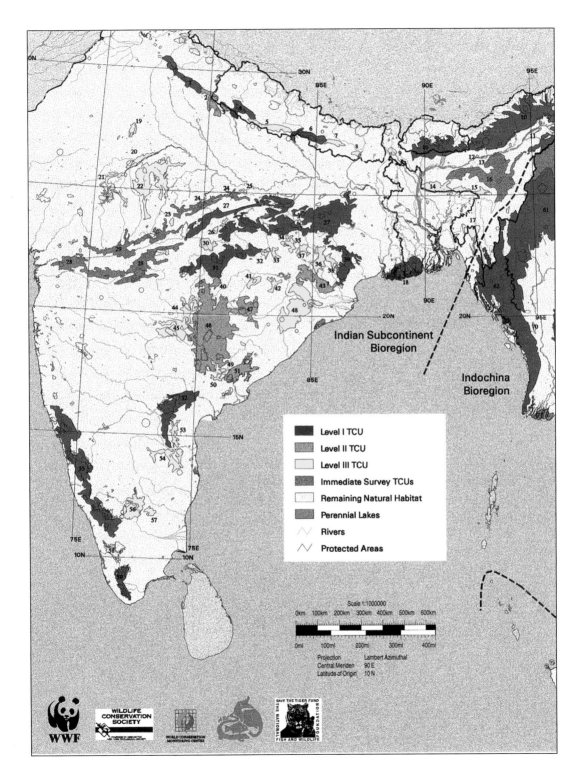

FIGURE 18.2
Map of the Indian subcontinent bioregion showing tiger conservation units (TCUs). The TCUs are numbered, and the numbers correspond to those in Table 18.3.

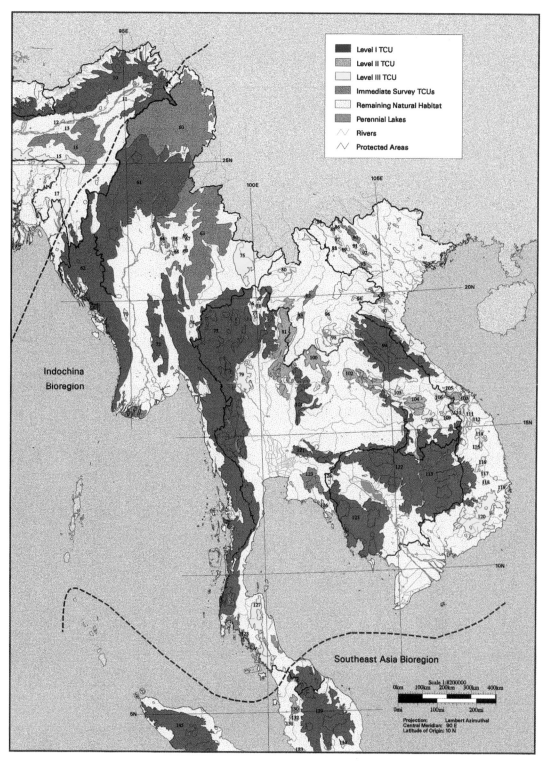

FIGURE 18.3
Map of Indochina bioregion showing tiger conservation units (TCUs). The TCUs are numbered, and the numbers correspond to those in Table 18.4.

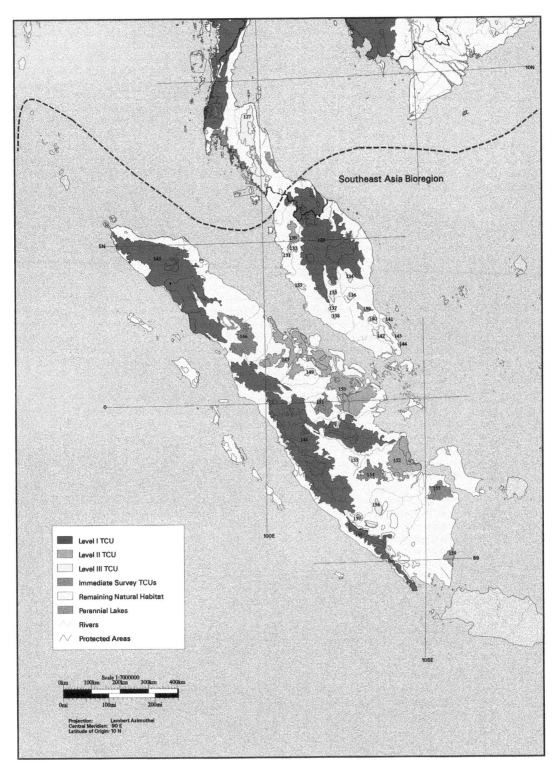

FIGURE 18.4
Map of Southeast Asia bioregion showing tiger conservation units (TCUs). The TCUs are numbered, and the numbers correspond to those in Table 18.5.

Table 18.3. *TCUs for the Indian subcontinent bioregion, with tiger habitat type (THT), priority category level, rank scores, TCU area, and extent of protected areas (PA) within the TCU*

TCU	THT	TCU Name	Level	Rank score	TCU area (km²)	PA area (km²)
6	AGD	Chitwan-Parsa-Valmiki	I	57	3549	2075
4	AGD	Bardia-Banke	I	54	2231	1437
1	AGD	Rajaji-Corbett	I	48	4357	1498
3	AGD	Dudhwa-Kailali	II	36	567	567
2	AGD	Sukla Phanta-Kishanpur	II	34	1897	439
11	AGD	Dibru	III	30	609	491
5	AGD	Dang-churia	III	28	4277	0
9	AGD	Mahananda	III	22	472	0
8	AGD	Trijuga-Koshi-Tappu	III	21	258	135
12	AGD	Orang	III	12	115	115
13	AGD	Laokhowa	III	12	170	170
18	MAN	Sundarbans	I	65	6624	873
10	SUF	Manas-Namdapha	I	55	59901	13181
7	SUF	Bara	III	30	230	0
27	TDF	Bagdara-Hazaribagh	I	55	61172	6042
52	TDF	Nagarajunasagar-Srisailam	I	51	13127	3862
47	TDF	Sitapani-Udanti	II	32	5743	1521
40	TDF	Kanha-Indravati Corridor	II	32	1377	0
43	TDF	Orissa Dry Forests	II	32	6763	0
56	TDF	Cauvery	III	29	7243	1003
44	TDF	Tadoba-Andhari	III	29	4387	578
53	TDF	Central Andhra Pradesh Dry Forests	III	25	2439	0
20	TDF	Ranthambhore	III	25	2494	338
54	TDF	South Andhra Pradesh Dry Forests	III	25	5495	0
42	TDF	East Madhya Pradesh Dry Forests	III	21	1303	0
21	TDF	Gandhi Sagar	III	19	5709	884
22	TDF	Madhar-Palpur Kuno	III	19	18170	1029
36	TDF	Bihar Dry Forests, block 1	III	17	1491	0
57	TDF	Tamil Nadu Dry Forests, block 1	III	17	899	0
23	TDF	Central Madhya Pradesh Dry Forests, block 1	III	16	505	0
38	TDF	Bihar Dry Forests, block 2	III	13	1409	82
25	TDF	South Uttar Pradesh Dry Forests	III	9	324	0
26	TDF	Central Madhya Pradesh Dry Forests, block 2	III	9	433	0
19	TDF	Sariska	III	9	1023	549
24	TDF	Panna-Son Gharial	S		8599	733
28	TDF	Melghat	S		25287	2415
29	TDF	Ratapani-Singhori	S		14089	406
31	TMD	Kanha-Pench	I	54	13223	2138
39	TMD	Simlipal-Kotgarh	I	46	7709	2980
46	TMD	Indravati-Navegaon	II	42	31413	4678
51	TMD	Papikonda	II	39	7293	770
14	TMD	Siju-Balphakaram	III	25	2677	37
45	TMD	Kawal	III	25	4307	1057
48	TMD	Satkosia Gorge-Kotgarh	III	25	11550	1169
37	TMD	East Madhya Pradesh TMD Forests, block 1	III	25	2893	0
15	TMD	Barail	III	25	1622	78
41	TMD	Barnawapara	III	20	1074	239

Table 18.3 (*cont.*)

TCU	THT	TCU Name	Level	Rank score	TCU area (km²)	PA area (km²)
32	TMD	East Madhya Pradesh TMD Forests, block 2	III	17	824	0
34	TMD	East Madhya Pradesh TMD Forests, block 3	III	9	214	0
35	TMD	Badalkhol	III	9	922	102
33	TMD	East Madhya Pradesh TMD Forests, block 4	III	9	408	0
49	TMD	Sabari-Sileru	III	9	377	0
50	TMD	Godavari	III	9	513	0
30	TMD	Central Madhya Pradesh TMD Forests, block 1	III	8	989	0
16	TMD	Kaziranga-Meghalaya	S		18984	488
59	TMF	Periyar-Kalakad-Mundunthurai	I	55	5440	1493
55	TMF	Dandeli-Bandipur	I	55	23881	7013
58	TMF	Parambikulam	II	29	2349	336
17	TMF	Pablakhali	III	21	1664	201

The THT acronyms are: TMF = tropical moist evergreen forests; TMD = tropical moist deciduous forests; TDF = tropical dry forests; SUF = subtropical and temperate upland forests; MAN = mangroves; AGD = alluvial grasslands and subtropical moist deciduous forests.

forest cover, nor the status of tigers in South China for the analysis, but suggest immediate surveys in this bioregion.

We also identified 16 TCUs (10%) as Immediate Surveys. The potential for long-term tiger conservation in these TCUs may be on a par with some Level I or II TCUs, but we lack data to confidently rank them. We urge surveys in these TCUs immediately to better rank these TCUs and to better determine their contribution to a regional tiger conservation strategy.

Some of the Level I, II and Immediate Survey TCUs we have identified are extremely large in size (e.g. TCU 73 in Indochina, Fig. 18.3). Ground surveys may well break these big units into smaller TCUs, but, using our criteria for habitat integrity (see Appendix 4A), any fragment of good habitat larger than 5000 km² would automatically receive the same score as the existing, much larger, TCU. Furthermore, according to our definition of a TCU, areas that are even narrowly connected should still be considered as a single TCU if at least one tiger per generation successfully disperses to and breeds in an adjacent block of habitat. If there is tiger dispersal across these large TCUs, they represent

the last remaining places in Asia where there is still an opportunity to plan land-use options based on landscape conservation principles before these large forest tracts succumb to fragmentation. We therefore prefer, at this juncture, to present these large TCUs to emphasise the large spatial requirements for tigers and to urge that any land-use options in these areas be planned with tiger conservation in mind (i.e. establishing a network of core areas and corridors, encouraging eco-tourism in restored buffer zones as a revenue source, using tigers as umbrella species for conservation plans, etc.).

Indian subcontinent bioregion

Priorities within tiger habitat types

We identified 11 Level I TCUs and seven Level II TCUs among the six tiger habitat types (Table 18.3, Fig. 18.2) in the Indian subcontinent. Three of the six tiger habitat types have adequate replication of a combination of three Level I or II TCUs: alluvial grasslands, tropical moist deciduous forests, and tropical dry forests. In tropical moist evergreen forests, we elevated to Level II a Level III TCU (58 Parambikulam) to achieve adequate representation.

There was only one TCU each in the mangroves and subtropical upland forests; thus, increased representation was not possible in these tiger habitat types.

The alluvial grasslands /subtropical moist deciduous forests tiger habitat type is represented by three Level I and two Level II TCUs (Table 18.3). This tiger habitat type once covered a huge swath of grasslands and riverine and moist semi-deciduous forests along the major river systems of the Gangetic and Brahmaputra plains, but has now been largely converted to agriculture or severely degraded. Today the best examples of this tiger habitat type are limited to a few blocks of habitat at the base of the outer foothills of the Himalaya. Tiger densities are high, in part a response to the extraordinary biomass of ungulates that constitute the tiger's prey base. The Chitwan-Parsa-Valmiki and Bardia-Banke TCUs are prime examples of this habitat. Kaziranga, which is part of the Kaziranga-Meghalaya TCU (16), also contains some prime alluvial grassland habitat, but was placed in the tropical moist deciduous forest habitat types because most of the TCU extends into the Assam Hills. We suggest elevation of the Kaziranga-Meghalaya TCU to Level I, with necessary TCU modifications if ground surveys warrant such a decision.

The best example of the mangroves tiger habitat type in this bioregion (and one of the best in the world) is the Sundarbans TCU (18). Although not often thought of as typical tiger habitat, tigers survive in this area, swimming between islands in the delta to hunt prey. The Sundarbans is under increasing pressure from woodcutters. Because extensive tracts of mangroves have largely disappeared or tigers disappeared from them across tropical Asia, the Sundarbans TCU stands as the last best chance to conserve mangrove ecosystems containing tigers.

The subtropical and temperate upland forests tiger habitat type is represented by one Level I TCU (Table 18.3). This large TCU (10), which extends from the Manas reserve complex in India and Bhutan to the temperate hardwood and conifer forests in north-eastern India, represents the best chance to conserve tigers that live close to the timberline and prey on montane ungulates, and also offers the last opportunity to conserve an intact example of temperate Himalayan forests. This TCU allows for tiger movement from the alluvial grasslands to forests bordering on alpine areas; thus protection of the slivers of land that contain alluvial grasslands on the lowland edges of the TCU is also critical. The steep terrain and heavy precipitation over much of the TCU precludes large-scale development in core areas, but the low-lying grasslands are susceptible to anthropogenic changes. The conservation of these altitudinal corridors is also vital to many other species of vertebrates that move seasonally along elevational zones.

The tropical dry forests are represented by two Level I, three Level II and three Immediate Survey TCUs (Table 18.3). This tiger habitat type, which includes thorn scrub, once covered large tracts of central and western India. Although much of it is now reduced, some of the larger blocks of habitat remaining in India are in this tiger habitat type. Tigers are distributed across the extreme edges of the dry forests and thorn scrub, from Sariska and Ranthambhore in Rajasthan, across to Bihar, and south to the Nagarjuna Sagar TCU.

Tropical moist deciduous forests are represented by two Level I, two Level II and one Immediate Survey TCUs. These are probably some of the most productive habitats for tigers and their prey in the subcontinent. The extensive loss of this habitat type across India calls for vigilant protection of the remaining blocks. The forests of this tiger habitat type and the species assemblages they support are similar to subtropical moist semi-deciduous forests but are not driven by the dynamics of flooding as are the subtropical forests to the north. This tiger habitat type contains some of the most famous tiger reserves in Asia, most notably Kanha National Park.

Tropical moist evergreen forests are represented by two Level I TCUs, one level III, and one Level III which was elevated to Level II (Table 18.3). This tiger habitat type represents one of the less common tiger habitat types in the Indian subcontinent, being largely limited to the upland areas and wetter parts of the Western Ghats. Tigers once ranged along the entire range of the Western Ghats, but the fragments north of the 16° Parallel no longer contain tigers. Tropical moist evergreen forests support a different prey base than moist deciduous forests, and support lower densities of tigers. Alternatively, tigers in moist evergreen forests serve as umbrella species for an area of the subcontinent with high endemism of plants, small vertebrates and invertebrates.

Landscape features and degree of protection
The mean size (±SD) of all TCUs in this bioregion was 7035±12 060 km², and the mean area of priority TCUs, (i.e. Level I, II and Immediate Surveys) was 14 692±16 604 km². Eight of the priority TCUs were >10 000 km², including two that were >50 000 km² (Table 18.3).

Of the 59 TCUs in this bioregion, 20 did not contain any protected areas (IUCN categories I–IV). The mean size of protected areas for all TCUs that had any protected areas was 1621±2458 km², representing 17% of the TCU area. Only two of the 22 priority TCUs did not contain strict protected areas (Table 18.3). The mean area protected within the priority TCUs was 2482±2984 km².

Indochina bioregion
We identified 10 Level I and eight Level II TCUs among the five tiger habitat types in this bioregion (Table 18.4, Fig. 18.3). Three of the five tiger habitat types have the minimum number of TCUs to achieve adequate representation. Of the other two tiger habitat types, the subtropical and temperate upland forests tiger habitat type is represented by two Level II TCUs, whereas the mangrove tiger habitat type does not have any at Levels I and II.

Priorities within tiger habitat types
Little of the once-extensive mangroves of the Irrawaddy and Mekong deltas remains. Thus, it may be impossible to achieve good representation of tiger populations in the mangrove tiger habitat type in the Indochina bioregion. We identified three mangrove TCUs, but none achieve a Level I or II (Table 18.4). Thus, we encourage immediate surveys in two – TCUs 71 and 128 – to determine if they can be classified as Level I or II TCUs.

The subtropical and temperate upland forests in this bioregion are represented by two Level II TCUs, but three other large TCUs have been identified for immediate surveys (Table 18.4), and elevation to Level I or Level II as deemed appropriate. Although these forests along the southern slopes of the Himalaya are not prime habitat for tigers compared with the lowland forests because of the different ecological role of tigers in these montane ecosystems and the different prey assemblages, these habitats have to be represented in a portfolio of conservation areas.

The tropical dry forest tiger habitat type is represented by three Level I and two Level II TCUs (Table 18.4). These forests that once covered much of central and eastern Thailand and parts of central Myanmar have now been degraded or cleared. But fairly extensive areas of tropical dry forest still exist in Cambodia with the two large Level I TCUs covering much of this country; however, with the granting of extensive logging concessions to foreign corporations these forests are now under severe threat. One TCU (113) is particularly significant to the recent trans-boundary conservation and protected areas management initiatives underway in the region, as it extends over Cambodia, Lao PDR and Vietnam. This TCU can be used to develop a truly regional conservation model with multinational collaboration on conservation issues.

The tropical moist deciduous forests are represented by two Level I and one Level II TCU (Table 18.4). These forests were extensive throughout much of central Myanmar, with smaller tracts of forests in northern Thailand and Lao PDR. Today, only fragments of these forests remain. The Chin Hills and Pegu Yoma in Myanmar now represent the best opportunities for conservation of this habitat.

The tropical moist evergreen forests tiger habitat type is represented by five Level I, three Level II, and one Immediate Survey TCUs. Several longitudinal bands of these forests once extended throughout the Indochina bioregion. Although now fragmented, relatively large extents of this tiger habitat type still exist, especially along the Arakan Yomas in western Myanmar, the Myanmar-Thailand border, the

Table 18.4. *TCUs for the Indochina bioregion, with tiger habitat type (THT), priority category level, rank scores, TCU area, and extent of protected areas (PA) within the TCU*

TCU	THT	TCU Name	Level	Rank score	TCU area (km²)	PA area (km²)
126	MAN	Namtok Phlui	III	16	449	89
71	MAN	Irrawaddy Delta	S		1733	122
128	MAN	Hat Chao Mai	S		1971	226
64	SUF	Maymo	II	42	1114	28
95	SUF	Bu Huong-Nam Xam	II	32	1121	316
75	SUF	North Indochina SUF, block 1	III	30	8984	0
80	SUF	North Indochina SUF, block 2	III	28	1017	94
92	SUF	North Indochina SUF, block 3	III	28	581	0
97	SUF	Northeast Indochina Montane Forests, block 1	III	20	512	0
65	SUF	North Central Myanmar Upland Forest, block 1	III	16	287	0
84	SUF	North Indochina SUF, block 4	III	16	496	0
66	SUF	North Central Myanmar Upland Forest, block 2	III	16	541	0
91	SUF	North Indochina SUF, block 5	III	15	137	0
88	SUF	North Indochina SUF, block 6	III	15	145	0
69	SUF	North Central Myanmar Upland Forest, block 3	III	15	152	0
67	SUF	North Central Myanmar Upland Forest, block 4	III	12	72	0
68	SUF	North Central Myanmar Upland Forest, block 5	III	12	33	0
60	SUF	Northern Triangle	S		33884	109
90	SUF	Nui Hoang Lien	S		1550	395
63	SUF	Shan Plateau	S		41075	153
113	TDF	Virachay-Xe Piane-Yok Don	I	54	52643	13897
101	TDF	Phu Khieo-Nam Nao	I	52	5702	3325
122	TDF	Kulen Promtep-Thap Lan	I	51	45880	11311
102	TDF	Phu Pha Man	II	42	4384	716
100	TDF	Phou Khao-Phu Kham	II	32	3579	299
79	TDF	Ramkhamhaeng	III	28	3784	409
127	TDF	Khao Luang-Khao Banthat	III	26	5641	2886
120	TDF	Nam Bai Cat Tien	III	25	7930	1599
103	TDF	Central Laos TDF	III	20	473	0
78	TDF	North Thailand TDF, block 1	III	17	159	0
76	TDF	North Thailand TDF, block 2	III	17	110	0
77	TDF	North Thailand TDF, block 3	III	17	212	0
118	TDF	Dalat Plateau, block 1	III	16	325	0
109	TDF	Bolovens Plateau, block 1	III	16	305	0
61	TMD	Chin Hills	I	66	82464	3078
72	TMD	Pegu Yoma	I	64	12600	4600
81	TMD	Thung Salaeng-Nam Poui	II	32	13823	2977
116	TMD	Dalat Plateau, block 2	III	20	982	0
94	TMD	Northeast Indochina Montane Forests, block 2	III	20	475	0
96	TMD	Northeast Indochina Montane Forests, block 3	III	16	246	0
117	TMD	Dalat Plateau, block 2 (?3)	III	16	341	0
70	TMD	Irrawaddy Plain	III	12	325	0
82	TMD	Luangprabang	S		1067	0
83	TMD	Muang Xaignabouri	S		689	0
73	TMF	Huai Kha Khaeng-Thung Yai Naresuan	I	58	155829	32459
62	TMF	Arakan Yomas	I	57	52353	446
121	TMF	Khao Yai	I	56	1945	1852

Table 18.4 (*cont.*)

TCU	THT	TCU Name	Level	Rank score	TCU area (km²)	PA area (km²)
125	TMF	Phnom Bokor-Aural	I	51	31715	10833
99	TMF	Nakai Nam Theu -Vu Quang	I	46	24626	6830
107	TMF	Bach Ma-Nui Thanh	II	38	2081	188
104	TMF	Xe Bang Nouan	II	32	2196	1260
93	TMF	Song Da Forest	II	32	1079	0
114	TMF	Kon Kai Kinh	III	30	1138	221
98	TMF	Phou Khao Khouay	III	30	1924	1493
124	TMF	Southeast Thailand TMF	III	28	678	745
110	TMF	Kontum TMF, block 1	III	28	1726	0
108	TMF	Bolovens Plateau, block 2	III	23	1129	0
106	TMF	Annamite Range, block 1	III	20	441	0
111	TMF	Ngoc Linh	III	19	1486	201
105	TMF	Annamite Range, block 2	III	16	442	0
74	TMF	East Myanmar TMF	III	16	325	0
112	TMF	Kontum TMF, block 2	III	16	275	0
119	TMF	Dalat, block 3	III	16	404	33
89	TMF	North Indochina SUF, block 7	III	16	248	0
86	TMF	North Indochina SUF, block 8	III	15	165	0
87	TMF	North Indochina SUF, block 9	III	15	354	0
85	TMF	North Indochina SUF, block 10	III	15	153	0
115	TMF	Dalat Plateau, block 4	III	15	150	0
123	TMF	Khao Ang Ru Nai-Khao Soi Dao	S		4751	1798

The THT acronyms are: TMF = tropical moist evergreen forests; TMD = tropical moist deciduous forests; TDF = tropical dry forests; SUF = subtropical and temperate upland; MAN = mangroves.

Lao-Vietnamese border, and in south-western Cambodia. Many of these forests lie along national boundaries, and trans-boundary conservation measures are needed if they are to retain their integrity as TCUs.

Landscape features and degree of protection

The mean size of all TCUs in the Indochina bioregion is 9906±23 502 km², and the mean size of the priority TCUs was 22 379±34 301 km². There were 11 TCUs that were >10 000 km², including three that were >50 000 km² and one that was >150 000 km² (Table 18.4).

Thirty-six of the 69 TCUs in the Indochina bioregion did not contain any strict protected areas. The mean size of protected areas in those TCUs which had any protected areas was 3 182±6225 km², or 17%

of the TCU area. Three of the 26 priority TCUs did not contain strictly protected areas (Table 18.4), and the mean area protected within the priority TCUs was 3884±6892 km².

Southeast Asia bioregion

Southeast Asia covers a much larger region than Peninsular Malaysia and Sumatra, but, with the loss of tiger populations in Java and Bali, these are the only two geographic areas that still contain tigers. Much of the lowlands throughout Peninsular Malaysia and Sumatra that were once prime tiger habitat have been lost. What remains is often peat swamp forest, which harbours interesting biodiversity but is poor tiger habitat, or upland evergreen forests. Because other forest types (mangroves, tropical moist deciduous forests) are

Table 18.5. *TCUs for the Southeast Asia bioregion, with tiger habitat type (THT), priority category level, rank scores, TCU area, and extent of protected areas (PA) within the TCU*

TCU	THT	TCU Name	Level	Rank score	TCU area (km²)	PA area (km²)
148	TMF	Kerinci Seblat-Seberida	I	60	50884	16605
145	TMF	Gunung Leuser-Lingga Isaq	I	60	36530	11423
129	TMF	Taman Negara-Belum-Halabala	I	56	27469	7135
158	TMF	Bukit Barisan Selatan-Bukit Hitam	I	48	6594	4784
130	TMF	Selama	II	38	1684	0
159	TMF	Way Kambas	II	34	1300	1300
147	TMF	Siak Kecil-Padang Lawas	II	34	2235	1995
150	TMF	Kerumutan-Istana Sultan Siak	II	34	11816	1742
152	TMF	Berbak-Sembilang	II	34	6671	2196
139	TMF	Endau-Rompin	II	32	788	0
140	TMF	Endau North	III	30	496	0
133	TMF	Sungai Dusun	III	27	428	0
142	TMF	Endau-Kota Tinggi	III	23	1108	46
153	TMF	South Central Sumatra TMF	III	18	444	0
156	TMF	Benakat	III	18	599	28
157	TMF	Bukit Raja Mandara-Gumai Pasemah	III	17	154	133
149	TMF	East Central Sumatra TMF	III	17	112	0
136	TMF	Tasek Bera	III	16	348	1
134	TMF	Phang, block 1	III	15	242	0
143	TMF	Johor Coast, block 1	III	14	84	0
135	TMF	Phang, block 2	III	14	364	0
137	TMF	Negeri Sembilan, block 1	III	14	382	0
144	TMF	Johor Coast, block 2	III	14	43	0
141	TMF	Mersing	III	14	190	89
138	TMF	Negeri Sembilan, block 2	III	13	147	0
131	TMF	Perak, block 1	III	13	181	0
132	TMF	Perak, block 2	III	13	150	0
155	TMF	Padang Sugihan	S		2505	652
151	TMF	Air Sawan	S		2444	605
146	TMF	Sibolga-Dolok Surungan	S		4685	594
154	TMF	Dangku	S		3431	106

The THT acronym is TMF = tropical moist evergreen forests.

so limited, we classified all remaining habitat as tropical moist evergreen forests. Some of the TCUs are quite extensive in this bioregion but, in general, tropical moist evergreen forests tend to be poorer quality habitat for tigers than other forest types. We identified four Level I and six Level II TCUs in this bioregion (Table 18.5, Fig. 18.4). Of these, two Level II and one Immediate Survey TCUs – 150, 152, and 155 – consist mostly of peat swamp. Although these three TCUs are large the habitat integrity was scored in the >50% degraded category. Tigers are becoming

increasingly restricted to these upland tropical moist evergreen forests because the lowlands are being converted to oil palm plantations and other forms of land use.

Landscape features and degree of protection
The mean size of TCUs in the Southeast Asia bioregion is 4654±11 505 km². The mean size of the priority TCUs was 11 360±15 013 km². There were four TCUs that were >10 000 km², including one that was >50 000 km² (Table 18.5).

Fourteen of the 31 TCUs in this bioregion did not contain any strictly protected areas. The mean size of protected areas in those TCUs which had any protected areas was 2908±4544 km², or 36% of the TCU area. Two of the 14 priority TCUs did not contain strictly protected areas (Table 18.5). The mean area protected within the priority TCUs was 3510±4798 km².

Russian Far East bioregion

With the exception of two smaller blocks of habitat on the Russian-Chinese border, the rest of the Russian Far East remains a single Level I TCU and is the only unit in its bioregion. It is the only area containing tigers where the connectivity still exists for tigers to move across a large landscape. Detailed plans for tiger conservation are available for this bioregion (WWF 1995; D.G. Miquelle *et al.* this volume Chapter 19).

South China bioregion

The lack of knowledge regarding the status of free-ranging tigers in the central and southern China bioregion hinders conservation efforts. Koehler (1991) provides a status report of tigers in South China. We recommend that surveys to gather information begin immediately.

Discussion

Where to invest first: the representative portfolio

Conservation biology has often been referred to as a crisis discipline. Practitioners are required to set priorities that will make the best use of scarce resources while relying at times on incomplete data. Determining where to invest first to save tigers is a classic example of working in a crisis mode. In the ideal world, we would be able to wait until all 162 TCUs were systematically surveyed regarding the status of tiger habitat, the natural dynamics of tiger and prey populations, poaching pressures, and the severity of other threats operating across TCUs. The recent crisis of poaching across the range of the tiger precludes the luxury of waiting until all of these data are collected and analysed.

Does the framework we have presented help us make hard choices? Using our hierarchical method, we were able to prioritise among the potential tiger conservation areas across Asia, and identify just 16% as Level I and 13% as Level II TCUs, representing all tiger habitats in the three bioregions (Tables 18.3–18.5). Sixteen other TCUs have been identified for immediate surveys to confirm tiger presence and population trends, verify poaching pressure, or to ground-truth habitat integrity. Because tiger habitat is a limiting resource, we urge that surveys in these TCUs be conducted immediately since these have the potential to become Level I or II TCUs. However, even though this is a broad-scale, rapid assessment, we are confident of our ability to designate TCUs as Level I and II, and feel that even after extensive field surveys of the TCUs, few, if any, of the Level III TCUs will be elevated to Level I or II.

Thus, the framework allowed us to objectively identify tiger areas with the greatest conservation potential, where investments in tiger conservation will have the most effect. But we offer two caveats. First, the analysis represents a regional priority-setting exercise, and is not meant to be used as a local or site-specific conservation planning tool. Secondly, while we present the Level I and Level II TCUs as worthy of global or regional priority areas, some Level III TCUs may be important for national or local-level planning, and this analysis is not meant to replace these initiatives.

We highlight several important findings from this analysis:

Strict protected areas in Asia are relatively small (Dinerstein & Wikramanayake 1993), and typically cover only a small fraction of TCUs. Some TCUs do not have any protected areas. But protected areas provide essential refuges for tigers (Karanth 1991, 1995) and may serve as the essential breeding populations from which other, more degraded, parts of the TCU can be repopulated. All high-priority TCUs should have strict protected areas, so an obvious target for conservation donors is to help finance the creation and recurrent costs of new or expanded protected areas.

The indices used in the analysis also emphasise

the need for a landscape approach to tiger conservation that extends beyond the boundaries of
protected areas and the importance of prey species
for tiger conservation (Seidensticker 1986; Karanth
1991; Rabinowitz 1993). Several Level I and II TCUs
are very large and do not, and likely will not, receive
complete protection. An essential part of the landscape-level planning in these TCUs would be to
create a linked core protected areas system, with
buffer zones, set in an overall matrix of other land-
use options. By highlighting these large areas as
significant tiger conservation areas we emphasise
the need for such appropriate land-use planning
before these large habitat blocks become fragmented, thus losing much of their conservation
potential. This goal is likely to require increased
co-operation among multiple sectors of national and
state governments, and its pursuit will promote
better land-use planning.

We have identified some Level I and II TCUs that
are very large, exceeding 50 000 km² in area.
Ground-truthing surveys may break these up into
smaller TCUs but, using our criteria for habitat
integrity (see Appendix 4A), any fragment of good
habitat larger than 5000 km² would receive the
same score as the existing, larger TCU. According to
our definition of a TCU, areas that are even narrowly
connected to allow for tiger dispersal should still be
considered a single TCU, and the maps that we used
to demarcate the TCUs indicate that such connectivity is present. These large TCUs will not receive
complete protection, but if there is tiger dispersal
among adjacent blocks of habitat they represent the
last remaining places where there is still an opportunity to plan land-use before these large forest
tracts succumb to fragmentation.

Protected areas typically cover only a small part of
the large TCUs, and tiger conservation efforts
should move beyond their boundaries. The tiger can
be used as a key species to design landscape-level
land-use options that include conservation, restoration of degraded lands and sustainable natural
resource use plans to meet the needs of the local
people (Panwar 1987). Several initiatives in which
conservation of tigers directly or indirectly benefits
local people through integrated conservation and

development programmes are already underway in
the region, especially in India (Panwar 1987) and
Nepal (Mishra *et al.* 1987; E. Dinerstein *et al.* this
volume) and can be used as models to adapt and
replicate in other areas.

The analysis also indicates that many Level I and
II TCUs straddle international borders (Figs. 18.2,
18.3). Over the past few years, there have been
several fora where trans-boundary conservation
strategies have been discussed (Anon. 1995a, b,
1996a; Dinerstein *et al.* 1995; Ji & Rabinowitz
1995; Rabinowitz 1995a; Dillon & Wikramanayake
1997). A subregional biodiversity conservation
strategy (Sub-regional Biodiversity Forum Project,
RAS/93/102) for Indochina (Cambodia, Lao PDR,
Thailand and Vietnam) is now being coordinated
by WWF's Indochina programme. A similar programme for the Himalayas extending across India,
China, Nepal, Bhutan and Myanmar is also being
planned by UNDP/GEF (Himalayan Regional
Biodiversity Conservation Initiative; proposed). The
tiger can be used as a 'flagship' species to promote
such strategies (see Rabinowitz 1995a), and as an
umbrella species to plan such strategies, and organisations such as the Global Tiger Forum should give
priority to these regional conservation initiatives.

Finally, our approach provides a viable alternative to the taxonomy-based approach that uses
putative subspecies to maintain representation.
Tiger subspecies are based on weak evidence
(Hemmer 1987; Herrington 1987; A. C. Kitchener
this volume; J. Wentzel *et al.* this volume). We feel
that an ecology-based approach is a stronger, more
powerful way to maintain representation of ecological phenomena associated with tigers as the top
predators in a variety of ecosystems.

Overall, this assessment provides new perspectives relevant to tiger conservation. It provides a
viable, ecology-based method for selecting areas for
tiger conservation, while maintaining genetic,
demographic, behavioural and ecological representation of tigers across their range within the
portfolio of priority sites. We also encourage the
adoption of this approach as an alternative to the
taxonomy-based method for other wide-ranging,
endangered species.

19

A habitat protection plan for the Amur tiger: developing political and ecological criteria for a viable land-use plan

Dale G. Miquelle, Troy W. Merrill, Yuri M. Dunishenko,
Evgeny N. Smirnov, Howard B. Quigley, Dimitriy G. Pikunov
and Maurice G. Hornocker

Introduction

As the twentieth century comes to a close, conservation of the remaining five subspecies of tigers is one of the greatest challenges facing range countries and the world conservation community. Tigers have an inherent charisma and value to human societies (P. Jackson this volume), and because, like other large carnivores, viable populations of tigers require large tracts of land, conservation of tigers can provide a means for conserving native biodiversity in many range states (Noss *et al.* 1996). However, numerous challenges to the survival of tigers exist in the face of ever greater pressures from an increasing human population for decreasing land resources. Anti-poaching programmes and management of the prey base will be critical short-term conservation strategies (S. R. Galster & K. V. Eliot this volume; K. U. Karanth & B. M. Stith this volume; D. G. Miquelle *et al.* this volume chapter 6; M. Sunquist *et al.* this volume); but, long-term, the ultimate threat is loss of the large, intact natural ecosystems upon which tigers depend for survival. Tigers can co-exist with people and have for a long time survived in proximity to Man, but co-existence usually occurs along boundary zones between areas of low human disturbance (i.e. tiger habitat) and landscapes severely impacted by Man. For the tiger to survive, dramatic steps must be taken to secure large blocks of land where tigers can live in natural conditions, minimising the environmental and genetic challenges generally faced by small populations, and, as much as possible, retaining the

evolutionary forces that have shaped each of the subspecies.

The Amur tiger is presently threatened with extinction in the wild. Formerly found throughout the southern Russian Far East (primarily within the Amur River Basin) to the Sea of Japan, northeast China and the Korean Peninsula, its range has collapsed to its eastern perimeter, with the only remaining viable population now occurring in Primorski and southern Khabarovski Krais (Provinces) in the Russian Far East. The last census of tigers in this region, carried out in 1996, estimated that 330–371 adult animals range between the two Krais, and may have some contact with a remnant population in Heilongjiang and Jilin Provinces in northeast China (Matyushkin *et al.* 1996). Dramatic increases in poaching activity have had significant impacts on this population (Galster *et al.* 1996). Therefore, immediate actions must be taken to ensure that this subspecies is not lost from the wild.

On the 7th of August 1995, Prime Minister Chernomyrdin signed decree no. 795 of the Government of the Russian Federation 'for conservation of the Amur tiger and other rare and endangered species of wild fauna and flora in the territories of Primorski and Khabarovski Krais'. This decree called for, among other things, the development of a national strategy for tiger conservation in the Russian Federation. This recently completed strategy identifies many management needs, including anti-poaching, environmental education, policy decisions, management of prey base, minimisation of conflicts between people and tigers, and development of a habitat protection plan (State Committee of the Russian Federation for the Protection of the Environment 1996).

We focus on this last component: development of a habitat protection plan for the Amur tiger in the Russian Far East. Our efforts are based on a long-term, intensive research project studying the ecology of the Amur tiger (Miquelle *et al.* 1993, 1995, 1996b; E. N. Smirnov & D. G. Miquelle this volume) and our understanding of the present political situation surrounding environmental protection and tiger conservation in the Russian Far East (e.g. Yelyakov *et al.* 1993; Newell & Wilson 1996). We develop a set of goals and key components for population and habitat protection, establish both political and biological criteria to guide development of the plan, and then present a stepwise, incremental process to reach pre-established goals. We primarily focus on presentation of a core network of protected areas, but emphasise that management of non-protected lands will be just as essential to survival of the Amur tiger. We do not present this process as a blueprint for all tiger range countries, but there may be components applicable to other areas.

This work was made possible with support from The National Geographic Society, The Exxon Corporation, The National Fish and Wildlife Foundation, The Save the Tiger Fund, The National Wildlife Federation and The Wildlife Conservation Society. Although this work does not necessarily represent their views, these proposals and our work would not be possible without the knowledge, input and friendship provided by our colleagues, including: V. V. Aramilev, J. Goodrich, L. Kerley, A. N. Kulikov, I. G. Nikolaev and G. P. Salkina. Our field team in Sikhote-Alin, including B. Schleyer, I. G. Nikolaev, N. N. Reebin, and A. V. Kostirya, deserves special recognition for their dedication to the Siberian Tiger Project; all our efforts are ultimately dependent on their devotion to the daily grind of field work. Work in Sikhote-Alin Zapovednik would not be possible without the co-operative support of Director A. A. Astafiev, and Science Coordinator M. N. Gromyko. The text benefited substantially from comments from J. Goodrich and L. Kerley.

Study area

Our 266 349 km² planning region included all of Primorski Krai (165 900 km²) and the southern seven Raions (Districts) of Khabarovski Krai (Bikinski, Vyazemski, Lazo, Khabarovski, Sovganski, Nanaiski and Komsomolski Raions). This region, situated between 42° and 50° 5' latitude in the southeastern corner of the Russian Far East, includes all administrative units of Russia that have a resident population of tigers (Matyushkin *et al.* 1996) (Fig. 19.1). The area can be divided into four major ecoregions, the first of which is the Sikhote-Alin Mountains, a coastal range that parallels the coast of the Sea of Japan from Vladivostok 1000 km north into Khabarovski Krai and comprises the majority of our planning region. The second eco-region, the Amur River Basin and its tributary, the Ussuri River Basin, drains the Sikhote-Alin Mountains on the inland side, forms the border with China, and is a lowland area that has been subjected to the most intensive human development in the region. Lake Khanka, the third ecoregion, includes a massive shallow lake, but most of the surrounding steppe lands have been converted to agricultural fields. The fourth ecoregion, southwest Primorski Krai, is a narrow strip of land that borders China to the west and North Korea to the south and represents the eastern edge of the East Manchurian mountain system.

All these regions were formerly inhabited by tigers, but the highest densities probably occurred in regions now most severely impacted by Man (Baikov 1925; Kucherenko 1985). Today, due to human disturbance in the Amur Basin and Khanka regions, approximately 95% of the Amur tigers in Russia occur in the Sikhote-Alin ecosystem, and the remaining 5% in the East Manchurian Mountains in southwest Primorski Krai (Matyushkin *et al.* 1996). These two regions, formerly connected, have become separated due to a corridor of human development from Vladivostok to the city of Ussuriysk. The southwest population has become further fragmented into two components (Matyushkin *et al.* 1996).

Tiger habitat in Sikhote-Alin and the East

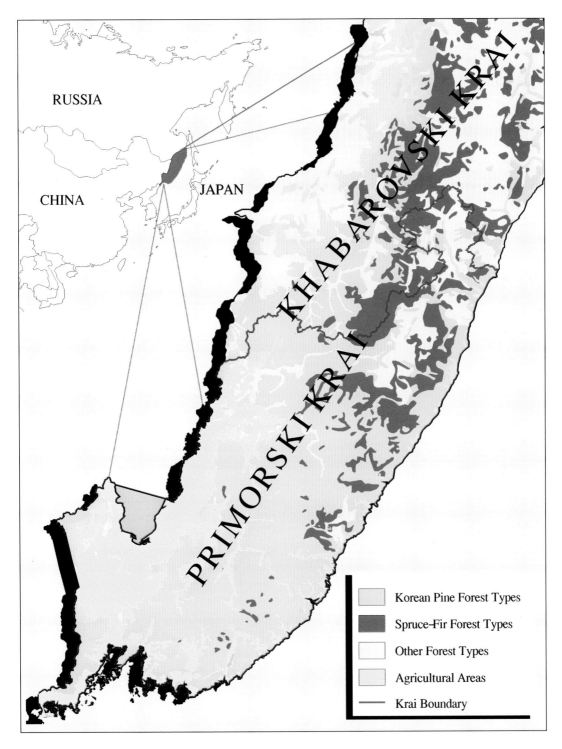

FIGURE 19.1
Study area, showing administrative boundaries of southern Khabarovski and Primorski Krais, Russian Far East, and dominant vegetative cover of the region (from Grebovoy *et al.* 1968). Potential tiger habitat is identified by suitable vegetative covers (see text) on Forest Service lands and zapovedniks.

Manchurian Mountains is nearly completely covered by forests. Mountains are relatively low (the highest peak is 2004 m), and are forested with a combination of conifer and broad-leafed species. Korean pine/broad-leaf forests were the defining forest community of this region, but the large majority of these forests have been selectively logged at various times in the past. A history of logging and fire has converted much of the region to secondary broad-leaf forests, especially oak and birch. Above 700–800 m, spruce-fir forests prevail in central Sikhote-Alin. This elevational boundary for a predominantly coniferous forest type increases to the south, and decreases northward until, at 47°20' latitude, coniferous forests occur at sea level.

Methods and results

We delineated the following steps as a process for developing a land-use plan for tigers in the Russian Far East:

1   Define the planning process:
    a.   Define the goal;
    b.   Define key components of the plan;
    c.   Determine political criteria to guide the planning process;
    d.   Determine biological criteria to define the adequacy of conservation units.
2   Implement the planning process:
    a.   Assess adequacy of existing protected areas;
    b.   Include regional and federal proposals for new protected areas that already have some legislative precedent;
    c.   Include multiple-use areas – proposed or existing – where tiger management will be a primary consideration;
    d.   Propose new protected areas if necessary;
    e.   Ensure connectivity;
    f.   Identify unprotected areas for inclusive tiger management plan.
3   Assess the potential effectiveness of the plan:

The following sections review these processes in detail.

Define the planning process

For a habitat protection plan to be successful, we propose that it should have well-defined goals, that key components must be delineated from the start, and that criteria must be delineated to guide the process. Furthermore, it must be politically viable, i.e. there must be a reasonable chance of such a plan being implemented, and it must be biologically sound, i.e. it must be based on an understanding of key ecological parameters of tiger biology that are relevant to land-use planning and population dynamics. We review definition of our planning goal, define key components and determine the criteria we employed as a basis for developing this plan.

Define the goal

We defined the goal of this habitat protection plan as protection of all existing tiger habitat, i.e. no further loss of Amur tiger habitat should occur. We focused on managing the Russian Far East Amur tiger population as the last viable wild population of this subspecies (see 'Determine political criteria to guide the planning process political criteria' below). However, we focused on habitat protection instead of population management as the key mechanism to long-term survival. Definition of conservation units of land is 'more concrete', more specific and less politically and biologically debatable than definition of population goals. We did not attempt to define a target tiger population for the Russia Far East because:

1. Existing estimates were of unknown accuracy, and surveys in general have low resolution. When we initially developed this plan in 1995 (Miquelle *et al.* 1995), the best estimates for remaining tigers in the Russian Far East ranged from 250 (Pikunov 1990) to 338–350 (Mescheryakov & Kucherenko 1990), but the true number of tigers was a topic of heated debate. Given the uncertainty associated with tiger surveys in general (Karanth 1987), we felt it inappropriate to focus on numbers of animals as an overall goal because there would be no way to

The goal of the Amur tiger habitat protection plan: no further loss of Amur tiger habitat should occur.

accurately determine when we had achieved our management goal.

2. Existing estimates may not adequately reflect the potential of existing habitat. Even if accurate population estimates existed, they might not indicate how many tigers could be supported on a given land base, if properly managed. Therefore, extrapolating the potential population based on existing information might result in an inaccurate estimate of the 'carrying capacity' of the habitat.

3. Our ultimate goal was the protection of habitat. Although the exact number of tigers was unknown, it was clear that the last remaining stronghold of habitat occurred in the Russian Far East, and that habitat degradation and fragmentation would be long-term threats. No matter what the present status of the population, an adequate habitat base will be the key to long-term survival. Therefore, we felt it critical to define as our goal no further loss of tiger habitat.

Define key components of the plan

To achieve this goal of no habitat loss, we propose that an effective habitat protection plan for Amur tigers should consist of three components.

1. *There must be a core area that consists of a network*

of conservation units. Amur tigers require large tracts of land (Yudakov & Nikolaev 1987, see below). Therefore, it is politically unfeasible to include all tiger habitat in protected areas (see 'Determine political criteria to guide the planning process political criteria' below). However, a core network of conservation units with a high level of legal protection should guard against catastrophic events and minimise the effects of long-term habitat and genetic erosion. The protected areas network should: (1) provide for a minimum target population of tigers (see 'Determine biological criteria to define the adequacy of conservation units' below); (2) provide a 'source population', or a protected 'reservoir' out of which tigers can emigrate to areas where localised extinction may occur; (3) mimic the shape of the existing habitat to ensure continuity, i.e. because the existing tiger habitat is somewhat elliptically shaped on a north/south axis (Fig. 19.1), an effective network of protected areas should extend throughout, also in a north/south projection.

2. *All potential tiger habitat must be identified and included in the planning process.* The core network by itself provides for only a minimum number of tigers. To ensure survival of the entire population, management must be extended to non-protected

areas outside the core area. To achieve this objective, all potential tiger habitat must be identified within the study area. Outside the core network, a management regime must be developed to ensure that unprotected habitats are not lost. Therefore, management zones must be delineated for all potential tiger habitat, and for each management zone a set of management guidelines should be established to guide land-use practices. This process sets priorities on which areas are most important, and acknowledges that there are some areas not suitable for occupation by tigers.

3. *All important tiger habitat must be interconnected.* A system of ecological corridors connecting both protected and unprotected areas is essential to ensure viability of the tiger population. Connectivity of the entire population will avoid the impact of genetic impoverishment, reduce the probability of fragmentation and reduce the chances of localised extirpation of small, isolated subpopulations, resulting, eventually, in loss of the entire metapopulation.

Determine political criteria to guide the planning process

For any land-use conservation plan to be successful, there must be a high probability that it will be politically viable. We attempted to assess the present political situation in the Russian Far East, and delineated the following political criteria.

1. *Proceed with efforts at international cooperation, but develop a plan that provides for a secure population of tigers within Russia, independent of an uncertain international alliance.* Available information suggests that there are perhaps 12 tigers remaining in fragmented patches of habitat in northeast China (Ma & Li 1996), and an unknown but small number remaining in North Korea. Given the small number of tigers in adjacent countries, and the variable and relatively 'cool' relations among North Korea, China and Russia, it would be unrealistic to rely on an international plan. Survival of the Amur tiger should not depend upon close political alliances between these three countries. A plan for tiger conservation in Russia should provide the opportunity for collaborative efforts, but should

be designed to be effective and independent of efforts outside the national borders. Although there are developing opportunities for co-operative conservation measures (Anon. 1996d), we do not address that issue here; our premise is that there is minimal habitat left in China and perhaps Korea, and that Russia must develop a sustainable plan for conservation of the tiger independently.

2. *Rely on existing legislative precedence and existing plans for protected areas wherever possible for creating a protected areas network.* Russia is reeling from political and economic chaos resulting from the dissolution of the Soviet Union and the gradual replacement of a state-driven, command economy by a free-market economy (Yergin & Gustafson 1995). In this political environment, it is difficult to develop the consensus necessary for the type of long-term planning essential for conserving sufficient habitat for tigers. Perhaps most importantly, during this period of financial crisis and political impasse, federal and regional administrations are unlikely to commit to establishment of any protected areas that do not already have some legislative precedent. Therefore, the ideal protected areas network would minimise the need for new legislation, and would use existing and proposed protected areas wherever possible.

3. *Much existing tiger habitat will remain outside strictly protected areas and may require some legislation for multiple-use lands to be managed for tiger conservation.* Tiger habitat represents about 75% of Primorski Krai (Matyushkin *et al.* 1996), and coincides with the area of highest human density in southern Khabarovski Krai. It would be politically impossible to achieve our goal – protection of all tiger habitat – by removing all tiger habitat from the resource base of the Krais and turning the lands into protected areas with limited or no human access. Therefore, a large percentage of tiger habitat will remain outside the core protected areas network. For this reason, a zoning process, as delineated above (key component 2), will have to be a political process. There is at present no adequate means of protecting tiger habitat outside conservation units, nor is there a legislative context to the concept of ecological corridors, which are a key component of

this plan. Therefore although we must attempt to minimise the need for legislative actions, some new legislation will be essential.

4. *Reduce reliance on state funds for maintaining tiger habitat*. Federal and regional funds for conservation efforts are minimal and many local politicians must generate hard currency income through international 'joint venture' business enterprises, many of which rely on natural resource utilisation and thereby conflict with natural resource protection. Politicians are less likely to support legislation for new protected areas without confirmed financial backing or a mechanism for reducing the need for federal and regional financial support. Therefore, multiple-use lands, which can generate their own funds, must be incorporated into land-use planning for tigers.

Determine biological criteria to define the adequacy of conservation units

A plan for preservation of tiger habitat must have a solid scientific basis. However, there is usually limited time and financial support for developing land-use plans for tigers. Therefore, it is usually necessary (as it was in our situation) to use available information at the time planning occurs. Four types of biological criteria were necessary for our planning process.

1. *Define potential tiger habitat*. To delineate tiger habitat, it was first necessary to define what tiger habitat is. We used two primary criteria for defining potential tiger habitat: land ownership and cover type. Nearly all lands in Primorski and Khabarovski Krais fall into one of four land ownership/use categories: settlements, agricultural production, forest service and zapovedniks, or reserves (forest resources on zakazniks, or wildlife refuges, and national parks are administered by the Forest Service). Although much land managed for agricultural production is forested, we did not include it as potential tiger habitat because its primary use is contrary to tiger conservation, and there would be little chance of affecting future land-use decisions. Therefore, we included only those lands managed by the Russian Forest Service, and zapovedniks, which are managed under the State Committee for

Protection of the Environment. We did not assume that all these lands were actually inhabited by tigers, although later analyses have demonstrated a close relationship between land ownership and tiger distribution (T. Merrill *et al.* unpubl. data). Rather, we assumed that these land categories had the potential to sustain tigers, and hence should be incorporated into the planning process. Within these regions, we used a 1968 forest-cover map (scale 1:2 500 000; Grebovoy *et al.* 1968) to define potential tiger habitat. The forest classification system for this map incorporates a variety of robust variables that define geobotanical potential and represent a description of enduring ecosystems, thus reducing the importance of the time interval between date of creation and the present. Using this cover map, we excluded all alpine and subalpine vegetative types, as well as high-elevation spruce-fir forests on Forest Service and zapovednik lands, because deep snows prohibit use by tigers (Matyushkin *et al.* 1996). All other forest types on these two land categories were included as potential tiger habitat.

2. *Define population goals within the protected areas network*. As noted above, we did not attempt to define a target tiger population for all tiger habitat. While the overall goal focuses on total land base, there has to be a rationale guiding the development of a protected areas network, and a decision base for determining how much land was required.

We defined the objective within the core network of protected areas as a population of 50 resident adult females. We focused on adult resident females as a basis for planning because they are usually the critical component of a population. Female reproductive parameters (litter size, age of first breeding, interval between breeding, breeding longevity) are often key demographic factors affecting viability of a population (e.g. see Beier 1993; Tilson *et al.* 1994), and an 'Allee effect', whereby animals at low density would have difficulty finding mates (e.g. Beier 1993), is not likely to be a problem with the present distribution of Amur tigers. We did not conduct a rigorous minimum viable population analysis, partly because many of the variables necessary for accurate projections are poorly known.

We used the value of 50 breeding females as a basis for planning with the following three justifications. First, conservation genetics suggests that an effective population size of 50 adult breeding animals is required as a minimum to prevent short-term genetic deterioration (Soulé 1980; Frankel & Soulé 1981). We did not have the necessary information to estimate effective population size, but data from Indian tigers suggest that effective population size is significantly smaller than actual population size (Smith & McDougal 1991); thus the need for a population of females considerably larger than 25 (half the suggested minimum adult population). Secondly, in general, genetic variation is related to population size (Frankham 1966), and therefore, 'the larger the population, the better'. However, thirdly, Amur tigers require large tracts of land (see below), and a population of even 50 females would require a massive network. Therefore, a qualitative 'middle' position for this core population was to ensure 50 resident adult females, thereby ensuring that it would be greater than the suggested minimum effective population size, but not so large as to be politically non-viable.

We emphasise that this core population is not considered to be a minimum viable population, and that this core population is not intended to be the only component of a habitat protection plan. This core network represents that segment of lands within existing Amur tiger habitat that would receive the highest level of protection, and should act as a buffer against catastrophic events (e.g. poaching episodes or extremely severe winters).

3. *Define criteria to estimate potential tiger population size within protected areas.* We propose that land-use planning for tigers should be based on defining the number of adult resident females that can be protected within existing or proposed conservation units. By knowing the area required by an average resident adult female tiger, it is possible to estimate the number of resident adult females that may occur within protected areas. This process provides a mechanism for determining the potential population size within any proposed conservation unit.

We propose that two key ecological parameters

must be known to employ this approach: (1) home range size of adult resident females; and (2) social structure (or land tenure system). Average home range size delineates the amount of land required by an individual resident female, and an understanding of the social structure will determine the amount of overlap among resident females. These two parameters can be used to estimate the number of resident females that can 'fit' into a prescribed conservation unit.

Results of our research efforts in Sikhote-Alin Zapovednik indicate that the average home range size of five resident females was approximately 470 km², based on a 100% minimum convex polygon (MCP) estimator, or 350 km² based on a 95% MCP estimator (Miquelle *et al.* in prep.). Overlap between adjacent resident females averaged less than 4%, using the 95% MCP estimator, and approximately 15% overlap using the 100% MCP (Miquelle *et al.* in prep.). Based on this analysis, we believe that adult female tigers in Sikhote-Alin Zapovednik are territorial (i.e. maintain exclusive home ranges). An intensive snow-tracking study elsewhere in Primorski Krai also suggested that resident female Amur tigers have exclusive home ranges (Yudakov & Nikolaev 1987).

We believe that while the 95% MCP estimator probably provides a better representation of the ecological requirements of adult female tigers, the 100% MCP is a more conservative and more appropriate estimator for large-scale planning. Although there is more overlap in adjacent home ranges using the 100% MCP estimator (suggesting less land per individual would be required), much land included in a large-scale planning project would be unsuitable tiger habitat. Areas with high human disturbance, high-elevation forests, alpine and subalpine communities, or any habitats that contain few prey are unlikely to hold resident females, but would be included in low-resolution, large-scale planning (the 1:2 500 000 map used to identify potential tiger habitat will not identify many small patches of unsuitable habitat). Using the larger estimator, and still assuming exclusive territoriality, we provide a conservative estimate of the number of females that could use a given landscape.

Our comparison of projected estimates based on these criteria with the estimated number of adult females in Sikhote-Alin Zapovednik (E. N. Smirnov & D. G. Miquelle this volume) indicates that this estimator is reasonable, at least within one conservation unit.

Not all protected areas included in a network would be of equal value. A protected areas network will include conservation units that vary in level of legal protection (and therefore the extent of human impact), and in inherent environmental parameters that affect habitat quality. To account for variation in habitat quality and extent of human impact, we adjusted the estimated potential number of resident females in a conservation unit by a crude index that compared the quality of habitat to the best existing habitat in the Russian Far East. Large, well-protected zapovedniks (e.g. Sikhote-Alin, Lazo and Ussuriysk) received the highest ranking. The relative density of resident females in other areas was adjusted in 25% increments that provided an index of their relative value in comparison to zapovedniks.

4. *Define ecological corridors.* In general, suitable data to adequately define what constitutes an ecological corridor is lacking (Shafer 1990), and represents an especially difficult issue for large carnivores (however, see Beier 1993, 1995). In the Russian Far East, along the 1000 km linear distance of tiger range, corridors will by necessity be long, as they must provide connections between conservation units that will be great distances apart. Therefore, corridors must not only provide for the opportunity for tigers to briefly pass through a region, but must be large enough to sustain prey populations. Tigers should be able to live for extended periods of time, if not indefinitely, within designated corridors. This is a more conservative, broader definition, including larger land areas, than has been proposed for cougars (Beier 1993), and reflects the large scale of planning required for Amur tigers.

We consider corridors to be a specially designated land zone (see 'Determine political criteria to guide the planning process political criteria'), and therefore they have a political as well as ecological context. Because they must be areas where tigers could live indefinitely, we delineated corridors to be, on average, at least equal to the minimum diameter of female tiger home ranges. In Sikhote-Alin, the average minimum width of the 100% MCP of adult resident females was 16.5 km (D.G. Miquelle unpubl. data.). We sought to designate corridors that maintained this width criteria.

Implement the planning process

We employed an incremental, six-step process in delineating a habitat protection plan. We first focused on developing a protected areas network using the goals, key components and political and biological criteria delineated above, and secondly suggest a zoning process for the remaining tiger habitat. This stepwise process uses the political criteria to delineate the sequence of inclusions to a protected areas network, uses the biological criteria to define how many tigers could be protected, and at the same time ensures that adequate linkages exist between conservation units. Finally, we used large-scale (low-resolution) mapping data to define all potential tiger habitat (see 'Determine biological criteria to define the adequacy of conservation units'), and suggest a zoning process for lands outside the protected areas network (see 'Determine political criteria to guide the planning process political criteria').

Step 1. Assess the adequacy of existing protected areas

Using the estimates of territory size for an adult tigress, and the known size of existing zapovedniks, we estimated the number of females occurring in existing protected areas (Table 19.1) to be at least 18 adult resident female tigers in approximately 10 500 km² of existing protected areas. This estimate fits well with separate estimates in existing protected areas (e.g. Matyushkin *et al.* 1996; E. N. Smirnov & D. G. Miquelle this volume). We rated two conservation units as having lower-quality habitat: Botchinski Zapovednik, which is at the northernmost boundary of tiger range along the coast in Khabarovski Krai; and Barsovy Zakaznik, which sustains severe human disturbance due to the presence of a military training site within its boundaries.

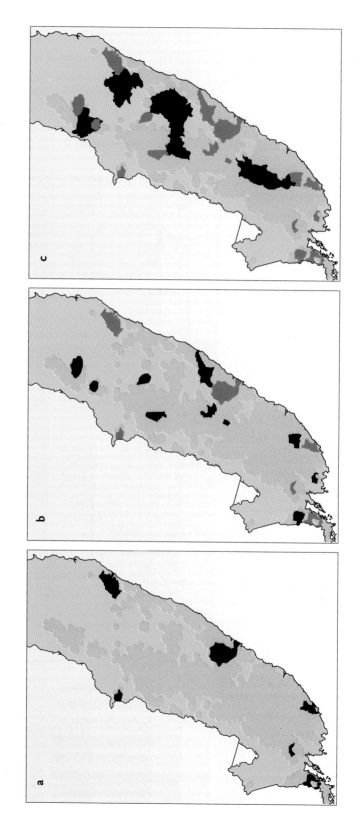

FIGURE 19.2

a. Step 1 in developing a protected areas network for the Amur tiger: assessment of existing protected areas (zapovedniks and key zakazniks) within potential tiger habitat, Russian Far East. **b.** Step 2 in developing a protected areas network for the Amur tiger: assessment of existing and proposed protected areas (zapovedniks and key zakazniks) within potential tiger habitat, Russian Far East. **c.** Step 3 in developing a protected areas network for the Amur tiger: assessment of existing and proposed protected areas (zapovedniks and key zakazniks), proposed and existing zones of traditional use, and multiple-use lands within potential tiger habitat, Russian Far East.

Not only does this estimate fall short of our planning goal for the number of females in a protected areas network (18 versus 50), but these conservation units are widely dispersed (Fig. 19.2a). Many are presently connected by existing forests, but there is no existing legal mechanism to ensure long-term connectivity. Therefore, we must increase the area of the network, and increase connectivity.

Step 2. Include regional and federal proposals for new protected areas that already have some legislative precedent

Incorporation of new lands into a protected areas network is substantially easier if there is some legislative precedent, or existing plans for protected areas (see 'Determine political criteria to guide the planning process political criteria'). Primorski Krai has developed a long-term plan for nature conservation sponsored by the Krai Administration (Yelyakov *et al.* 1993). Included are plans for three national parks, one nature park (Krai-level legislation), and one proposed zakaznik that would add significant lands to a tiger protected areas network. In Khabarovski Krai, one national park and three zakazniks have been proposed that are within tiger range. In total, creation of these conservation units would increase the total protected lands to over 26 000 km² (Fig. 19.2b). The relative gain in land area is not matched by relative gain in number of resident females, however, because the relative quality of these lands is less (Table 19.1). Nonetheless, this complex of lands would provide territory for an additional 23 resident females, bringing the total to approximately 42. However, there are still major gaps in the system that could lead to fragmentation in the future (Fig. 19.2b).

Step 3. Include multiple-use areas – proposed or existing – where tiger management will be a primary consideration

In an attempt to avoid the need for proposing new protected areas (see 'Determine political criteria to guide the planning process political criteria'), we assessed the feasibility of including existing and proposed multiple-use areas as tiger conservation units. Two types of multiple-use areas could be potentially included in the network. There exist two large tracts of land with proposed and/or existent zones of traditional use for indigenous people under the Primorski Krai plan for nature protection (Yelyakov *et al.* 1993). These two areas, the Bikin and Samarga, represent the last unlogged basins in Primorski Krai. The middle Bikin River Basin (already gazetted as a zone of traditional use) represents some of the best unprotected tiger habitat in northern Primorski Krai, while the upper Bikin (proposed zone of traditional use) is lower quality tiger habitat except close to the river bottoms (Matyushkin *et al.* 1996). The other proposed zone for traditional use, the Samarga River Basin, has only recently been recolonised by tigers (Smirnov & Miquelle 1995) and is close to the northern boundary of tiger distribution along the coast (Matyushkin *et al.* 1996). Nonetheless, there is evidence of breeding females there (Smirnov & Miquelle 1995; Matyushkin *et al.* 1996).

Traditional-use zones are open to hunting and potentially other extractive processes, thus decreasing their potential value for tiger conservation. However, indigenous people traditionally have a respect for tigers, and formerly did not hunt them (Arseniev 1941). Just as importantly, traditional-use zones represent large conservation units that protect habitat with little administrative cost associated with tiger conservation.

There are two international multiple-use planning projects within Amur tiger habitat. The United States Aid for International Development (USAID) Environmental Policy and Technology Project is developing a multiple-use plan for Chuguevski Raion, the centre of Primorski Krai, which has, as an explicit component of the plan, development of an 'ecological corridor' to link protected areas in the south (Lazovski Zapovednik) with those in the north (Sikhote-Alin Zapovednik) (USAID 1995). A series of protected zones and management of ungulates species have been incorporated into the planning effort. The Chuguevski Multiple Use Plan is important because: (1) it covers a large territory (7680 km²); (2) it is the first attempt in the Russian Far East to incorporate all user groups into a single planning effort; and (3) it incorporates endangered species and game species

Table 19.1. *Summary of proposed network of protected areas for Amur tigers in Primorski and southern Khabarovski Krais, and an estimate of potential density of adult female tigers in each conservation unit. See also Fig. 19.3*

Conservation unit	Status[a]	Area (km²)	Relative density[b]	No. adult female tigers
Khabarovski Krai:				
1. Manominski Ecological Corridor		2078	0.25	1.1
2. Annui National Park	proposed	3000	0.5	3.2
3. Gassinski Model Forest	active	3800	0.25	2.0
4. Tigrini Dom Zakaznik	proposed	1500	0.75	2.4
5. Khor-Mykhen Ecological Corridor		4684	0.25	2.5
6. Chuken Zakaznik	proposed	2000	0.5	2.1
7. Matei Zakaznik	proposed	2000	0.75	3.2
8. Botchinski Zapovednik	1995	674	0.5	2.8
9. Bolshe Khetkhsirski Zapovednik	1963	451	1	1.0
North Primorye:				
10. Samarga Traditional Use Zone	proposed	4000	0.25	2.1
11. Samarga Ecological Corridor			0.25	0.0
12. Upper Bikin Zone of Traditional Use	proposed	6500	0.25	3.5
13. Bikin Zone of Traditional Use	active	6000	0.5	6.4
14. Kema Ecological Corridor		52	0.5	0.1
15. Kema-Amgu National Park	proposed	3000	0.75	4.8
16. Sikhote-Alin State Reserve	1935	4000	1	8.5
17. Central Ussurka National Park	proposed	1059	0.75	1.7
18. Central Zakaznik		282	0.5	0.3
Central Primorye:				
19. Dalnyegorsk Ecological Corridor		7145	0.25	3.8
20. Chuguevski Planning Project	in process	7680	0.5	8.2
21. Upper Ussuri National Park	proposed	971	0.75	1.5
South Primorye:				
22. Lasovsky Ecological Corridor		390	1	0.8
23. Lasovsky State Reserve	1935	1200	1	2.6
24. Nature Park Ecological Corridor		834	0.75	1.3
25. Southern Primorye Nature Park	proposed	715	0.5	0.8
26. Southern Sikhote-Alin Ecological Corridor		1180	0.5	1.3
27. Ussuri State Reserve	1932	404	1	0.9
Southwest Primorye:				
28. Borisovkoe Plateau Zakaznik	1996	613	1	1.3
29. Kedrovya Pad	1916	180	0.75	0.3
30. Barsovy Zakaznik		974	0.5	1.0
Totals		69366		71.5

[a] date of creation, officially proposed, or, if blank, newly proposed.
[b] relative index of habitat quality for tigers.

into the planning process. Tiger densities will be considerably lower in this region than in protected areas, but the process represents an attempt to include endangered species such as tigers into a land-use plan, an important precedent for Amur tiger conservation.

At the northern fringe of tiger range, there exists a Canadian-financed Gassinski Model Forest

Table 19.2. *Summary of land use categories for proposed protected areas network, and percentage of total land base of each administrative unit*

Land use category	Primorye		Southern Khabarovsk[a]		Total		
	Area (km²)	% of area	Area (km²)	% of area	Area (km²)	% of admin. areas	% of network
Existing zapovedniks	5784	3.5	3125	3.1	8909	3.3	12.8
Existing zakazniks	1869	1.1	0	0.0	1869	0.7	2.7
Existing multiple-use zones	0	0.0	3800	3.8	3800	1.4	5.5
Existing ethnographic zones[b]	6000	3.6	0	0.0	6000	2.3	8.6
Proposed national parks	5745	3.5	3000	3.0	8745	3.3	12.6
Proposed zakazniks	0	0.0	5500	5.5	5500	2.1	7.9
Proposed ethnographic zones	10500	6.3	0	0.0	10500	3.9	15.1
Proposed multiple-use zones	7680	4.6	0	0.0	7680	2.9	11.1
Proposed ecological corridors	9601	5.8	6762	6.7	16363	6.1	23.6
Total	47179	28.4	22187	22.1	69366	26.0	100

[a] Southern Khabarovsk includes Bikinski, Vzyamski, Lazo, Khabarovski, Sovganski, Nanaiski and Komsomolski Raions.
[b] Ethnographic zones of Khabarovski Krai are not included in this assessment.

Project (Anon. 1996d). To a lesser extent, this project also includes endangered species in its management plan. Tigers do occur within the model forest, but, at the northern limits of tiger distribution, there are few breeding females in the area.

As multiple-use areas, these regions would not receive the level of protection allocated to other conservation units within the plan, and will not maintain the same density of tigers. However, inclusion comes with little additional political, financial, or administrative burden. Cumulatively, although the relative quality of the areas for tigers is less, inclusion of these lands means that the goal for a protected areas network is exceeded, with an estimated 59 resident females occurring on 52 000 km². Just as importantly, inclusion of these lands has filled many of the large 'gaps' between existing and proposed protected areas. Linkages, however, are still necessary (Fig. 19.2c).

Step 4. Propose new protected areas if necessary
Due to political inertia and the additional financial burden, we believe that proposals for new protected areas should be developed only when all existing and proposed protected areas and all multiple-use lands

have been assessed and included in the planning process. At this point in our planning process, 8% of Primorski Krai is allocated to existing or proposed protected areas (zapovedniks, zakazniks, and national parks), and nearly 10% to zones of traditional use (Table 19.2). In southern Khabarovski Krai, 11.6% is allocated to existing or proposed protected areas (Table 19.2). Ethnographic zones, common in southern Khabarovsk, are not included in the assessment due to differences in legislative context of this land-use designation. In total (i.e. including multiple-use zones), 26% of the land base has been proposed for incorporation into the conservation network for tigers. Although less than 10% would be strictly protected (i.e. the other 16% would support human activities), proposals for new protected areas for tigers are unlikely to be politically viable. At the same time, we have exceeded our goal of 50 resident females. Therefore, we make no new proposals for protected areas, but seek to ensure that connections exist between all conservation units.

Step 5. Ensure connectivity
To prevent fragmentation, we attempted to link all existing components of the proposed network with

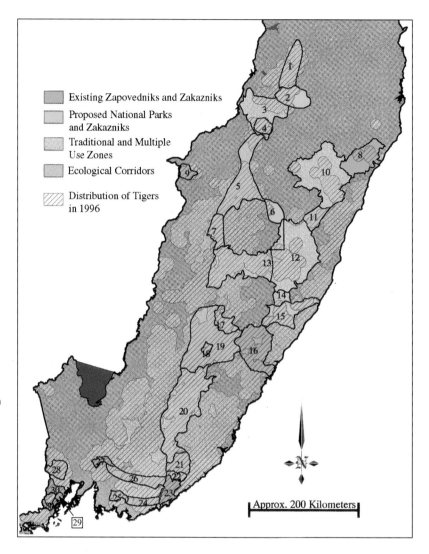

Existing Zapovedniks and Zakazniks

Proposed National Parks and Zakazniks

Traditional and Multiple Use Zones

Ecological Corridors

Distribution of Tigers in 1996

FIGURE 19.3
Step 5: A protected areas network of 30 conservation units connected by ecological corridors (except for southwest Primorye and Bolshekhetskhirski Zapovednik) for the Amur tiger in the Russian Far East. Numbers for conservation units are referenced in Table 19.1.

ecological corridors. This was possible in a north/south axis because forest cover is still intact throughout the Sikhote-Alin Mountain Range (Fig. 19.3). However, the development corridor between Vladivostok and the city of Ussuriysk makes restoration of a corridor linking the Sikhote-Alin and East Manchurian Mountains unlikely biologically and politically.

In total, 69 366 km², or 26% of the land base, is included in a network of conservation units (Table 19.2). This network would protect approximately 70 resident females, far exceeding our numeric goal,

but not quite meeting the criteria for connectivity. There still exists a fragmentation point separating southwest Primorye and the Sikhote-Alin system (Fig. 19.3), but the network does protect a continuous band of tiger habitat along 900 km in a north/south gradient.

Step 6. Identify unprotected areas for inclusive tiger management plan
Once the network of protected areas had been identified, we used the potential tiger habitat map to identify all regions that could sustain tigers that

Table 19.3. *Allocation of lands in Primorye and southern Khabarovski Krais, Russian Far East, to protected areas network, unprotected potential tiger habitat, and percentage of lands presently inhabited by tigers in protected areas network and unprotected potential habitat*

Administrative Region	Within administrative region			Within protected area network				Outside protected area network		
	Total area	Total forested area[a] (km²)	Potential tiger habitat[b] (km²)	Potential habitat with tigers[c] (%)	Total protected network (km²)	Portion of admin. unit (%)	Area with tigers[c] (%)	Potential tiger habitat (km²)	Portion of admin. unit (%)	Area with tigers[c] (%)
Primorye	165 900	118 803	91 265	82.0	47 179	28.4	75.6	70 589	42.5	73.3
Southern Khabarovski	100 449	73 156	36 428	48.0	22 187	22.1	64.7	31 012	30.9	37.0
Total	266 349	191 959	127 693	65.0	69 366	26.0	70.2	101 601	36.7	55.1

[a] GosLesFund (Forest Service) lands and zapovedniks.
[b] GosLesFund and zapovedniks minus unsuitable tiger habitat (alpine, subalpine, swamps, and spruce-fir forests).
[c] Based on 1996 winter census (Matyushkin et al. 1996).

were not incorporated in the network. A total of 91 265 km² of Forest Service and zapovednik lands representing 55% of Primorski Krai, and 36 428 km² of forest lands in southern Khabarovski Krai (36%) were identified as potential tiger habitat. A significant portion of tiger habitat is outside the protected areas network (Table 19.3).

All tiger habitat outside the protected areas network must be included in a habitat protection plan for tigers. It is beyond the scope of this chapter to fully develop this component, but we offer a brief outline of key components of a plan for non-protected lands.

We recommend development of a zoning system for all of Primorski Krai and southern Khabarovski Krai. Zoning systems have been applied successfully for other large carnivores (e.g. Mech 1977). Because tiger habitat represents such a large percentage of these administrative units, it is important to incorporate the entire region, even areas considered unimportant to tigers. The objective of a zoning plan is to identify all areas important to tiger conservation, and then apply specific guidelines for managing those areas. Such a zoning process would accomplish several critical management tasks:

1 Identify all important tiger habitat, and provide an index of relative importance with a zoning system.
2 Identify all regions considered unacceptable for tiger habitat, and develop applicable management actions when tigers are located in such areas.
3 Provide management guidelines for each zone as minimum standards for tiger conservation.

We recommend that specialists familiar with specific regions of tiger habitat develop recommendations for a zoning plan, based on their knowledge of the region, and existing information on past and present tiger distribution (Yudakov & Nikolaev 1973; Pikunov & Bragin 1987; Matyushkin et al. 1996). We propose that three to five zones be created, ranging from areas deemed unacceptable for tigers (e.g. forest patches within village and city limits), to regions where protection of tigers is one of the highest priorities of land use. Each zone that includes tiger habitat should have an associated set of management guidelines. We suggest that management guidelines include recommendations for:

1   Type and extent of human activity allowed.
2   Type and extent of logging activity allowed.
3   Amount of hunting allowed, or target densities of key ungulate species.
4   Restrictions on road development, and road closures where possible.
5   Restrictions on development activities.

The importance of some of these variables to the Amur tiger population has been discussed elsewhere (Miquelle *et al.* 1993, 1995, D. G. Miquelle *et al.* this volume chapter 6). We believe that these human activities all have important impacts on tiger densities within potential tiger habitat and, at the same time, are manageable activities on Forest Service lands.

Assess the potential effectiveness of the plan

Ideally, an assessment of the potential effectiveness of a land-use plan should be conducted. Our land-use plan is based on the assumption that existing protected areas are quality tiger habitat, but that most forested lands have the capacity to sustain tigers. However, it was not clear what percentage of the proposed protected areas network actually contained tigers, or how distribution of tigers within the network compared to the total distribution of tigers, i.e. was the proposed network of protected areas better than a random selection of potential tiger habitat?

A complete survey of existing tiger habitat in 1996 (Matyushkin *et al.* 1996), initiated shortly after development of this plan, provided an opportunity to assess its effectiveness in comparison to the known distribution of tigers.

For approximately three months in the 1995–1996 winter census, tracks of tigers were reported on 652 count units distributed throughout approximately 90% of potential tiger habitat. Field counters plotted out the location of each track on 1:100 000 maps. A geographic information systems (GIS) database was created with this information, and we generated a distribution map of tigers by encompassing each track with a 10-km radius circle, representing the approximate minimum diameter (see above) of female home ranges (Yudakov &

Nikolaev 1987; D. G. Miquelle *et al.* unpubl. data).

The results of this process suggested that tigers occurred on 82% of potential habitat in Primorski Krai and only 48% of Khabarovski Krai. Overall, tigers occurred on 70% of lands inside the protected areas network (including lands that were not classified as potential habitat), but only on 55% of potential habitat outside the network (Fig. 19.4; Table 19.3). These results suggest that the protected areas network is more effective than a random selection of potential tiger habitat, but the differences are not great. There appear to be two reasons for the relatively minor differences:

1   Much of the land within the network is composed of proposed protected areas, or proposed multiple-use/ethnographic zones. Without adequate protection currently in place, it is not expected that these lands would presently have a tiger density higher than average. The recent survey clearly indicated that protected lands had higher densities of tigers (Matyushkin *et al.* 1996). We predict that if proposed protected areas received an adequate level of protection, density of tigers would increase.
2   Some areas included in the network are unsuitable habitat for tigers. This is particularly true of the ethnographic zones, which include large tracts of spruce-fir and larch forests, which are poor quality habitat for Amur tigers. These tracts are retained in the plan because they maintain connectivity of habitat, and there is little political/ administrative cost in incorporating them into a tiger conservation plan. However, it must be recognised that such lands will hold few resident tigers.

Discussion

We initiated our research with the objective of obtaining the necessary information for formulation of a conservation plan, and were fortunate to

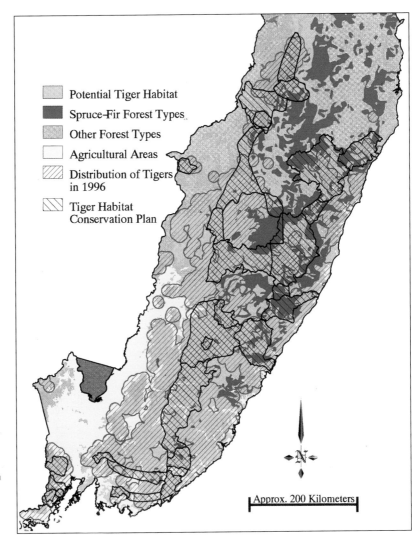

FIGURE 19.4
Distribution of tigers, based on a 10-km radius of tracks located during the 1996 winter census (Matyushkin *et al.* 1996), in comparison to the proposed protected areas network, and identified potential tiger habitat.

have a sufficient database when a federal decree ordered development of just such a plan. We were able to use the 'high-resolution' data on ecological parameters of Amur tigers, combined with 'low-resolution', large-scale data for forest cover, tiger census data and administrative units to develop a biologically and politically defensible land-use conservation plan for tigers. During a workshop in September 1995 to formulate a national strategy for tigers, we were able to present our proposal to the federal government for consideration (Miquelle *et al.* 1995). As a result, the basic concept for a protected areas network and a zoning plan for tigers was incorporated into the national strategy document (State Committee of the Russian Federation for the Protection of the Environment 1996). We believe that this is a valuable lesson of the need for research that addresses conservation issues, and of the value of applying ecological data to the formulation of federal policies for natural resource management.

Planning is a process, and often it is necessary to initiate that process with inadequate information. We are moving forward with continued research on

**Box 19.1   People and tigers in the Russian Far East: searching for the
'co-existence recipe'**
Dale G. Miquelle and Evgeny N. Smirnov

There is approximately 156 000 km² of tiger habitat remaining in the Russian
Far East (Matyushkin *et al.* 1996). Presently, only 7% of this area is protected
as zapovedniks (reserves) or zakazniks (wildlife refuges). Even under the
most optimistic habitat protection plan (D. G. Miquelle *et al.* this volume
Chapter 19), no more than 16% of tiger habitat would be strictly protected. If
the Amur tiger is to survive in the wild, tiger management cannot be restricted
to nature reserves.

Perhaps the greatest hope for tiger conservation in Russia is the low
human population density; one of the lowest in all range countries at between
1 and 14/km². Nonetheless, there are approximately 3.9 million people living
in or adjacent to tiger habitat in the Russian Far East. While 80% of these are
urban dwellers, villagers throughout tiger habitat are dependent on forest
resources. Virtually all tiger habitat outside protected areas is exploited for
timber, wildlife and non-timber forest products. Given the present economic
instability in Russia, in the foreseeable future forest resources will be viewed
by administrations as a source of hard currency, and by local villagers as a
source of food, raw materials and livelihood. Halting natural resource
exploitation is not an acceptable alternative.

It is clear, therefore, that a successful plan for tiger conservation in the
Russian Far East will combine a system of protected areas with a
management regime in unprotected lands that gives high priority to tiger
conservation. Thus the need for a co-existence recipe on unprotected lands.
The search for this co-existence recipe must first identify potential points of
conflict – situations in which tigers adversely impact humans and arouse
animosity, and human activities that threaten the tiger's existence – and then
seek solutions. Some conflicts, such as competition for prey, impose mutual
adverse impacts.

Adverse impacts of tigers on humans
Loss of human life   Although early accounts suggest that attacks on Man in
Manchuria and Russia were not uncommon (Prezhewalski 1870; Baikov
1925), since 1970 attacks have been exceedingly rare (Nikolaev & Yudin 1992).
Although meetings with tigers are common (e.g. Smirnov 1997), there have
been only six cases of unprovoked attacks and man-eating since 1970, and in
only one case have there been repeated attacks (two) by one tiger. Provoked
attacks (usually the result of hunters wounding tigers) have occurred more
frequently. Provoked or not, all interactions with tigers resulting in human
death receive inordinate attention from the local press, while the daily

shootings in Vladivostok go virtually ignored. Presently, all such incidents are investigated by state-employed specialists; permits to kill culprit tigers are usually issued to state officials where evidence suggests unprovoked attacks, or where severely wounded tigers pose future threats. Some education programmes may reduce the number of such incidences, but as long as tigers and Man co-exist this conflict will never be completely eliminated. Man-killing tigers, though rare in Russia, are bad publicity and hurt tiger conservation efforts. Long-term success may depend on demonstration to local people that their voice is also heard: problem animals must removed from the wild quickly and professionally.

Livestock depredation   Depredation on livestock has increased throughout the last 50 years, coincident with an increasing tiger population; 30 depredation incidences (cattle and horses) were reported in Russia between 1920 and 1940, 102 during 1941–1960, and 362 from 1961 to 1970 (Yudakov & Nikolaev 1973). Detailed recent records are unavailable, but evidence suggests that this trend is decreasing (Miquelle *et al.* 1996b). Depredation is often episodic and short-lived; nonetheless, local small-time farmers cannot bear the financial burden. Resolution of this situation will depend on an adequate natural prey base as well as compensation. Locally financed, self-sustainable 'insurance' programmes for farmers may help alleviate the economic loss and reduce animosity.

Competition for prey   In a survey of hunters, over 55% were either neutral or agreed with the statement that tigers are a problem because they kill too many ungulates. The wildlife management system in Russia allocates specific territories to an individual or group, which is then responsible for proper management with harvests at a level appropriate for the game numbers within that unit. Hunters often complain that tigers are removing prey that could have been included in their quota. While scientific research may demonstrate that tigers do not regulate prey or impact prey densities, personal convictions are often stronger than any hunter education programme. Working with local hunting groups to reduce poaching, to improve habitat conditions for ungulates, to improve hunters' economic status and capacity for self-management, and to increase their sense of ownership of and responsibility for wildlife resources – all activities that should increase prey populations – is beneficial both to hunters and to tigers. Model programmes based on this premise are promising. Ultimately, hunters must see some benefit in 'carrying the burden' of the tigers inhabiting their hunting unit; reduced license fees or other state-sponsored assistance must be forthcoming.

Adverse impacts of humans on tigers

Poaching   Poaching is considered in detail elsewhere in this volume (S. R. Galster & K. V. Eliot). Ultimately, positive and negative reinforcements will both be necessary. Stiff penalties, high conviction rates of poachers, and a belief by potential poachers that capture is a distinct possibility must come in conjunction with long-term education programmes.

Timber harvest   Logging is not incompatible with tiger conservation. Selective cutting of appropriate species and age-classes can actually improve tiger habitat in some situations. But not all logging is beneficial. A clear understanding of the complex relationship between logging and its impact on tiger habitat is yet to be fully delineated, but a concise description of logging regimes compatible with tiger conservation must be developed and incorporated into Russian Forest Service policy and enforcement guidelines.

Habitat destruction   To provide for a viable population of tigers in the Russian Far East, all remaining habitat must be retained (D.G. Miquelle *et al.* this volume chapter 19). Any type of development that reduces tiger habitat in the region reduces the long-term chances of survival of this population. Existing state forest lands (GosLesFund) must be managed for long-term ecosystem integrity; and, if possible, easements to secure forested lands not included in GosLesFund should be developed.

Selective logging is compatible with tiger conservation, but the roads it leaves behind can make illegal access easier if steps are not taken to close them.

Road access   Road construction is the Achilles heel of selective cutting, which in comparison to clear-cutting requires vast areas (and therefore a vast road network) to secure equal harvest volumes. A high-density road network provides easy access for all kinds of forest exploitation, and has severe impact on tigers and their prey through both legal and illegal hunting. Experimental road closures are now being developed in conjunction with programmes of assistance for hunting societies and agreements with local Forest Service officials. A large-scale programme for road closures is of the utmost importance to increase the extent of secure habitat for tigers and prey.

Many components of the 'co-existence recipe' in Russia have been identified in the national 'Strategy for Conservation of the Amur tiger in Russia'. Management regimes for other large carnivores (e.g. brown bears, wolves and mountain lions) from other countries can provide clues in seeking appropriate resolution mechanisms. However, ultimately realisation of the goals of the national strategy will depend on forthcoming results in two arenas. Within the federal and regional political arena, policy decisions must be made to enhance the potential for tiger survival. At a local level, 'grassroots' efforts must be focused on components of the 'co-existence recipe' that demonstrate feasible and positive benefits to local inhabitants as well as protection of tigers.

tiger ecology, and development of more detailed cover maps on land ownership, land use, road networks and forest cover. This information will be critical to refinement of a land-use plan for tigers. We believe it better to initiate the planning process even when perceived information needs are inadequate. Planning will always occur with some degree of uncertainty; therefore, it must be acknowledged from the start that planning is a process and that plans will change as experience and more information become available.

Acquiring 'high-resolution' data on tiger ecology requires intensive fieldwork and substantial financial investment, and will not be possible in all tiger range countries. Therefore, the process of using ecological parameters to determine the adequacy of protected areas in other range countries may be limited, unless reasonable estimates of those parameters can be inferred. 'Low-resolution' information on present tiger distribution, existing forest habitats, land ownership, land use and road networks are often available, or can be developed fairly rapidly. These geographic data layers are the key starting point to developing protected area networks, and can provide invaluable information in formulating and adjusting plans. For instance, we are presently using land ownership, land-use maps and forest-cover maps as indicators to identify potential fragmentation points in the Sikhote-Alin tiger population, and to search for potential linkages amongst Chinese and Russian habitat patches.

This proposed plan, which includes nearly 70 000 km² , would represent one of the largest protected areas networks in the world, if implemented. For instance, it is comparable in size and scope to any of the Wildlands Projects (e.g. Noss 1993). The size of the network makes the process of implementation a daunting one. More daunting, perhaps, is the political and economic context in which implementation must occur. Russia's political process is unstable and evolving on a daily basis, and the economic environment forces businesses to focus on maximising short-term returns. This is not a conducive atmosphere for long-term natural resource planning.

Nonetheless, there are four points that favour the process.

1   Perhaps most importantly, the human population density in the region is very low: 3.3 million people live in Primorski Krai, but 78% live in urban centres (Kungurova undated), while 80% of the 1.6 million people in Khabarovski Krai are urban dwellers (Pensen *et al.* 1995). The intense pressure on tiger habitat by the burgeoning populations of the Indian subcontinent, for instance, are not present in the Russian Far East; the density of people in tiger habitat (outside the urban centres) ranges from 0.1 to 3.5/km² (Kungurova undated; Pensen *et al.* 1995).

2   The Sikhote-Alin ecosystem represents a largely intact, continuous habitat for tigers. The highest-quality habitat for Amur tigers has already been lost to human development in China, Korea and Russia during colonisation phases and rapid human growth over the last century (Stephan 1994; Janhunen 1996). The remaining habitat is largely mountainous and unsuitable for intensive human exploitation. To date mineral exploitation has been localised, and the selective logging regime most commonly practised does not destroy tiger habitat (Miquelle *et al.* 1993).

3   Many components of the plan have already been proposed. We are not proposing a totally new set of recommendations to the federal and local administrations, but are mostly asking that the existing recommendations be implemented. A large percentage (46%) of the proposed network is comprised of proposed protected areas and proposed ethnographic territories. However, many of these proposed protected areas are in the

process of being established (USAID Environmental Policy and Technology Project 1995; V. K. Berseniev pers. comm.).

4   Finally, tiger conservation can be compatible with other land uses. There needs to be a core area with a high level of protection, but well-managed resource extraction regimes can, in some instances, improve tiger habitat and, in many others, minimise potential impact.

There are many components of this plan that will require adjustments in the future. For instance, ecological corridors comprise 25% of the protected areas network, and are exceptionally long in some places, e.g. linking Ussuriyski Zapovednik and the proposed southern Primorye Nature Park (75 km); and linking Matei Zakaznik and Tigrini Dom Zakaznik in Khabarovski Krai (160 km). This concept of ecological corridors is quite different from the typical concept of a short linkage between distinct habitat patches (e.g. Beier 1993). Moreover, the whole concept of ecological corridors is new in Russian land-use planning. At present there is no legislative precedence to provide guidelines that would legally define what constitutes an ecological corridor, nor is it clear what exactly would be the management criteria for ecological corridors. Even within the proposal presented here, not all biological criteria were always met, e.g. minimum width criteria for ecological corridors were not maintained in all instances. Exact delineation of what constitutes corridors, and exactly where they will occur, awaits a detailed planning process conducted within each administrative region.

At the same time there are a myriad of legislative and political issues that must be addressed for full implementation. A large percentage (38%) of the network is comprised of proposed protected areas and proposed zones of traditional use. All these lands must undergo a tedious bureaucratic process for gazetting.

The value of the southwest Primorye lands to tiger conservation is debatable. At present, the population of tigers in this region is believed to be less than 10 (Matyushkin *et al.* 1966). By themselves, these protected areas are insufficient to maintain a viable population of tigers. Therefore, two critical steps need to be taken in the near future: (1) assess the feasibility of reconnecting this fragment of habitat to the Sikhote-Alin ecosystem; and (2) assess the feasibility of creating international protected areas with China (Anon. 1996b). The potential exists in trans-boundary lands on both sides of the Sino-Russian border to connect southwest Primorye with another fragment of habitat in Pogranichny Raion near Lake Khanka, thus greatly expanding the potential habitat for tigers. This region also represents the last habitat for the Far Eastern leopard, a large cat subspecies even more endangered than the Amur tiger (Miquelle *et al.* 1996a). Therefore, focus on protecting this block of habitat is doubly important.

Although not presented in detail here, the implementation of a zoning process outside the protected areas network is just as important a step as the creation of a network itself. Here, challenges will be faced in delineating a zoning process for tigers that does not impede sustainable resource use, does not commit federal and regional administrations to undue financial burdens, and that can be accepted by the local populace. Finding an acceptable balance between tiger conservation, logging interests and hunting rights will be crucial tests in this arena.

The process of developing and implementing such a plan is long-term. We invested over five years to acquire an adequate database for development of a defensible habitat protection plan, which was then incorporated into the national strategy for tiger conservation. Now, there must be extensive and intensive efforts to implement the plan.

20

The tragedy of the Indian tiger: starting from scratch

Valmik Thapar

Introduction

India today is a country of nearly a billion people. Fragmented and deeply scarred, this is the land of the tiger and is home to 50% of the world's wild tiger population. Comparing past and present forest maps of India, one can see that large chunks of forest have now ceased to exist in the west and north. We are ending up with little islands that may or may not survive the pressures of time. Whether it is the green floodplains of Kaziranga, the monsoon and rainforests of peninsular India, the mangrove swamps of the Sundarbans, the lower Himalaya in Manas or Corbett, the problems of these areas mount and we enter an era where the tragedy of the tiger overwhelms. It is time to start from scratch.

If we look at our premier tiger habitats – the 23 Project Tiger Reserves that encompass 33 000 km² – we face virtual disaster in 12 000 km² and an uncertain future in 11 000 km². If we are lucky and effective, the possibility exists of saving the remaining 11 000 km² (Fig. 20.1). The fate of the tiger's habitat outside Project Tiger reserves is even more precarious. We have reached this state because of the endless pressure on habitat, incessant poaching and, of course, unplanned large-scale 'development' of areas by dams, industry, mines and so much more (Fig. 20.2).

'Political will' to protect the tiger is presently lacking across India, and the reasons for this need to be clearly understood in a historical perspective. When former Prime Ministers Indira Gandhi and Rajiv Gandhi – convinced conservationists – spear-headed the Congress Party in the 1970s and 1980s, most of the 'Tiger States' were ruled by the same political party (Fig. 20.3a). This resulted in effective political will because the leaders of the party were sympathetic to forests and wildlife, and the decisions they took quickly reached State Governments and were translated into field action. In the last decade the entire picture has changed. Now, in 1997, 11 different political parties rule the Tiger States and the Federal Government has power in only a couple of them (Fig. 20.3b). Political will has been totally dissipated, and the process of decision making has taken a hammering. To look at the prevailing situation let us first analyse the processing of funds and budgets that are allocated to save the tiger.

Financial mechanisms

Project Tiger's budget is approved by the Federal Government every five years. The Director of Project Tiger clears proposals with the Additional Inspector General of Forests (Wildlife) and the Secretary in the Ministry of Environment and Forests. They are then examined by a committee of experts that look at proposals for India's Five Year Plans. With everyone's consent, the proposal for funds goes to the Planning Commission, and after approval ends up with the Expenditure Finance Committee in the Ministry of Finance, which then releases the budget for Project Tiger in Delhi. But how does the Field Director of a reserve tap this money?

Each year the Field Directors complete their utilisation certificates for the previous year's funds and formulate plans of operation for the coming year. These are then approved by the Chief Wildlife Warden and Forest Secretary of each state and sent to the Director of Project Tiger in Delhi. If there are any glaring mistakes they are sent back to the state. If accepted, they go to the Additional Inspector

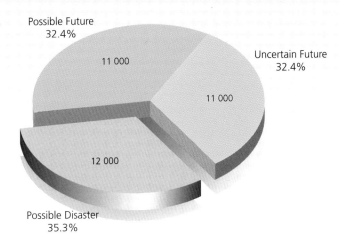

STATE OF THE TIGER RESERVES

(23 Project Tiger Reserves - 33 000 km^2 area)

Possible Future
32.4%

11 000

Uncertain Future
32.4%

11 000

12 000

Possible Disaster
35.3%

FIGURE 20.1
State of the Project Tiger
Reserves.

General of Forests and, after travelling through various departments and offices, end up with the Internal Finance Department of the Ministry of Environment and Forests. If cleared, they travel back to the Director of Project Tiger. Through all this travel, budgets, allocations and even heads of expenditure could have undergone drastic change. Release orders are then sent to the State Government but, believe it or not, the route is through a small city in Maharashtra called Nagpur, where the Accountant General's office is located. All the individual budgets of each tiger reserve go to Nagpur and it is from here that the funds are transferred to State Governments.

But it does not even stop there. In the State of Bihar, yet another file moves to the Forest Secretary and then to the empowered committee in order to route the release of funds to the planning and financial departments. Then the file drops into the lap of the State Chief Minister. If the budget is below 250 000 rupees (about $7000) they can clear it; if above, it goes to the cabinet and waits in a queue for the cabinet to find the time to approve it. It finally ends up with the administrative department, which sends the money to the Field Director. The tragedy for the tiger is obvious. I have never been so shocked by the system of wildlife governance, and I realise that if we are to save the tiger we require immediate reform and change in these antiquated mechanisms that only hinder rapid action. Look at how long it can take for money to travel; in the year 1996–1997, two of our premier Tiger Reserves in Bihar (Palamau and Valmiki), for which over half a million dollars were sanctioned, did not receive their money until the end of the financial year, at which stage the money had to be returned for revalidation in the next financial year. Such fatal delays make it nearly impossible for Field Directors to plan for the future or manage their area effectively.

Attitudes and decision making

It is not just the mechanisms for funding, but also the very attitudes associated with our wilderness that have led to the current complete lack of

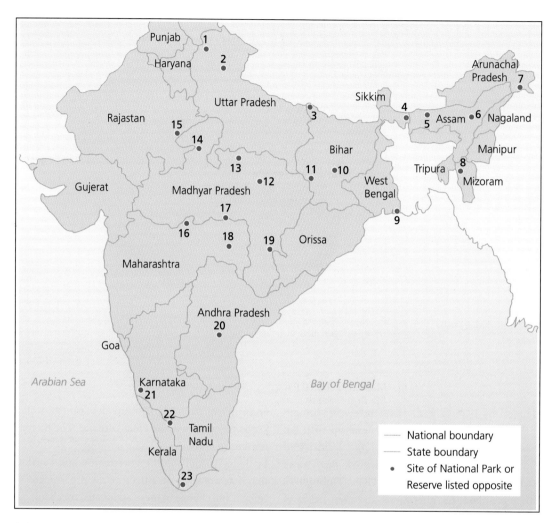

FIGURE 20.2
Land-use conflicts in tiger
habitats in India. (Based
on: *Fifty Indian Tragedies in
the making*, by Bittu Sahgal.
Sanctuary Asia volume
XVII(5), October 1997.)

Location	Threat
1. Rajaji National Park, Uttar Pradesh	Monoculture plantations, army ammunition dump and canal
2. Corbett Tiger Reserve, Uttar Pradesh	Encroachment and excessive tourist development
3. Valmiki Tiger Reserve, Bihar	Railway embankment causing flooding, roads and mining
4. Buxa Tiger Reserve, West Bengal	Major canal project will bisect reserve and damage Jaldapara and Gorumara Sanctuaries
5. Manas Tiger Reserve, Assam (World Heritage Site)	Insurgency, encroachment and road project
6. Kaziranga National Park	Encroachment and pesticide pollution
7. Namdapha Tiger Reserve, Arunachal Pradesh	Road to Myanmar
8. Dampa Tiger Reserve, Mizoram	Hydro-electric project
9. Sundarbans Tiger Reserve, West Bengal	National Waterway proposed for deep draft vessels from West Bengal to Bangladesh
10. Hazaribagh National Park, Bihar	Mining
11. Palamau Tiger Reserve, Bihar	Mining and forest submergence
12. Bandhavgarh National Park, Madhya Pradesh	River pollution from thermal power plant, dam and excessive tourism
13. Panna Tiger Reserve, Madhya Pradesh	Mining, river pollution and road development
14. Madhav National Park, Madhya Pradesh	Mining and dam project
15. Ranthambhore Tiger Reserve, Rajasthan	Mining and road development, and river pollution
16. Melghat Tiger Reserve, Maharashtra	Irrigation projects threatening forests while State Govt has denotified large area for logging
17. Pench Tiger Reserve, Madhya Pradesh	Commercial fishing in core area
18. Tadoba Tiger Reserve, Maharashtra	Mining and road projects
19. Sitanadi Sanctuary, Madhya Pradesh	Irrigation project threatening forests
20. Nallamalai Forests, Andhra Pradesh	Highway project
21. Kudremukh National Park, Karnataka	Mining
22. Nilgiri Biosphere Reserve, Tamilnadu	Hydro-electric and road projects
23. Kalakad-Mundanthurai Tiger Reserve, Tamilnadu	Tea, coffee and cardamom plantations

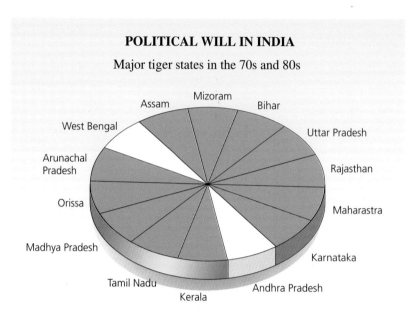

POLITICAL WILL IN INDIA

Major tiger states in the 70s and 80s

All ruled by Congress Party Govts except the two in white

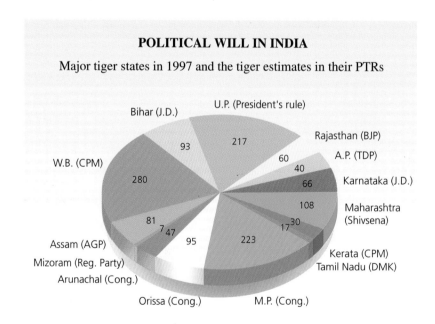

POLITICAL WILL IN INDIA

Major tiger states in 1997 and the tiger estimates in their PTRs

FIGURE 20.3
a. Political will in India – control of major tiger states in the 70s and 80s.
b Political will in India – major tiger states in 1997 and the tiger estimates in their Project Tiger Reserves.

understanding, knowledge and concern for the problems on hand. I quote a series of different people in the hierarchies of wildlife administration in India and the way they look at tigers, their habitat, and the so-called development of an area. It is

shocking indeed that understanding of wildlife has reach such a low ebb.

In changing more than just the patterns of use of tiger habitats, a senior forest officer stated of his Tiger Reserve: 'Though an area of 81 ha. of Tiger

Reserve would be lost under submergence, a new area of about 530 ha. will be turned into an ideal wetland ground for fish populations, which in turn will attract diverse aquatic bird species in large numbers.'

When denotifying more than 500 km² of the Melghat Tiger Reserve, the State Chief Wildlife Warden declared: 'This will help the tiger in the long run.' He added: 'Felling operations will also provide employment for "poor adivasis" (tribal people).' When asked whether the area was significant from the wildlife point of view, a Principal Chief Conservator of Forests replied: 'No. But tiger and panther are present in the Sualkadu Block Reserved Forest,' – an area separate from Melghat.

On the matter of mining in a National Park, there are two examples of the stand of the Ministry of Environment and Forests (MOEF): 'Forest clearance issued by the Forestry Division of MOEF for 3106 ha. of forest land is not valid. About 16 km² belongs to Madhav National Park, for which permission for forest clearance cannot be given – seven red sandstone mines are in operation inside the Madhav National Park, which is illegal under Wildlife (Protection) Act (1972).' The same Ministry, in another letter, stated: 'I am directed to refer to the letter of the Chief Minister, Madhya Pradesh . . . addressed to the Minister (Environment & Forests) regarding above mentioned subject and to say that, *as a very special case, this Ministry has decided to grant permission up to 31st March 1996 only, for removal of existing material and completion of mining operations in already broken up areas in respect of seven mines.*' (author's italics). Despite the Ministry's declarations, mining continues in Madhav National Park, which has also lost 16 km² illegally taken for an irrigation project.

On irrigation projects in another protected tiger area, the Ministry wrote: 'The proposal was approved after receiving the advice of the Chief Wildlife Warden, Madhya Pradesh, that impounding of water would in fact improve the quality of the habitat for the animals.' A senior official of the MOEF told the State Government, on the instructions of a Minister in whose constituency the lake fell: 'This is not an *in situ* lake. So if any use of

bio-resource of the pond [72 km²] ultimately helps in the improvement and better management of the Reserve . . . this may be explored as envisioned in Section 35(6) of the Wildlife (Protection) Act.' Nineteen months later, the same official, under a new Minister, wrote to the State questioning why it had followed what were his own instructions. The State replied: ' . . . the State Government was in favour of stopping fishing in Pench National Park area and was acting accordingly. But the State was compelled to change this stand due to the Government of India's instructions.' So basically there is no policy that is followed; even the Wildlife Act is interpreted on Ministers' whims as they come and go.

When refusing permission for a biologist with a foreign passport to work in India, a senior official of the MOEF said: 'The Government of India considers conservation of tigers a sensitive issue,' and this was because of the 'chance dissemination of sensitive information on tiger conservation'. This quote reveals the complete ignorance that can exist about both tiger conservation and public interest.

A senior secretary to a State Government said to me: 'That man (a Forest Service official) has too much experience in wildlife. We must now transfer him, even if against his wishes. He must learn commercial forestry operations.' We must ensure that people such as this senior secretary never serve the tiger.

These attitudes have also percolated through to the non-governmental organisation (NGO) sector. 'Alarmist reports about one tiger being poached every 18 or 24 hours cannot be called authentic or justified. Wild conjectures or mere guesstimates, which cannot stand scrutiny, will not do. . . . The only agency which has the authority . . . is the Forest Department,' said the Indian branch of the giant WWF. NGOs like WWF could not even assess the facts and figures. Their position was compromised.

An analysis of government files, its interpretation of laws, and its decisions reveal a vast mass of contradictions that have littered the corridors of our Federal and State governments. Few decisions are taken in the interest of the tiger. I find it incredible that the tiger has managed to survive at all with the ignorance that surrounds both the intervention of NGOs and some key government officers. This ignorance, combined with India's other problems, could end up being fatal for the tiger.

The crisis

A sharp decline of political will began in 1987–1988, when not only were there more diverse political parties ruling the States (Fig. 20.3b), but at the same time India entered the era of the free market economy and spiralling consumerism. Cable television invaded city homes and, quite suddenly, overnight, all our traditional gods, that rode on vehicles like the tiger, lion, and wild boar were transformed into 'Quick Money Gods' as more and more joined the race for cash, tearing at the very heart of the wilderness of the country. I have witnessed several cases where animals have been poached and trees cut in order to obtain cash to buy television sets and video recorders. Even activist groups have been sucked into the fashions of the day and they endlessly debate the rights of tribal people over our forests, even when the group of people they fight for are actually in the business of commerce and have no connection with traditional beliefs. Today in every corner of India there is a 'right' that is being fought over in some forest or other. The forest slowly dies, as there are few who fight for the right of a tree or an animal to survive.

We should be very clear at the outset that if the tiger is to be saved it has to be done with a combination of both guns and guards, and love and affection. One has to work hard with the local people, but also enforce the law in the strictest possible way. One has to deal firmly with big business interests that mine vital habitats of the tiger, and the timber mafias that rip great scars in the vital corridors that link habitats together. At the same time, the mass tribal hunts or *akhand shikar* that occur every year in places like Simlipal Tiger Reserve in Orissa to kill large numbers of wild animals must be banned forever without people attempting to take refuge in the concept of 'tradition' or 'tribal culture'; because, as you step

into this rhetoric, the forest ends up being choked to death on the very excuse of 'tradition'.

Solutions

So how is India going to save its tigers? Only by emergency action, reform and a total change in the system of wildlife governance.

Of the roughly 120 forest tracts with tigers, at least 50 with over 65 000 km² of area must be targeted. To do this requires new and innovative measures.

The Prime Minister of India with the support of all State Chief Ministers must empower a special Tiger Task Force with finance and the necessary power to save the tiger and its habitat. Our system of governance needs reform and change, be it in funding, administration or enforcement. We can no longer govern India's wilderness using an infrastructure 30 years old, created without a clue that the crisis would intensify to the extent it has. It is only the Prime Minister and the Chief Ministers who can initiate reform and change in the antiquated system that exists, and at the same time launch an action plan, so as to provide immediate relief for the tiger.

An Action Plan for 1000 days for the Tiger Task Force

1   Identify 50 tiger sites in India.
2   Post the right person to the right job. Select, train and empower 100 forest officers to administer the 40–50 sites for a minimum tenure of five years. This will be a vital first step to bring change and reform into the Indian Forest Service by creating within it an essential wildlife cadre. Without this, wildlife will remain relegated to insignificance and wildlife postings will continue to be looked on as 'punishment posting'.
3   Fully equip all forest guards and provide complete backup infrastructure and legal support. Until the people who guard our natural treasures are taken care of there is

Equipment needed by forest guards to effectively perform their jobs has been increasingly provided by NGOs: these motorbikes were supplied by Care for the Wild (UK).

little hope for the tiger. A special wage incentive could also be considered.

4   Encourage field research in at least 25 locations. Good information is the only way to find solutions for the future. Our research institutions need to be specially geared up to deal with a crisis.

5   Start 50 community-based activities. The local people must feel proud to protect their natural heritage and their innovative interventions must be a part of the collective management of an area. Local people and local communities cannot be ignored. It is their home, it is their life, and we have to work together. We have to find links, so that the forest guard you are arming – it may be with a stick, it may be with a gun, it may be with a woollen sweater – also has a link then with the village. We cannot just have armed guards looking after a place with no connection with communities. We must have in each of these selected areas one small but effective community conservation project. It could be a family planning centre; it could be something to do with replanting trees; it could be entirely based on what that village or community wants to do, but we have to have a place where people feel that they are also participating in the conservation of natural heritage. It is vital for those forest guards who protect the forest to know that they have trustworthy people in their village, who share their concern for wildlife and on whom they can lean. There are people who still dress up like tigers and dance. There still is worship of this incredible animal. We have got to work with such people because they believe in nature, a belief that is alien to most people today. So community conservation combined with strong grassroots protection is a recipe for effective tiger conservation.

6   Focus on education and awareness programmes, especially on local language television and newspapers.

7   Encourage NGOs to help and provide support for field initiatives. In the last two years it is the NGOs who have played critical roles in strengthening the protection infrastructure in the field. At least half a million dollars of inputs have been given to vital tiger habitats from Assam in the east to Karnataka in the south, and it is because of such efforts that some areas have shown an increasing commitment to resolve prevailing problems.

8   Monitor and evaluate, from 1999 to 2000, the impact of all the above activities.

Box 20.1 A boost for Project Tiger?
Valmik Thapar

'We are bound by a national commitment to protect wildlife and, in particular, the tiger. Project Tiger will never suffer and we shall take stringent action against poachers.'

This statement by India's Prime Minister, Inder Kumar Gujral, on 3 October 1997 confirmed growing hopes during 1997 that political will, lacking for several years while the tiger situation deterioriated, had returned at the highest level of government.

Early in the year, Mr Devi Gowda, Mr Gujral's predecessor as Prime Minister, had convened a meeting of the Indian Board for Wild Life, which had not met for nine years. After hearing reports on the situation he sent a message to Chief Ministers of all States telling them that the ecological and environmental security of the nation was affected. He called on the Chief Ministers to review the situation in their areas, adding : 'The concerned departments and authorities will have to be geared up to meet the challenge of protecting our wildlife resources.'

Mr Gujral became Prime Minister in April and in a message to the nation on World Environment Day (5 June) he declared: 'The tiger is India's national animal, the peacock India's national bird, and a host of other wildlife form the rich natural diversity of this country. Their protection must have the highest priority across the length and breadth of this country.'

1997 was the 50th anniversary of India's independence. It was marked by 320 Members of Parliament from 22 parties appealing to the Prime Minister 'to initiate new and immediate reforms in the mechanisms of administration, funding and enforcement in order that the tiger and its habitat can be saved.'

They added: 'Prime Minister, the future of the tiger lies in your hands. Its extinction will herald a downward spiral of ecological decay which can only escalate the misery of millions. There is not a moment to lose.'

I had presented the signatures to the Prime Minister. In response he stated: 'It is indeed heartening that an historic move made by our Parliamentary members to save the tiger has cut across all the Parties. We are already looking into the issue of reform and change in the mechanism that governs wildlife and I hope that we will shortly convene a meeting to initiate some measures in order to ensure the safety of the tiger and the natural heritage of our nation, which are on the highest national agenda of my government. I hope we can work closely together on these issues.'

A new and dynamic Director of Project Tiger, the government's conservation programme, declared that he believed that poaching had caused a drop in the tiger population to below 3000, compared with a 1993 census report of 3750. At a meeting of the project's Steering Committee, proposals to make revolutionary changes in management and funding of tiger reserves were discussed, and the Minister for Environment and Forests, Saifuddin Soz, was asked to consult with the Home Minister, Inderjit Gupta, on measures to bring under control several tiger reserves affected by insurgents and illegal activities.

The Supreme Court, which had earlier displayed its 'green credentials' by ordering the cessation of illegal felling and sawmills in government forests, now banned State administrations from denotifying protected areas without permission from the Indian Board for Wild Life. Tackling the lengthy delays in formally establishing national parks and sanctuaries, which have sometimes run into many years, the Court called on State governments to complete all legal formalities within one year. It also gave orders that forest guards should be given modern arms and communication facilities.

Mr Gujral made his public commitment to protect wildlife and the tiger at a function on 3 October at which scores of children representing 'The Children of the World' presented an appeal signed by 7000 British children on behalf of the tiger. He told the media that Project Tiger would never suffer from lack of funds, because once the species was destroyed it could never come back or be replaced, but money could.

Things have started moving again in India. We await the outcome with great anticipation. Will some of these words be translated into field action?

Conclusion

It is only the Prime Minister of India, with all the 17 Chief Ministers of Tiger States, who can activate a special Task Force with the finance and necessary power to save the tiger and its habitat. One of the vital roles of such a Task Force will be to initiate all the necessary changes in wildlife administration and policy so that by the year 2000 all the inherent weaknesses in the management of these areas are resolved and reformed. This is the most important long-term objective of the Task Force. There are over 100 forest tracts in India today with tigers. If some of the initiatives suggested above are translated into practice we could save as many as 40–50 of these. If little field action results then fewer than 15 will survive.

The battle to save the tiger is on; it is up to all of us to engage in it. The tiger is a symbol of the natural heritage of our planet. None of us want it to end up as a bag of bones, or its home as furniture for our homes. Neither do we want it caged in a zoo. It is a moment for the entire conservation community, in and out of government, to join hands, to find a path to drop our differences and ensure that the tiger has a fighting chance for survival.

Recently I went to Namo Buddha in Nepal, over 2000 m up in the Himalaya. This is where the Buddha is depicted on painting and stone giving up his life, blood and flesh to feed a starving tigress and her cubs. Today people pray and worship the tiger at this spot. It is how religions embraced nature, and it is a lesson for all of us about sacrifice in the interest of nature and for its greatest symbol, the tiger.

Dancers made up as leopards and tigers performing a ritual in Karnataka, India.

21

Reconciling the needs of conservation and local communities: Global Environment Facility support for tiger conservation in India

Kathy MacKinnon, Hemanta Mishra and Jessica Mott

The India ecodevelopment project

India is one of the world's 12 megadiversity countries (Mc Neely *et al.* 1990). The country's biodiversity is coming under increasing pressure from a high, and growing, human population, with many communities living in extreme poverty. Competing land uses and conversion and degradation of natural habitats for agriculture and infrastructure, industrial and commercial activities all place increasing pressure on India's forests, grasslands, wetlands and coastal and marine ecosystems. Human activities such as hunting, cattle grazing, cutting of trees for timber and fuelwood, collection of non-timber forest products, and uncontrolled fires, deliberately or accidentally started, put further pressure on natural ecosystems and their native species. The fate of the tiger, a large predator at the head of the food chain, is a good indicator of the conservation status of India's natural habitats and wildlife (Panwar 1987; Nowell & Jackson 1996).

More than 4.3% of India's land area is designated as protected areas; together 75 parks and 421 sanctuaries cover 14 million hectares across 10 biogeographical zones (World Bank 1996b). Selected high-priority parks and sanctuaries and contiguous reserve forest areas are managed as tiger and elephant reserves. Project Tiger began in 1973, covered nine reserves by 1975, and brought considerable attention and extra resources for tiger conservation (Panwar 1984) but, as government priorities have changed and the human population has increased, resources outside protected areas

have been destroyed and so pressure on the protected areas and their wildlife has increased. Some protected areas in India have few or no human inhabitants, but many more include or abut small villages with significant human populations. These village communities, both within and adjacent to protected areas, subsist on long-established sedentary agricultural systems. The need for land for crops and livestock grazing, losses of cattle to tigers from protected areas, and crop damage from deer and wild pig have led to increasing conflicts between humans and wildlife, and between villages and protected areas.

By the early 1980s, India was already experimenting with special ecodevelopment options and benefits for adjacent villages to reduce pressure on natural resources in selected protected areas (MacKinnon *et al.* 1986). In 1994, the Indian government requested World Bank assistance to begin preparation of the India Ecodevelopment project. The project seeks to promote conservation by addressing the impact of local people on the protected areas and their wildlife and by mitigating the impact of protected areas, and limitations on resource use, on the local people. If this pilot model proves effective in reducing pressures on the protected areas, the Indian government intends to expand ecodevelopment to another 100–200 protected areas nationally. Five of the seven reserves receiving support under the project are designated Tiger Reserves (Buxa, Palamau, Pench, Periyar and Ranthambhore) while Nagarahole is an Elephant Reserve with important tiger populations and Gir supports the world's only surviving population of Asiatic lions.

Together these protected areas cover a range of different habitats (wet evergreen forests to semi-arid grasslands and mountain ecosystems) and protect important plant communities and populations of large ungulates and other wildlife, as well as the charismatic large fauna for which they are most famed. All of these areas have been recognised as globally important for biodiversity conservation (MacKinnon & MacKinnon 1986). An eighth proposed site, Simlipal Tiger Reserve in Orissa, was excluded from the project after World Bank concern over involuntary resettlement of local people (World Bank 1996b). At all sites biodiversity is under considerable threat from competing land uses. At several sites poaching pressure on tigers and their prey species is severe. Ranthambhore, for instance, has witnessed a dramatic decline in its resident tiger population over the last decade due to poaching to supply the medicinal trade (V. Thapar pers. comm.).

At all six sites within the tiger's range the project area covers both the protected areas and the villages peripheral to them, but within a 2-km radius of the protected area boundaries. This is consistent with a landscape approach to conservation that acknowledges that reserves and their wildlife are affected by surrounding land uses. The tiger reserves generally comprise gazetted and proposed parks and sanctuaries, plus specified areas of reserve forest and tourism zones. Nagarahole, Gir, Palamau, Periyar, Pench and Ranthambhore are National Parks. The parks and associated lands within the project areas are zoned for different kinds of conservation and human use. Individual protected areas range in size from 643 to 1412 km² with core areas ranging from 192 to 393 km² (Table 21.1). Generally the core areas have little or no human settlement.

The tiger reserves cover some of the last remaining blocks of natural habitat in India, but these are not uninhabited areas of wilderness. About 427 000 villagers will participate in the project; of these more than 89 000 live within the protected areas, mainly in enclaves and other non-core parts of the tiger reserves where human settlement is allowed. Thirty-nine per cent of these project beneficiaries are tribal people, some of the poorest of the poor and most dependent on biological resources. In some areas these residents depend on harvesting non-timber forest products from within the protected area, an activity that can conflict with conservation objectives. The number of village participants ranges from 36 000 to 75 000 per protected area (Table 21.2). In this densely populated landscape this figure includes some, but usually not all, of the people living within 2 km of the protected area boundary. Thus for Periyar the project will work with 62 000 villagers, less than a third of the estimated 225 000 people living within 2 km of the reserve. Activities for changing behaviour and livelihoods will focus on those villages having the greatest impact on the protected area. Most eco-development investments will take place along the immediate boundaries of the protected areas.

The $67 million project will cover a five-year period from October 1996 to September 2001 and is financed with government of India funding of $19 million; a concessional loan of $28 million from the International Development Association; and a $20m grant from the Global Environment Facility (GEF) which provides grant funding to developing countries to support the incremental costs of conservation and sustainable use of biodiversity that is recognised as globally important. The government's willingness to borrow funds for conservation demonstrates a clear commitment to meet its national obligations under the Convention on Biological Diversity. From the total, $60 million will flow directly to the seven protected areas and will focus on three main components: (1) improved protected area management; (2) village ecodevelopment to take pressure off, and encourage support for, the conservation areas; and (3) education, visitor management, monitoring and research. The project has been designed to be participatory in both preparation and implementation to reflect the needs of affected populations and the concerns of major stake-holder groups. Project preparation was coordinated by the Ministry of Environment and Forestry in partnership with a national coordinating NGO, the Indian Institute of Public Administration,

Table 21.1. *Protected areas (PA): biological importance*

Protected area and state	Area (km²)[e]	Biogeographic zone and province	Major vegetation	Biodiversity values
Buxa Tiger Reserve, West Bengal	761[a] 315[b] 55[c] 391[d]	Central Himalaya Lower Gangetic Plains	Tropical moist and subtropical hill forests	High floral and faunal diversity, regional PA connectivity, clouded leopard, elephants
Nagarahole (Rajiv Gandhi) National Park, Karnataka	643 192 451 0	Deccan Peninsula: Deccan Plateau South	Tropical semi-evergreen; southern tropical moist and dry deciduous and bamboo forests	Nilgiri Biosphere Reserve, large prey–predator system, regional PA connectivity
Palamau Tiger Reserve, Bihar	1026 213 766 0	Deccan Peninsula: Chota Nagpur	Sal-dominated and mixed deciduous and bamboo forests	Large prey–predator system, regional PA connectivity, largest forested area in State
Pench Tiger Reserve, Madhya Pradesh	758 293 157 308	Deccan Peninsula: Central Highlands	Southern tropical dry deciduous forest; teak dominated and mixed	Large prey–predator system, regional PA connectivity
Periyar Tiger Reserve, Kerala	777 350 427 0	Western Ghats Mountain	Tropical wet and semi-evergreen and moist deciduous forests; montane grasslands	Very high diversity and endemism in plants, invertebrates and lower vertebrates
Ranthambhore Tiger Reserve, Rajasthan	1335 393 801 0	Gujarat Rajwara Semi-Arid	Tropical dry and northern dry deciduous forest; dry deciduous scrub; man-made wetlands	Large prey-predator system, wetlands, waterfowl

[a] Total area.
[b] Core area.
[c] Non-core sanctuary area.
[d] Forest reserve and enclave.
[e] NB: Project area even greater.
Source: World Bank 1996b.

and has involved a wide range of organisations, including local government agencies, local NGOs and community groups.

Even with this fully participatory approach, the project has excited considerable controversy, both within India and from the international NGO community. Accusations range from those from the conservation community that the project is too development-and-people-orientated, to those from advocacy groups that accuse the Government of India and World Bank of putting the interests of wildlife before the needs of people. The difficult challenge faced by the Government of India is to try to reconcile the legitimate needs of local communities with the conservation objectives of protected areas and to link conservation with development and poverty alleviation for the mutual benefit of both wildlife and people. The ecodevelopment approach has evolved as a response to conservation of protected areas in a context where enforcement alone cannot succeed but appropriate development is perceived as a tool to further conservation goals. Strengthened protection and management of the protected areas is an integral part of the strategy. Tigers, and their population trends within the protected areas, will be key indicators in assessing the success of this conservation approach.

Table 21.2. *Project beneficiaries and people living within tiger reserves*

Protected area (PA)	Total beneficiaries[a]	Population inside PA[b]	Comments
Buxa	36 000	15 600 (0)[c]	No settlement in NP, 37 Forest Villages in Reserve
Nagarahole	70 000	7100 (0)	54 settlements in notified NP; including landless labourers
Palamau	75 000	39 000 (630)	Three Forest Villages in intended NP (pop. 630). 102 villages in enclaves in sanctuary
Pench	48 000	12 000 (0)	No settlements in proposed NP; already resettled
Periyar	62 000	3900 (0)	Three tribal settlements (pop. 2036) and one agricultural settlement (pop. 1820) in sanctuary fringe. 225 000 people live within two km radius of park
Ranthambhore	64 000	4300 (1210)	Four villages on boundary of declared NP (pop. 1210); 25 villages in sanctuaries (pop. 3067).

[a] Some but not always *all* of the people living within 2 km of PA boundary.
[b] Inside PA but usually outside core conservation area.
[c] Population inside core area. N.B. if all Nagarahole NP (2465 km²) is regarded as core then there are 8940 people in the park.
Source: World Bank 1996b.

Protected area management

The project's main objectives are to strengthen protected area management to conserve biodiversity and to increase the opportunities for local communities to participate in, and realise benefits from, protected area management. As a tool to this end the project will promote ecodevelopment opportunities to reduce local pressures for land and resources from the protected areas and to mitigate the negative impacts of the protected areas on local communities. It aims to increase the collaboration of local people in conservation efforts.

Approximately $14 million, one-fifth of project resources, is allocated to the strengthening of protected area management through better management planning for the seven protected areas (six reserves with tigers). This includes preparation and implementation of management plans, ecosystem management and habitat restoration, and improved control of fire, poaching and problem animals. Protected area boundaries will be rationalised and management plans will be prepared using an ecosystem approach that integrates buffer-zone forest management with the conservation objectives of the adjoining protected area. The project explicitly recognises that the staff are the 'front line troops' in protected area management, enforcement and anti-poaching activities, and allocates resources for upgrading staff facilities and provision of field kits as well as training for career development.

The project will also seek to integrate protected area and conservation concerns into regional planning and regulation to ensure that regional development programmes support, rather than threaten, protected area integrity and viability. As part of the management strategy there will be greater consultation and involvement of local communities, both to increase local support and to develop benefit-sharing arrangements for protected area by-products and enterprises such as eco-tourism.

India's tiger reserves are the main strongholds for long-term protection of this endangered cat. Any strategy to protect the tiger depends on reliable and adequate financial resources for protection and management of the reserves. Yet India's protected areas suffer from the same budgetary shortfalls and delays as do many other protected areas in the developing world; too often conservation activities are at the bottom of the list and among the first items to be cut when governments are balancing their budgets. To address this issue strategies will be developed for each protected area to ensure sustainable financing to cover the recurrent costs of management activities.

Village ecodevelopment

More than half of all project funding ($34 million) is allocated to village ecodevelopment activities. Root causes of biodiversity loss within the protected areas include rural poverty, grazing pressures, over-exploitation of natural resources and poaching. The ecodevelopment resources address some of these by providing alternative livelihoods and promoting rural development for communities in and around the protected areas, but only development that is appropriate and consistent with conservation goals.

Teams comprising reserve staff and local NGOs will work with local communities and, through participatory rural appraisal techniques, determine village dependency and impacts on the protected areas. Detailed village-level planning will identify potential investments and activities that would mitigate negative reserve/people interactions and propose reciprocal ecodevelopment commitments and responsibilities. Investments will be provided to local villages to foster alternative resource use and livelihoods, while communities will commit to measurable actions to improve conservation efforts. Ecodevelopment investment categories could include alternative fuels, ecotourism, woodlot and forestry programmes, agriculture and watershed management, small-scale irrigation, bee-keeping and honey collection, sustainable harvesting or horticulture of non-timber forest products and medicinal plants, agro-processing, and artisanship to increase value of local products. Site-specific investments will be decided according to local appropriateness and the needs and desires of local

communities. Actions to enhance conservation efforts might include commitment to curtailed grazing of livestock, curtailed fuelwood collection within the protected area, and increased participation of local communities in anti-poaching efforts.

Village ecodevelopment investments will be allocated according to specific criteria in addition to village conservation commitments; investments associated with the reciprocal commitments must meet all of the following criteria. These investments must conserve biodiversity, either directly or indirectly by creating sufficient incentives for a village consensus that commits the local community to specific measurable actions that support conservation. The investments will also mitigate the negative impacts of reserve establishment (e.g. loss of access to resources) on vulnerable groups and ensure equitable distribution of benefits to groups currently dependent on the protected areas, especially tribal people, women and other disadvantaged groups. Investment must be supplemental and not replace current government or other sources of development funding. Proposed investment activities must be technically, financially, socially and institutionally feasible as well as environmentally sustainable. Investments will be selected by the communities themselves to ensure 'ownership' and would not exceed Rs10 000 ($285) base costs per family plus Rs500 ($14) during the initial micro-planning phase to establish credibility and early rewards for conservation action. The local communities are expected to contribute at least 25% of the village ecodevelopment investment themselves in cash, kind, or labour. The project will provide resources for capacity building and training in specific skills to ensure effective micro-planning and investment success. Several communities have already prepared micro-enterprise plans.

In addition to the ecodevelopment investments the project provides financing for special programmes that will mostly take place outside the national park and sanctuary boundaries but inside the 2-km periphery of the protected areas. The programmes would support local community involvement in joint forest management in reserve

Table 21.3. *Lessons learned from other integrated conservation and development projects*

Lesson	Impact on project design
Need for common understanding of project objectives.	Project objectives defined and tested through series of workshops involving local people and NGOs.
	Orientation workshops to project staff, local stakeholders and concerned NGOs.
Need to incorporate PA concerns into regional planning and regulation	Project design includes specific PA component.
Need for linkage between conservation and development objectives.	Village ecodevelopment funds dependent on reciprocal agreement that specifies conservation actions of local people
	Priority development interventions that directly depend upon or enhance biodiversity conservation.
Need for active participation of beneficiaries.	Participatory rural appraisal.
	Participatory microplanning.
Need to define project's scope.	Analysis of implementation capacity and human pressures used to select project areas and to define project scope for each PA.
Risk of project investments acting as magnet to draw migrants into area	Project design incorporates (1) voluntary relocation; (2) not providing social services; (3) strong encroachment control; (4) alternative livelihoods; (5) limiting investments; (6) monitoring of migration activity; (7) project areas with relatively sedentary populations.
Need to identify parties to disputes.	Provide mechanisms for conflict resolution.

Source: World Bank 1996a.

forests adjacent to protected area boundaries, using models tried and tested under other joint forest management projects in several states in India (World Bank 1996b; Table 21.3). Supplementary investments through a discretionary fund allow managers of protected areas to allocate resources to areas with special needs (e.g. special watershed management priorities and communities dependent on forest resources).

Several of the protected areas have substantial human populations within their boundaries whose agricultural, livestock farming and collecting activities bring villagers into competition and conflict with the needs of wildlife and conservation. The World Bank has strict guidelines on resettlement but recognises that human use and conservation objectives may often conflict within protected areas.

A primary concern is that any resettlement should be voluntary and that those resettled should not be worse off after the move. Under the project a small amount of project funds, less than 1% of total project cost, is allocated for voluntary resettlement of people currently living within the protected areas, including resources to aid households in the transition period, investment funds for alternative livelihoods, and support services.

Resettlement remains a complex and contentious issue. At Buxa, one village is actively seeking support to enable it to move out of the reserve. Forestry officials there intend to initiate a planning process soon that would conform with the project and World Bank guidelines and address the wishes of these local people. At Nagarahole, several thousand tribal people live within the boundaries of the new national park. Under the project there may be a variety of ways to address the issue of people living within the current park boundaries. For example, some of the people within Nagarahole live on a periphery that might be excluded from the park by redefinition of boundaries, and the protected area management and research activities of the project could consider the environmental feasibility of this option. The project could also provide support, in accordance with guidelines agreed with the World Bank, for those people who choose to relocate, after fully participating in a new 'bottom-up' planning process.

Education and visitor management

The project will develop an environmental education and awareness strategy with help from professional educators, NGOs, park and reserve staff, and scientists to increase awareness of conservation issues and needs in all sectors of society. This will include a mass-media campaign, the expansion of educational programmes for school children, and development and support of alternative media, such as dance troops, for promoting conservation messages. Visitor management and ecotourism strategies will be developed for individual areas as well as reserve-specific visitor information and interpretative services. As a flagship species and major attraction for visitors, the tiger is likely to feature extensively in educational, interpretative and promotional materials about the reserves.

Monitoring and research

Monitoring programmes and management-oriented research will improve understanding of issues and solutions relevant to protected area management, meeting species' conservation needs and the interactions between protected areas and local communities. The project will support both ecological and socio-economic monitoring to measure the effectiveness of the project in reducing pressures on the reserves by monitoring protected

area integrity, disturbance indicators, population trends of key species and changes in resource use by local communities. It will also support ecological and socio-economic research relevant to protected area management, species' conservation and sustainable use.

Specific resources are allocated for research on the ecology and population dynamics of key and endangered species such as tigers, for surveys on status, distribution and foraging patterns, and through radiotelemetry studies. Monitoring and research will be contracted to institutions and individuals with proven expertise. Local people will be involved in participatory monitoring as much as possible, especially in relation to human use of protected areas and adjacent lands for harvesting of natural resources, especially non-timber forest products and medicinal plants, and for cultural needs. There will be training and dissemination programmes to build local capacity for monitoring and to facilitate communication of findings on useful experiences and demonstration experiments.

To support all the above activities there will be a strong project management component, aimed at strengthening management at all levels from the individual protected area to national administration. This will include national policy studies and dissemination of lessons learned to other states. If this model of integrated conservation and development proves effective, a second ecodevelopment project will be prepared to build on, and further develop, ecodevelopment management strategies and apply them to other protected areas throughout India.

Ecodevelopment and tiger conservation

The India Ecodevelopment Project will contribute directly to tiger conservation through *in situ* protection of tigers and their habitats. It will strengthen management of six protected areas within the tigers' range, including provision of resources and training for field staff in ecological monitoring and anti-poaching activities. Through ecodevelopment investments and reciprocal conservation agreements with local villagers, the project will reduce and mitigate threats to the protected areas and native wildlife by providing alternative livelihood and land-use options to communities living within, or adjacent to, tiger reserves. Specifically these activities can be expected to minimise human/wildlife conflicts and to promote those that emphasise the benefits of tiger conservation (e.g. ecotourism).

Of the six reserves with tigers that will benefit under the project, Pench, Periyar and Nagarahole were identified as top priority Tiger Conservation Units (TCUs) in a recent strategy document for tiger conservation (Dinerstein *et al.* 1997; E. D. Wikramanayake this volume). These areas have been identified as conservation priorities for tigers as they have large blocks of suitable tiger habitat, adequate core areas, low poaching pressure and the best chance of retaining viable tiger populations over the long term. Nagarahole supports a large and relatively secure tiger population (K. U. Karanth this volume Chapter 8) and is part of the Nilgiri Biosphere Reserve, one of the largest conservation areas in India. Apart from its importance for tigers and its high ungulate population densities, Nagarahole is a central link in the seasonal migrations of elephants to Bandipur NP in the southeast and Wynad Wildlife Sanctuary in the southwest. Pench and Periyar are also both part of regionally connected habitats that are crucial to the long-term survival of large predator/prey systems.

Buxa, Palamau and Ranthambhore are given a lower conservation rating because of their isolation from other habitat blocks with tigers, fragmentation and high poaching pressure. However, it is recognised that they can harbour small populations of tigers and are important in national and state strategies for tiger conservation. Moreover, the tiger is a resilient and fast-breeding species and tiger populations can quickly recover with adequate protection of the big cats and their prey base. Improved protection and strengthened management, including strenuous anti-poaching measures at these smaller reserves, should therefore contribute quickly and significantly to tiger conservation.

The India Ecodevelopment Project is consistent with many of the priorities identified by Dinerstein *et al.* (1997) in that it promotes:

1   *In situ* conservation by protecting remaining habitats and strengthening protection to reduce hunting pressure on tigers and their prey.

2   A landscape approach to influence land use outside protected areas and in buffer zones, consistent with maintaining large areas of natural habitat needed by wide-ranging predators such as tigers.

3   Financing and support for anti-poaching networks, adequate staffing and training in protected areas and reserve forests, and integrated conservation and development programmes focused in and around protected areas.

4   Awareness and education to emphasise the importance of tiger conservation and to combat poaching for tiger bones and other body parts for the medicinal trade.

5   Monitoring of habitat integrity, poaching pressure and the population status of tigers and their prey populations. These indicators are powerful predictors of the long-term viability of tiger populations.

Linking conservation to development: opportunities and risks

Projects that attempt to integrate the dual objectives of conservation and development are often doomed to only limited success, or even failure, because their proponents fail to recognise that they are often trying to marry conflicting agendas (Wells & Brandon 1992; Brandon 1997; MacKinnon 1997; Wells *et al.* 1998). The India Ecodevelopment Project benefited from lessons learned from other conservation projects throughout Asia in integrating conservation and development and involving local communities in natural resource management, and from joint forest management, rural development and multi-state projects in India (Table 21.3).

While the protected areas selected for inclusion in the project were chosen according to their biological values, the long-term survival of those areas and their constituent species, including the tiger, will depend on a combination of political, social and economic factors. For any biodiversity conservation project to succeed it therefore needs to have clear project objectives and a common understanding of those objectives among all stake-holders. The active participation and support of all beneficiaries and stake-holders, and the mechanisms for identifying and resolving conflicts between them, will be crucial to project success. Considerable progress has been made in establishing these processes around individual protected areas in India.

Protected areas are the cornerstones of biodiversity protection, but protected areas alone will not conserve wide-ranging species like the tiger, especially if those protected areas are small, fragmented and resented as areas set aside from mainstream development. The India Ecodevelopment Project attempts to incorporate protected area and conservation concerns into regional planning and regulation and to ensure that protected areas are recognised as a legitimate and valuable form of land use in a densely populated landscape. In this context, small-scale development opportunities are presented to local communities as incentives for support of protected areas and conservation, with clear and explicit linkages between conservation and the economic investments.

One concern is that the development opportunities provided might act as a magnet to draw poor villagers into the project area, thereby further exacerbating pressure on the protected areas. In general the project areas have relatively sedentary populations rather than being frontier regions and there is limited financing for enterprise investments, so that the development resources are unlikely to attract new migrants. In addition, the project is not providing social services in core areas. Nevertheless, migration will be monitored to ensure that the project development activities are not acting as a draw for immigrants. Enforcement capacity will be strengthened to stop encroachment and poaching. Under regional development programmes the states could further reduce population pressure on the reserves and remaining tiger habitats by promoting employment and economic growth away from the protected areas. However, this is a strategic development planning issue mostly beyond the scope of the present project.

This is a critical time for the tiger in India (see other chapters in this volume). This chapter has focused much more on people and rural development issues than on tiger ecology. Tiger populations are adaptable, fast-breeding and resilient, but ultimately the tiger's fate in India will depend on how well its needs for habitat and prey can be accommodated in a country with the second highest human population on earth. Protected areas, and the staff and resources to protect them effectively, are essential ingredients in any tiger conservation plan. However, the tiger reserves are not remote wilderness areas, but small islands in a densely populated landscape of rural poor. The ecodevelopment model tries to address some of these social issues and root causes of biodiversity loss by providing alternative livelihoods and incentives for changed behaviour to local communities to take pressure off the protected areas and their wildlife. A major indicator of project success will be how well tiger populations fare in the targeted reserves.

22

Tigers as neighbours: efforts to promote local guardianship of endangered species in lowland Nepal

Eric Dinerstein, Arun Rijal, Marnie Bookbinder, Bijaya Kattel and Arup Rajuria

Introduction

Over the past five years the number of free-ranging tigers has declined across Asia, yet only meagre donor funding has been allocated to managing or expanding core tiger reserves, anti-poaching information networks and other more traditional responses to endangered species conservation (Nowell & Jackson 1996). The majority of funding has flowed into expensive integrated conservation and development projects (henceforth ecodevelopment projects) that attempt to integrate biodiversity conservation with economic development for poor villagers living around tiger reserves. The World Bank alone is pouring millions of dollars into such projects in ten reserves in India and others across Asia (K. MacKinnon et al. this volume).

Why have ecodevelopment projects become so popular among the multi-lateral and bi-lateral agencies? First, these projects have elements more familiar to the development-oriented agendas of the agencies. Secondly, traditional approaches to conservation – often dismissively described as the 'guns and fences' or exclusionary approach – have acquired a reputation for being ineffective. But the near-abandonment of traditional approaches and the heavy lean on ecodevelopment projects is also failing to save tigers. In short, the situation is not either/or.

In the case study presented here, an ecodevelopment project was built into an existing strict preservationist project producing favourable results for tigers and their human neighbours. Nepal's Royal Chitwan National Park (henceforth Chitwan)

supports the highest density of tigers in the world and perhaps the highest or second highest density of greater one-horned rhinoceroses (Dinerstein & Price 1991; Dinerstein et al. 1997; J. L. D. Smith et al. this volume Chapter 13). The 932-km² park, located in the relatively flat, low-lying Terai zone, encompasses an important mosaic of alluvial grass-lands and riverine forests that once dominated the Gangetic and Brahmaputra plains (Fig. 22.1; Dinerstein & McCracken 1990). Today, the popula-tion of tigers in Chitwan and the adjacent Parsa Wildlife Reserve has recovered to about 118 indi-viduals according to the 1996 census (Nepalese Department of National Parks & Wildlife Conser-vation 1996), and populations of rhinos are at the highest level in decades (Yonzon 1994; E. Dinerstein, pers. obs.).

These encouraging trends result from strict pro-tection inside the park, an effective anti-poaching information network, the presence of the Nepalese Army inside the park, and careful monitoring by His Majesty's government and such non-governmental organisations (NGOs) as the King Mahendra Trust for Nature Conservation (KMTNC). Others familiar with Chitwan attribute at least part of the recovery to a well-established ecotourism industry that has created jobs for local people, and the long-standing policy of allowing villagers access to Chitwan's grasslands to cut thatch grass and canes.

But a study of the economic impact of the privately based ecotourism industry on local house-hold income (Bookbinder et al. in press) refutes some of these observations. The study found that: only a tiny fraction of about $4–5 million per year

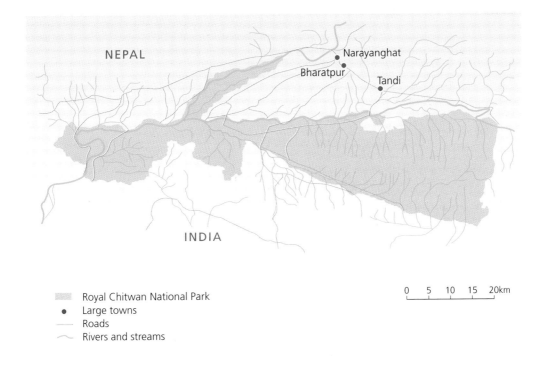

FIGURE 22.1
Map of Royal Chitwan National Park showing locations of towns, roads, rivers and streams.

earned by the hotels is recycled into local economies; only 1100 people are directly employed in the eco-tourism industry (out of 87 000 in the workforce); and indirect employment of local people is small with few household incomes affected by eco-tourism. The thatch grass effect is also a myth; plant succession in Chitwan's grasslands has greatly reduced the amount of thatch species available, and many people are more interested in concealing firewood in their grass bundles during the two-week free access period than in collecting the grass (J. Lehmkuhl pers. comm.).

What appears irrefutable is that the rebounding wildlife populations in Chitwan are the result of strict protection efforts against poaching and encroachment – not the benefits accruing to the local people living near the park. For species like tigers that are sensitive to high levels of human disturbance, there is no substitute for large core areas with strict protection as the centrepiece of landscape management. Without this element ecodevelopment experiments in park buffer areas are building on quicksand.

But another truth is also becoming clear; the park is too small to maintain viable populations of tigers and rhinos, and the prime habitat in the buffer zones and corridors attached to the park is being severely degraded.

How can degraded habitats and buffer zones be restored, dispersal corridors maintained, and core areas remain inviolate without collaboration of local villagers? They cannot. Here again, the situation is not either/or. In Chitwan, an ecodevelopment component layered over the strict preservationist project has begun to expand habitat for endangered species beyond the park boundaries to ensure their survival.

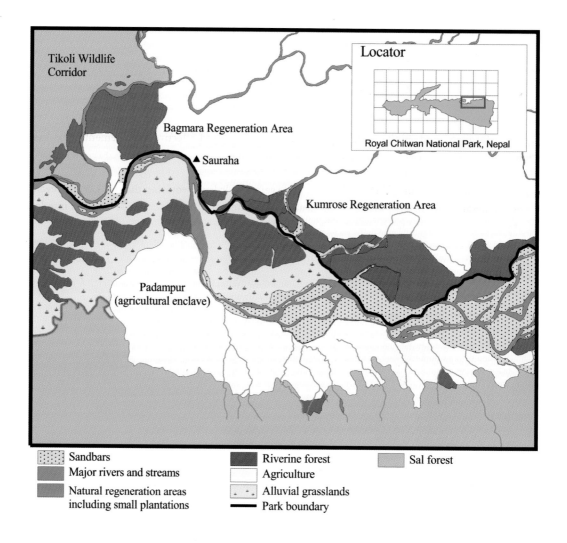

FIGURE 22.2

Buffer zone regeneration forests adjacent to Royal Chitwan National Park and the two regeneration areas mentioned in the text.

To examine the lessons learned in Chitwan, and to make recommendations for improving ecodevelopment projects so that they meet wildlife conservation goals, this chapter is divided into three parts. Part I describes the evolution of the development component of the project. Part II demonstrates that ecodevelopment projects can and must be measured for their biodiversity conservation value; we introduce a method that is doable, rigorous, shows results quickly and closely links investments to fundamental goals of biodiversity conservation (Olson & Dinerstein unpubl. data). Part III lays out a model to help predict where ecodevelopment projects in tiger habitats have the best chance for success.

We thank the staff at the Nepal Conservation and

Research Centre, particularly Shankar Choudhury, Bishnu Bahadur Lama, Harka Man Lama, Kapil Podhrel and Bul Bahadur Lama for their help in data collection. Binot Adhikari, Deepok Adhikari, Ramesh Choudhary, Rupendra Karmacharya, Maya Pandit, Chudamani Poudel, Om Rijal, Yogendra Tamang, and Binita Wagle provided essential help in conducting the questionnaires. The Department of National Parks and Wildlife Conservation supported our efforts and we thank Director Generals T. M. Maskey and U. R. Sharma. The manuscript benefited from comments by Janis Alcorn, Andrea Brunholzl, Hank Cauley, Robert Murray, David Olson, Kent Redford, John Robinson, Eric Wikramanayake and Anita van Breda. Funding for this project was provided by the Biodiversity Conservation Network, the Save The Tiger Fund, World Wildlife Fund-US, and from a donation by Jeffrey Berenson. The Biodiversity Conservation Network is funded by the United States-Asia Environmental Partnership led by the United States Agency for International Development. E. Dinerstein was supported by an Armand K. Erpf Fellowship and a grant from J. Berenson.

Part I: The evolution of the ecodevelopment component

Our project was conducted in the buffer zones adjacent to Royal Chitwan National Park (Fig. 22.2). Bordering three sides of Chitwan are 36 Village Development Committees supporting a total population of over 260 000 people. The annual per capita income is estimated at $150, and more than half of the population lives in absolute poverty, earning less than $100 annually (Keiter 1995). The demand for firewood, fodder and grazing areas puts pressure on the habitats of Chitwan and its buffer zone. One of the initial goals of the project was to take the pressure off these habitats by meeting the natural resource needs of local people through better management of the buffer zone forests.

Beginning in the early 1980s, Nepalese conservationists lobbied, without success, to allow local village committees to take over management of degraded Forest Department lands adjacent to protected reserves. But in 1993, a major reform in national policy allowed legal buffer zones to be created around existing protected areas. Management of these zones would be handed over to local User Group Committees (UGCs), providing that they developed effective management plans based on rational use of resources. Additional landmark legislation came in 1995, when Parliament ratified a series of bylaws requiring that 50% of the revenue generated by protected areas be recycled into local development programmes in the buffer zone surrounding national parks, instead of returning to the Ministry of Finance. Although not yet fully operational, these two initiatives paved the way for establishing legal economic incentives to reduce pressures on core reserves, and to conserve wildlife habitats outside parks. Our project was fortuitously timed to take advantage of the new legislation; without these policy changes, the project would have been impossible.

Starting small

A $10 000 award from USAID in 1988 financed the creation of a native tree nursery on the private land of a KMTNC forest ranger. The following year a degraded 32-ha plot of government land in the buffer zone (henceforth Bagmara Community Forest) was fenced and established as a locally managed tree plantation. Native rosewood and khair were planted with three other native tree species to provide timber and firewood. Some of the tree species, particularly rosewood, were also planted as an incentive for local people to support habitat regeneration long term; the valuable wood would sell for large sums in 20 years.

Fencing off of non-plantation areas was also established to allow natural regeneration and to discourage grazing and resource extraction activities. An additional 20 ha of plantations were added in 1989. During this initial phase, the local village group did not technically have ownership of this land.

The first key piece of legislation in 1993 sanctioned the organisation of UGCs to manage these buffer zone forests. Members of these committees,

Regeneration of vegetation after six months in the Bagmara buffer zone, Chitwan Valley, Nepal, when protected by fencing from domestic livestock grazing.

elected by local people, represent various community interests. Once established, a UGC petitions the local District Forest Officer for management rights to buffer zone land. Since 1997, UGCs petition the Department of National Parks and Wildlife Conservation – the agency that now manages development activities in the buffer zone – for authority to manage the land. Previously, buffer zone forests were managed strictly by the Department of Forestry – a centralised and financially limited agency unable to stem the degradation of important forested wildlife habitats and critical forest corridors adjacent to Nepal's parks and reserves.

Between 1994 and 1996, the Bagmara Community Forest expanded its regeneration area to 460 ha. Over the same interval, our project received major funding from the Biodiversity Conservation Network ($647 000 over three years) to finance, among other activities, the planting of native trees and the natural regeneration of another 1050 ha of degraded riverine forest and alluvial grassland in

collaboration with the Kumrose UGC (Fig. 22.2). The project's goals were to meet the firewood, fodder and grass needs of over 10 000 villagers in the two areas and to develop micro-enterprise activities in the second phase of the project. The rapid regeneration of the forest was the important precursor to achieving these goals. A recent grant from the Save The Tiger Fund has allowed 150 ha to be fenced for natural regeneration in the Kothar area in the summer of 1997 (not shown on map). Plans for further extensions are underway in several other areas of the buffer zone.

Community-based ecotourism and biodiversity conservation

Once the regeneration areas were fenced and protected, the same wildlife species that inhabit Chitwan began to recolonise the buffer zone. Simultaneously, many conservationists, and certainly many local villagers, recognised that tourism in Chitwan was predominantly a private enterprise controlled by a few. Both of these groups also suspected that the amount of profits recycled to the surrounding villages was marginal; our investigation of the privately owned ecotourism industry confirmed this perception (Bookbinder *et al.* in press).

To return more of the profits to local people, the Bagmara User Committee (representing over 584 households in the Chitwan buffer zone), WWF and the Biodiversity Conservation Network proceeded with plans for community-based ecotourism in the regenerating forest. Nature trails were established for elephant-back safaris, and guards were hired from the local community to protect the wildlife from poachers and the habitat from trespassers illegally collecting firewood. A wildlife viewing tower was also constructed, enabling tourists to enjoy overnight stays in the jungle and to observe rhinos using wallows in the viewing area. The User Committee also undertook management of the wildlife viewing tower.

The result has been a new sense of empowerment of local people living adjacent to the buffer zone forests. The plantations have begun to meet the natural resource needs for fodder grasses, and partly offset the needs for firewood. The biggest gains have been in the development of local people as entrepreneurs in a community-based ecotourism programme (see later) and the desire for replication of these activities in many other parts of the buffer zone of Chitwan.

Part II: Measuring the conservation impact of ecodevelopment projects

Biological components

Perhaps the weakest component of ecodevelopment projects is their lack of a framework, funds, and staff to measure their impact on endangered species and habitat conservation. The impact of the ecodevelopment component in our project was measured using a Biodiversity Conservation Framework (Olson & Dinerstein unpubl. data). This framework measures impacts at three levels of biodiversity: the status of sensitive indicator species; habitat integrity; and maintenance of critical ecological

processes. To apply this method to our study of the impact on wildlife and habitats of Chitwan and the regeneration areas, we identified the following indicators for each variable:

> Sensitive species: (1) densities of tigers; (2) relative abundance of tiger prey; (3) densities of rhinos; (4) recruitment of these species in regeneration areas of the buffer zone; and (5) poaching levels of rhinos and tigers. We also assessed the use of regeneration areas by rare forest birds (Crosby in press), and all forest birds (these data not included in this chapter).

> Habitat and landscape integrity: (1) regeneration of critical habitats preferred by endangered species; and (2) local guardianship and improved management of habitats restored by User Committee activities in the buffer zone.

> Ecological processes: (1) restored integrity of wildlife corridors as a proxy for measuring dispersal of tigers; (2) predation by tigers in regeneration areas; and (3) maintenance of early successional habitats containing the full suite of native species.

We set up a monitoring programme to estimate the numbers of target species that recolonised the regenerating (previously degraded) forests. We relied on photographic identification for rhinos (Dinerstein & Price 1991), pug mark and scat analyses for tigers and tiger prey, and direct observations for rhinos, tigers and tiger prey. Recruitment was monitored by direct observation (rhinos) and pug marks (tigers). Censuses were conducted twice per month from elephant back in the two main regeneration areas; each census used up to four elephants per census and the same observers. We also included a control plot in the census; Icharni Island, an adjacent block of habitat inside Chitwan that supports one of the highest recorded densities of rhinos anywhere (Dinerstein & Price 1991). The relative abundance of tiger prey in regeneration areas is also being monitored (A. Rijal pers. comm.).

We include rhinos in this analysis for several reasons. Calves under one year of age are preyed on by tigers (Dinerstein & Price 1991). Rhinos share the same habitat preferences as tigers and are also endangered. But most importantly, poachers are likely to take both tigers and rhinos, so information gathered from Chitwan's anti-poaching information network is useful for both species.

Collection of data on the poaching of rhinos and tigers is ongoing but only data from January 1990 to March 1997 are presented in this chapter. For rhinos, the data were taken mostly from field notes or conversations with the senior and assistant park wardens, Mr. Ram Pritt Yadav and Mr. Tika Ram Adikhari. Both officials lived in Chitwan during much of the monitoring period, and Mr. Yadav has been a warden for the past 25 years. We also took satellite location readings of each poaching incident and used a Geographic Information Systems (GIS) programme (Ecological Consulting, Inc. 1993) to display poaching incidents on a yearly basis in and around Chitwan. To analyse the distribution of poaching events and identify poaching hotspots, we used a harmonic mean isopleth analysis programme (Ecological Consulting, Inc. 1993) for the 45 poaching incidents occurring from 1990 to 1997.

To assess how project interventions led to the regeneration of critical habitats preferred by endangered species, we mapped the project area using GIS and calculated the size of restored blocks of habitat. We also monitored the extent to which User Committees followed recommendations for restoration. These recommendations included: excavating wallows for rhinos, fencing tall grassland habitat adjacent to riverine forests (Kumrose only), stopping the collection of firewood from inside the regeneration areas, carefully managing tourism in the wildlife viewing areas, maintaining short-grass clearings for tiger prey (Bagmara), and leaving behind standing dead trees for cavity-nesting birds.

Direct measurement of the maintenance or restoration of ecological processes is often difficult. For example, monitoring dispersal of tigers, an important ecological process, is best done by intensive radiotelemetry studies. We chose a surrogate measure, the restoration of habitat integrity in wildlife corridors, a prerequisite for effective dispersal. We also assessed predation by tigers on native herbivores in regeneration areas by counting kills. Taken together, the presence of tigers, their prey and breeding rhinos was used to indicate the maintenance of early successional habitats containing the full suite of native large mammal species.

Assessing threats to ecodevelopment projects

One concern about ecodevelopment projects is that raising local living standards or increasing the availability of resources may stimulate migration to the project site. Increased population means greater demands on existing resources – exacerbating the threats to biodiversity conservation and potentially reducing the impact of economic incentives. This concern was addressed by conducting a survey of 996 households chosen at random among six Village Development Committees (the methods used to conduct this survey are described in detail in Bookbinder *et al.* in press). We included a set of questions in the survey to identify land tenure patterns and immigration trends for the study area. The questions were designed to separate trends resulting from ecodevelopment project investments from other trends, such as the need for flat, fertile agricultural land or people's attraction to growing markets for commerce.

Rapid population growth – resulting from immigration or other demographic fluctuations – can dilute the benefits of project investments. To determine the demographics of the population in our study area, we identified and collected the data necessary to calculate eight key parameters: population density, population structure by gender and age class, growth rate, natural rate of increase, total fertility rate, population doubling time, infant mortality rate and average family size.

Conservation impact on sensitive species

A census of tigers conducted in Chitwan and the adjacent Parsa Wildlife Reserve in 1996 estimated a population of 118 tigers (Nepalese Department of National Parks & Wildlife Conservation 1996). The largest number of tigers in Chitwan Valley live adjacent to the buffer zone areas now, or soon to be, under regeneration by this project. Within the

Greater one-horned rhinoceros are increasingly utilising regenerating buffer zone forests in Nepal's Chitwan Valley.

regeneration areas there was no tiger use in Bagmara in 1994, nor was it likely in the years before management of the area began (B. Lama pers. comm.). In 1995 an adult male became a transient visitor in Bagmara. From 1996 to March 1997, five tigers were regularly using the two areas: one adult male, two subadults, and a female and cub.

Densities of rhinos have increased steadily in the regeneration area since the inception of the project. Between 1984 and 1988, when a census of rhinos was taken using a photo-ID technique, they were relatively common (see Dinerstein & Price 1991). But habitat degradation resulting from intensive human use and cattle grazing reduced rhino numbers between 1988 and 1994 (B. Lama, pers.

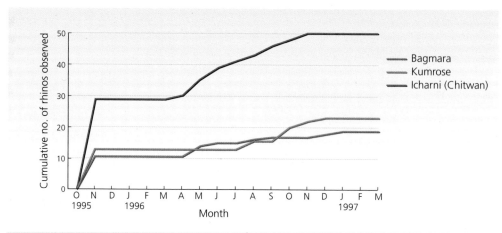

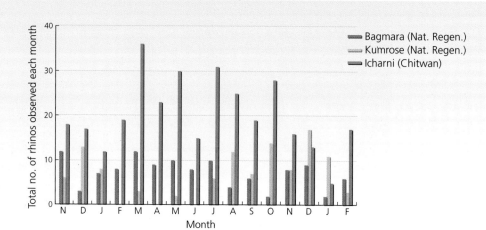

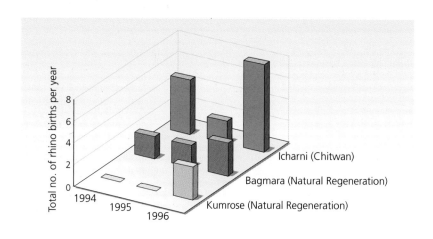

FIGURE 22.3a–c

Recovery of rhinos in regeneration (treatment) areas compared to a control plot (Icharni, Chitwan). **a.** Cumulative number of rhinos observed by month; **b.** total number of rhinos observed by month; **c.** total number of rhino calf births by year over a three-year period.

comm.). Individual rhinos were actively photographed again beginning in 1995 as part of this study. Up until January 1997, rhino numbers have increased in Bagmara and Kumrose and reached a plateau since then (Fig. 22.3a). From December 1996 to January 1997, more rhinos were using the Kumrose regeneration area than the control plot inside Chitwan (Icharni). The monthly average number of rhinos has been 7.3 in Bagmara, 6.9 in Kumrose and 20.3 in Icharni since November 1995, when census-taking began. These numbers correspond to an average density of 7.3 individuals/ 4.0 km² (or 1.8 individuals/km²) in Bagmara, 6.9 individuals/6.0 km² (or 1.2 individuals/km²) in Kumrose and 20.3 individuals/km² (or 6.8 individuals/km²) in Icharni.

As of March 1997, 23 individual rhinos use Kumrose and 19 use Bagmara, compared with 50 in the control site (Icharni) (Fig. 22.3b). Recruitment of rhinos in the regeneration areas has been excellent; 11 calves have been born and survived in the regeneration areas over the three-year period compared with 14 in Icharni (Fig. 22.3c).

Poaching data for rhinos and tigers are the best indicator of the impact of illegal hunting in Chitwan. Data from the past three years are consistent with the hypothesis that interventions of the Project add strength to the anti-poaching programme while moving beyond strict protection measures alone.

Tiger poaching data recorded in and around the park from 1990 to 1997 totalled 11 known incidents and concentrated in two areas in the eastern part of Chitwan (Fig. 22.4a). At least one tiger has been poached in the past year. Between 1990 and March of 1997, 45 rhinos were poached in and around Chitwan. Poaching of rhinos peaked in 1992 (24 incidences) during a breakdown of law-and-order in Chitwan District that weakened the anti-poaching information network. After restoring strict control and reinvigorating the anti-poaching information network, poaching has been insignificant – typically one to two individuals poached per year – since 1994.

Surprisingly, the distribution of tiger and rhino poaching incidents over a seven-year period show little overlap (Figs. 22.4a, b). Harmonic isopleth analysis of seven years of poaching data for rhinos (Fig. 22.4b) overlaid on the location of army guard posts and the park boundary revealed that:

1   Eighty per cent of the poaching was concentrated in two relatively small areas, despite rhinos ranging widely along the Rapti River.
2   Fifty per cent of all poaching occurred in two areas at the northern edge of the park totalling an area <54 km².
3   Most poaching incidents occurred north of major rivers where the Nepalese Army is not responsible for controlling poaching (outside of park) and where there are no guard posts.
4   No poaching occurred between 1994 and 1997 in or around the regeneration areas.

Conservation impact on habitat integrity and landscape management

By July of 1997, project interventions had led to the recovery of 16.5 km² of critical riverine forest habitat in the Chitwan buffer zone (partly shown in Fig. 22.2). The early success of local management and guardianship of these habitats is striking. First, in 1998 not a single dead tree was cut in the regeneration areas, while dead trees in our bird census plots inside Chitwan have been (E. Dinerstein pers. obs.). Secondly, in both Kumrose and Bagmara the local User Committees, using volunteer labour, have dug out extensive oxbows. These are now used as rhino wallows, making rhino censusing much easier. Thirdly, the Bagmara oxbow was also stocked with mugger crocodiles, and it is now home to many species of wading birds uncommon in the area before habitat management began. Fourthly, a critically important tall grassland used by tigers and rhinos has been fenced in Kumrose to keep out cattle, and is being expanded greatly in 1998; the mosaic of *Saccharum spontaneum* tall grasslands and riverine forests are where tigers reach highest densities. Fifthly, illegal firewood collection in the regeneration areas has ceased, as these areas are patrolled by local village watchmen paid for by the User Committees. Sixthly, wildlife ecotourism is carefully managed in both areas so as not to degrade

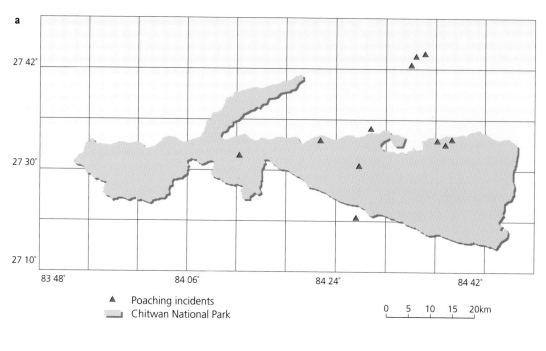

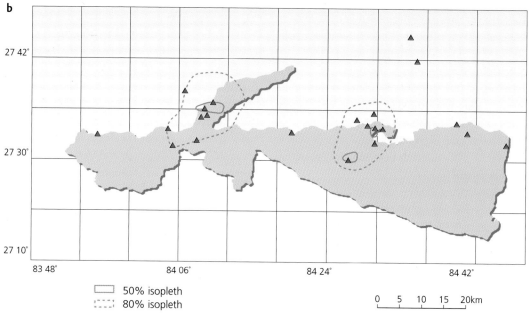

FIGURE 22.4a,b
Poaching of endangered species in Chitwan between 1990 and 1997. **a.** Poaching of tigers has been more prevalent in the eastern half of Chitwan; **b.** Degree of clustering of 45 incidences of poaching of rhinos over a seven-year period. Three small areas totalling about 54 km² encompass 50% of all poaching incidents (solid line). Eighty per cent of the poaching incidents occurred in the areas designated by the hashed line. All of the poaching incidents located near the regeneration areas occurred before the locally managed buffer zone programme began in 1994.

the habitat as is the case in parts of Chitwan. In Bagmara, elephant-back viewing is limited to one-hour rides, with a maximum of five elephants using the area at a time. Elephant drivers are required to stay on trails carefully designed for tourist wildlife viewing. The trails, the wildlife viewing towers and the short-grass clearings around them are maintained by the User Committees who pay for the upkeep using revenues from the ecotourism micro-enterprise.

Have community-based activities enhanced tiger conservation in the larger landscape of Chitwan? The answer appears to be a qualified yes. First, the User Committees and the local community members they represent view themselves as local guardians of endangered species and are now actively protecting the areas they manage from potential poachers and illegal wood cutters. Since the inception of our project, no rhinos or tigers have been poached in the vicinity over which local people have jurisdiction. Secondly, 16.5 km² of critical wildlife habitat is well on the way to recovery. Those unfamiliar with the extraordinary productivity of the Terai riverine forests and grasslands for tigers, rhinos and other endangered species should not view a total of 16.5 km² regenerated in three years as a trivial amount. Rhinos can reach densities of more than 10 individuals per square kilometre and tigers in these habitats achieve some of the highest densities recorded in Asia.

These regeneration areas and the profits generated in the first year have become renowned. Several other User Committees along the periphery of Chitwan now are determined to use the model provided by Bagmara and Kumrose. One of these newly planned regeneration areas will form part of the last remaining forested corridor connecting the deciduous hill forests of the Siwalik Range to the subtropical semi-deciduous forests of the low-lying Terai belt. This corridor to upland areas will have a large ecological impact on maintaining dispersal possibilities for tigers and many other species. Similar conservation investments initiated by local people are planned for five other village committees from 1998 onwards in the eastern part of Chitwan.

Shunting some of the ecotourism activities to the buffer zone has landscape management as well as financial benefits. First, shifting more tourists to the buffer zone – where they will see the same species as inside Chitwan – increases the dynamic core area of the park itself and reduces disturbance of the breeding population of tigers. Secondly, the buffer zone forest has a greater potential for ecotourism than Chitwan, because it is accessible even during the monsoon season when entry to the park proper is restricted by the flooding Rapti River. Finally, tourists who enter the wildlife viewing area in the regeneration forests must still pay a Chitwan entry fee, so the park benefits in revenue from this 'win-win' situation, even if a tourist never technically sets foot in the park itself.

Most Asian reserves are too small to maintain viable populations of large mammals over the long term, but auxiliary landscape features such as corridors, buffer zones and multiple-use areas may allow for the persistence of endangered species living in fragmented habitats. A recent analysis of remaining habitats containing tiger populations across Asia identifies 59 landscape units, called Tiger Conservation Units (TCUs) on the Indian sub-continent where protected areas occupy a small fraction of the total potential remaining block of habitat (Dinerstein et al. 1997; E. D. Wikramanayake et al. this volume). Bringing some fraction of these areas under conservation management before they are converted or further degraded is imperative to the recovery of tigers and other large mammals.

One concern about restoring tiger habitat in formerly degraded buffer zones is that it puts tigers up against the edge of villages. The proximity of tigers to villages sometimes leads to cattle predation and problems with local residents. However, tiger studies in Chitwan have already demonstrated that tigers preyed on domestic stock in the area now occupied by regeneration areas when they were highly degraded (D. Smith pers. comm.).

Still, reduction of cattle grazing must be part of any effort to improve landscape-scale conservation for tigers and their prey. A better solution (now being addressed in the third phase of our field project) is introducing improved breeds of livestock. If local villagers are provided access to veterinary

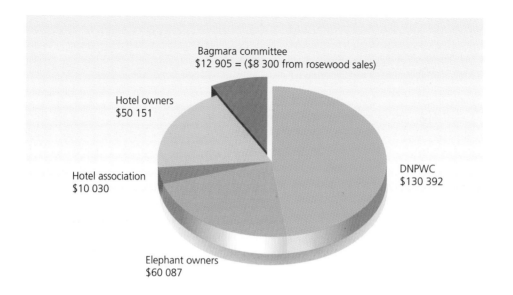

Bagmara committee
$12 905 = ($8 300 from rosewood sales)

Hotel owners
$50 151

Hotel association
$10 030

DNPWC
$130 392

Elephant owners
$60 087

FIGURE 22.5
Distribution of revenue generated from community-based ecotourism activities in the Bagmara area during the first year of operation. The fraction of the revenues allocated to the Bagmara User Group Committee will increase in year two of operations as they negotiate a larger percentage. Also in year two, the Hotel Association will no longer be awarded a percentage of the revenues. Half of all funds generated go to Chitwan via DNPWC for fees to enter the wildlife viewing area located in the buffer zone. Fifty per cent of this sum of $130 392 is recycled back to a local development fund, as mandated by the new federal legislation, to finance other buffer zone development projects. DNPWC = Department of National Parks and Wildlife Conservation.

care for livestock, they will accept improved breeds. Improved breeds are too valuable to be allowed to graze freely and are stall-fed, using fodder harvested from the adjacent plantations.

The promotion of stall-feeding of livestock has another ecological benefit. The dung, which is much easier to collect from stall-fed animals than from free-ranging cattle, is deposited in methane gas digesters to be used as a source of cooking fuel (biogas). In the Kumrose Village Committee alone, 47 methane gas digesters have been installed and another 90 are about to be installed in the coming year. Increasing use of biogas provides another incentive for stall-feeding livestock. Stall-feeding cattle also helps reduce predation by tigers living or dispersing along the borders of the buffer zone. The total population of domestic stock will also decline,

as farmers will sell off unproductive local cattle once they see the benefits of stall-feeding.

Where people live close to core areas, conservation planners must seriously consider moving villages to safer areas outside the TCU. In Chitwan and in our study area, the entire Padampur Panchayat (Fig. 22.2), an area of 14 villages, has been partly resettled and will be completely shifted within the next three years. The villagers of Padampur have been urging for resettlement closer to the major towns of Chitwan District and away from the edge of the park where their crops are destroyed by wildlife and their fields threatened by the inundation of silt deposited by monsoon floods. The flood of 1993 left several villages homeless and initiated the push for resettlement to a site about 20 km away that was set up previously for flood victims. The Padampur example is one that conservationists can point to where resettlement was the desire of the people. Once the transfer is complete, the area once occupied by the Panchayat could allow the tiger and rhino populations in Chitwan to expand by at least 5%. As long as people are willing to shift to equally valuable or superior sites, conservation planners should encourage resettlement away from core areas to the edge of TCUs.

Conservation impact on restoring ecological processes

Project interventions have fostered plant succession in the buffer zones which has led to the restoration of their native vertebrate communities. To date, no tigers have preyed on wild ungulates in the buffer zone, although in early 1997 an adult male tiger killed several domestic wild buffalo in Kumrose. Data are lacking to determine if activities have enhanced dispersal of tigers, but restoration along the buffer zone most likely has a positive effect. Restoring the part of the Tikoli forest that connects Chitwan to the outer range of the Himalaya – a known tiger dispersal corridor – is part of the 1997 workplan (see Fig. 22.2).

Evaluation of the community-based ecotourism project

During the first year of operation (November 1995 through October 1996), 10 632 tourists visited the Bagmara Community Forest, generating $276 432. Prior to November 1995, the income earned by local people and the National Park for use of this area was zero; local nature guides took tourists to the Bagmara Community Forest, but they had no legal right or mechanism to collect a user fee, nor did Chitwan staff.

Between November 1995 and October 1996, all visitors to the Bagmara Community Forest were charged $26, $13 of which is the Chitwan entry fee collected by the Department of National Parks and Wildlife Conservation (DNPWC) in support of the park. Approximately $12 905 – or 4% of the total revenue generated during the first year – was allocated to the Bagmara UGC, and $130 392 or 50% was generated for the DNPWC (Fig. 22.5). The

money earned by the Bagmara UGC was used to build and refurbish three schools and a health post; this money also covered yearly maintenance costs for the 450-ha regeneration area and the wildlife viewing tower.

The Bagmara UGC earned an additional $8300 (estimated) in 1996 from the prescribed thinning cycle for rosewood trees planted on 52 ha adjacent to the wildlife regeneration areas. It will do so every five years. In 13 years, when trees are ready for harvest, the amount earned will be substantially higher.

Half of the 1996 revenues turned over to the National Park from the local ecotourism project ($69 108) has been put into an account containing other funds to be recycled to local development. Because of federal legislation, half of the overall tourism revenue of Chitwan from April 1995 to April 1996 (the Nepalese calendar year) which totalled $852 333 (from park entry fees and hotel concessions) will also be added to this account. As a result, $491 362 is available from 1996 alone for local development activities. (This sum does not include revenues generated in the Kumrose viewing area, which began operating in late 1996.)

Assessing potential threats to ecodevelopment projects; demographics and land tenure

Population growth continues to be of concern, but immigration is not. Along the northern boundary of the park and its buffer zone, villagers live in relatively high densities (Table 22.1). The population is also young. Twenty-two per cent of the population is under the age of 15 and less than 6% are 50 years or older. Twenty-four per cent of the women in the study area are in their reproductive years (15–45 years), and the average number of children a woman will have, assuming that current age-specific rates remain constant throughout her child-bearing years, is between four and five. Within the study area, the immigration rate was only 0.17, but, with a natural increase rate of 2.7, the population will double in 24 years. The average household size is seven.

Land tenure is relatively stable around Chitwan; approximately 87% (865) of the villagers we

Table 22.1. *Comparison of demographic parameters for study area with national and regional statistics*

Demographic data	Chitwan	Nepal	South Central Asia
Birth rate	33	38	31
Death rate	5	14	10
Percentage of population under age 15	22	42	38
Natural rate of increase	2.7	2.1	2.4
Total fertility rate	4.5	5.8	3.8
Population doubling time	24	33	29
Population density (km²)	161	165	130
Infant mortality rate	190	102	79

Chitwan refers to the seven Village Development Committees in the study area. The statistics for Chitwan were calculated from data collected in the 1994 household survey. For Chitwan, the infant mortality rate is a general estimation based on births and deaths over time, not only for 1995. Population density for Chitwan is based on His Majesty's Government/National Planning Commission Secretariat 1991 population census, published in 1994. Nepal and South Central Asia 1995 statistics were compiled and published by the Population Reference Bureau, Inc. Infant mortality rate is based on the number of deaths per 1000 births in a given year.

interviewed in our household survey were landowners. The remaining 131 villagers were landless, with the greatest number (32%) living in Piple Village. Loss of property and farmland due to flooding in 1993 were the primary reasons for landlessness in this UGC. The shifting course of the monsoon-swollen rivers of the Terai often claims agricultural land or deposits alluvial sands, making farming impossible. In addition, 45% of the heads of households are originally from the Chitwan District. Ecotourism has not attracted immigration. The average number of years since immigration to the Terai among the non-native residents (55%) is approximately 22 years – most arrived before the development of a large ecotourism industry.

Part III: Where are ecodevelopment projects most likely to succeed in conserving and extending tiger habitat?

Previous failed attempts at implementing eco-development projects across Asia illustrate that they are not suitable everywhere. Based on the long-term project in Chitwan and observations in other sites, we propose a model that helps predict where eco-development projects are likely to succeed in meeting their overall conservation goals (Fig. 22.6). The three axes of the graph represent what we consider to be the determining variables of tiger conservation efforts: habitat integrity, including the relative resiliency of the main habitat type; poaching pressure on tigers and their prey and the ability to mitigate those pressures; and strength of economic incentives to encourage villagers who reside in the greater TCU to work with conservationists in developing landscape-scale conservation plans. The TCUs in the lower left of the graph experience high poaching pressure and are essentially isolated fragments of tiger habitat. Ecodevelopment projects targeted for these types of TCUs are essentially rural development projects and will have little benefit to long-term tiger conservation.

We argue that ecodevelopment will work best in situations that meet or potentially meet the following scenario characterised by the upper right hand corner of the graph: (1) the TCU consists of a highly resilient habitat type, such as flood plain

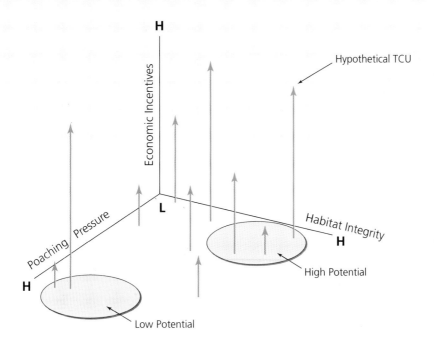

FIGURE 22.6
A model to predict where ecodevelopment projects in Tiger Conservation Units (TCUs) will have the greatest conservation impact on endangered species and habitats. Each vertical line indicates the location in a three-dimensional space created by the intersection of three axes: habitat integrity (including resiliency); poaching pressure on tigers and their prey and the ability to mitigate pressures; and economic incentives that are widely shared and powerful enough to offset threats to tigers and tiger habitat.

grasslands, savannahs, dry woodlands, thorn scrub, or mangroves; (2) there is low poaching pressure on tigers and their prey or a good chance of mitigating such pressures; and (3) there are powerful economic incentives to encourage locals to actively protect wildlife and wildlands.

Most tiger habitats are fairly resilient because they experience annual monsoons; even the driest parts of the tiger's range in India would rebound, we believe, if cattle grazing were removed from these areas. TCUs in more fragile habitat types (e.g. tropical moist forest) also have a high chance of success if they contain very large, unfragmented block(s) of habitat. Thus, outside of the tropical

moist forests, it is largely block size and spatial configuration (e.g. connectivity, width of dispersal corridors, size of core areas) that would contribute to the habitat integrity score. But, because large parts of the tiger's range are not likely to be restored in the near future, this variable is perhaps the least fluid.

The ability to mitigate poaching of tigers and their prey is perhaps the most fluid of the three variables because the historical demography of tiger populations shows that they are very resilient across their range (Sunquist 1981; Matyushkin et al. 1996). Tiger populations rebound quickly if protected and if the integrity of their habitat and prey base is intact.

Economic incentives, the third axis, may thus be the most critical aspect for determining the success of ecodevelopment projects, assuming that the goal is to achieve landscape-scale conservation of tiger populations. For incentives to work, they must be recognised as being of immediate benefit to stakeholders, shared by many, and of sufficient magnitude to offset the threats of habitat degradation and poaching across much of the TCU. We believe that in some of the high-priority TCUs interdigitated with heavily populated areas, nothing short of major revenue-sharing schemes between parks and communities will guarantee the persistence of tigers occupying the larger landscapes or using dispersal corridors between core areas. Without these incentives, strict protection remains the only effective tool for conserving tigers, but this approach will be limited to increasingly isolated reserves as surrounding landscapes become permanently altered.

Conclusion

Our project shows that short-term conservation gains can be achieved with strong protection efforts by the government, even if local residents do not share in the benefits. Over the long term, however, landscape management for tigers requires partnerships with local people. Continued monitoring is now needed to determine the extent to which the recycled revenue programme, regeneration of the buffer zone, and generation of income from micro-enterprise activities are increasing long-term local stewardship towards biodiversity in Chitwan while decreasing threats to the park. We urge conservationists in other tiger range states to lobby for recycling park revenues into local development, as is now the law in Nepal. Without adequate economic incentives, it will be hard to convince poor villagers to make room for tigers in an increasingly crowded Asia.

Box 22.1 Pre-requisites for adapting the Chitwan model to other Tiger Conservation Units in Asia
Eric Dinerstein

Why is the ecodevelopment project succeeding in Chitwan? The Chitwan project falls in the high-potential area of the model presented in Fig. 22.6 for several reasons. Chitwan is a very forgiving, dynamic landscape – a flood plain habitat with high resiliency and moderate to potentially high integrity. Poaching pressure on tigers and prey is low, and powerful economic incentives are already in place to potentially conserve larger landscapes for tigers.

But there are other reasons why the project has been a success. First, the virtual absence of sophisticated firearms among local people reduces poaching pressure. Secondly, the relatively law-abiding nature of Nepalese citizens works in the favour of endangered species conservation, reinforced to a degree by the presence of the Nepalese Army stationed in the reserve. Thirdly, the passionate commitment of a local villager and King Mahendra Trust for Nature Conservation staff member, Mr. Shankar Choudhury, illustrates that the efforts of a single individual on one small plot of land can start a process that conserves a larger landscape. Mr. Choudhury spearheaded the on-farm forestry project on his own property in 1988 and organised the village committees to experiment with plantations and regeneration areas. Few residents gave this project much chance of success a decade ago.

Mounting pressures on natural resources in developing nations in Asia make conserving lands adjacent to protected areas an important goal. Community-based ecotourism in regenerated buffer zones is one way to provide incentives. But as demonstrated in Bookbinder *et al.* (in press), the privately based ecotourism industry is essentially exploitative and unlikely to recycle significant funds to change local attitudes towards conservation. Clearly, local support for biodiversity conservation in the larger landscape requires co-ownership, co-management and policy change. The policy change has already occurred through the recycled park revenue law recently ratified by the Nepalese Parliament. Thus, the single most powerful tool for enhancing success of landscape management is federal legislation that changes the pre-eminent, exclusive character of the privately owned ecotourism industry by linking biodiversity conservation with community development.

Just as the privately owned ecotourism industry in Chitwan was not a panacea for tiger conservation, neither would we recommend the community-based ecotourism approach described in this chapter as a cure-all for conserving tigers in other high-priority areas in Asia. We have identified a distinct set of conditions essential for successful implementation of this approach. These include: a relatively accessible wildlife reserve with a well-protected core area containing a charismatic megafauna; a fraction of buffer zone (or strips of land between the protected area and the agricultural frontier) remaining for regeneration and micro-enterprise activities; a secure land tenure system to minimise immigration in response to the magnet effect of ecodevelopment projects; a stable, privately owned ecotourism industry that can serve as a precursor to a community-based approach and absorb some of the initial costs (or a benevolent entrepreneur willing to help local leaders); policies that enable local people to participate in enterprise activities in buffer zones adjacent to protected areas; a co-operative working relationship between local people and protected area officials; and strong local institutions that enforce local conservation rules, ensure the equitable distribution of benefits from joint activities, and respond to changing economic conditions and new opportunities with a community mindset.

Conservationists must also be careful not to place too much emphasis on economic incentives that promise to raise the standard of living for individual households. A better approach is to stress improving the quality of life for communities, as revenue generated from ecotourism may be more beneficial and easier to manage when spent on community goods.

Epilogue – vision and process in securing a future for wild tigers

John Seidensticker, Sarah Christie and Peter Jackson

In a remote Javanese village, a farmer went out one morning to find a tiger sound asleep beneath his rice barn. Even sleeping, this tiger was a problem the farmer knew was beyond his ability to solve. So the farmer hastened to consult with his village head. The village head accompanied the farmer back to the barn, where the tiger still lay sleeping. Agreeing that this problem was beyond both their abilities, they hurried a few miles to tell the subdistrict officer about the tiger. All three returned to the barn to view the sleeping tiger, then went off to enlist the help of the district officer. Progress up the bureaucratic chain to seek a solution to the sleeping tiger went on all day until finally a by-then large group of men reached the regional commander of the army. The commander marched out to the village and laid out a plan to deal with the sleeping tiger, but before it could be implemented the tiger woke up and moved away. So now they had a different, but still real problem; there was a tiger near the village but no one knew where it was. Suddenly leaving the barn was a very risky proposition.

This story was told to anthropologist Clifford Geertz in the 1950s in the Southern Mountains region of Java, an area where some of Java's last tigers lived. It may be a local joke about bureaucracy, as Geertz believed the story too well formed to be literally true (Geertz pers. comm.), but we believe it to reflect the central dilemma in saving tigers. Lurking unseen or asleep under a barn, a tiger is perceived to be a problem requiring a solution. In the preceding chapters our contributing authors have provided the vision and analytical tools to reach beyond our conventional problem-solving approaches in tiger conservation; a whole chestfull of options and processes with which tiger conservation can proceed in these turbulent times. The

vision that emerged from Tigers 2000 is that the tiger can be a star; not a fading star as a barometer of human intrusion and destruction of habitat, but a guiding star in consensus-building for sustainable relationships between people and resources, from which the tiger also benefits.

It is well understood that the metaphorical 'holocaust' for species on the brink of extinction, such as tigers, is the clearing of forests (Wilson 1993). The renewed tiger crisis of the early 1990s was the fear that the tiger could be lost because a new economic reality was driving demand for tiger parts. As described in the preceding chapters, there has been some success in reducing this demand and a process has been initiated for eliminating it entirely in time. The shock of this crisis for everyone who values wild tigers was that the tiger could be lost not only in the holocaust of species extinction but also to E. O. Wilson's metaphorical rifle shot. Could it be that with literally a few shots and snares tigers would be erased from ecosystems where they otherwise would persist?

Strong protective legislation and the means to strictly enforce it go hand-in-hand with the process of reducing and eventually eliminating the demand for tiger parts. Increasing the capacity of countries to control the illicit trade in tiger parts through strengthening law enforcement infrastructure should be a priority investment for the future of tigers. Needs include specific laws with meaningful penalties, government agencies with clearly defined responsibilities, trained people-power and intelligence-gathering networks. These are lacking or remarkably ineffectual in nearly every tiger range state. Several consumer states have enacted specific measures to control the tiger trade; we applaud these measures, but wait to judge their effectiveness. Discussions at Tigers 2000 emphasised that these critical short-term responses need to be linked to solutions that give living tigers and the places they live sufficient value for the people who also live near them. Good intelligence is the lifeline of effective enforcement and security, and an important insight from Tigers 2000 was the need to form partnerships with local people when establishing intelligence-gathering networks. But payment for information must be timely, as must payment be for lost livestock where compensation is practised; delays here erode credibility and the tiger loses. NGOs are usually more effective in the timely disbursal of funds than are governmental agencies.

The 1990s tiger crisis was a wake-up call and a time to re-examine the status of tiger habitats throughout tigerland. Even though it is generally known that many more tigers live outside reserved lands than in them, tiger habitat has usually been equated with reserved lands in conservationists' minds. This has trapped our thinking and remains a trap for many tiger conservationists. Surprisingly, no one had attempted to assess the extent and character of the entire suite of potential tiger habitats that still present opportunities to secure a future for wild tigers until Eric Dinerstein and his colleagues did so in 1997. Their report was subsequently revised by Eric Wikramanayake and associates and presented at Tigers 2000. The broad outline of this tiger conservation unit or TCU concept provides a firm foundation on which we can base more detailed work to advance the limits of our knowledge about where tigers can and do live. Contributors to Tigers 2000 suggested that geographical priorities for tiger conservation actions should be based on bio-regional distribution rather than on ill-defined subspecific designations and ranges for tigers. This shift moves tiger conservation from a taxonomic to an ecological emphasis, while still recognising the importance of the ecological, behavioural and demographic differences in tiger populations throughout their vast range.

Tigers 2000 was a chance to synthesise what new research was saying about the tiger's resilience and its ecological needs, and a more complete picture emerged than we have ever had before. While habitat loss, fragmentation and degradation remain as primary threats to the tiger's future, it is not the loss of habitat that has placed the tiger in its recent state of crisis. There are substantial blocks of potential tiger habitat remaining. The immediate problem facing tigers is that Wilson's metaphorical rifle shot in the forest has been all too literal and has substantially reduced prey populations in many of the remaining forest tracts that could otherwise harbour tigers. No large mammal prey, no tigers, is the crux of this newly highlighted threat. The advice that emerged from Tigers 2000 was to take the stress off tiger numbers and focus our resources on maintaining and monitoring habitat quality – especially an abundance of large ungulate prey and on identifying and maintaining the connective habitat linkages within tiger conservation units.

Many of us think of those few places where tiger still live as natural wonders – little Edens, if you will. Saving Eden at whatever cost and means has been thought to be the goal. The vision emerging from Tigers 2000 is that we should not limit our vision to just the Edens with their tigers, but think instead about recovering landscapes. We should put aside the notion that tiger reserves are self-contained islands, disconnected in a sea of non-compatible land-uses. Our vision should be of landscapes with their ecological and genetic processes intact. Protected areas are principal building blocks in this vision, but they cannot stand alone. We have emphasised the linkages between the needs of the tiger and the welfare of the people living near it. The future of wild tigers lies in a recovery process for tigers that includes, as essential elements, establishment and maintenance of sustainable relationships between people and their resources in habitats surrounding reserves, and maintenance of the connectivity within TCUs that is essential for ensuring long-term tiger population sustainability. The future of the tiger lies in recovering natural capital. Continued environmental deterioration is the alternative; along this road, the tiger and much else will be lost.

The specifics of shifting the boundaries of the problem domain and crafting effective solutions for 'the large carnivore problem' are dependent on the ecological and sociological contexts prevailing where remaining tigers live. Clark *et al.* (1996) point out that ecological factors are one of five variables in defining the 'problem' and seeking the solution. Cultural history, valuation, management system and policy process are all equally as important. These factors will interact very differently in different regions of the tiger's range, as the contributors to Tigers 2000 have shown.

Tigers 2000 anchored a vision, and initiated the beginning of a process to define and manage the flow of problems that emerges as we seek to secure a future for wild tigers and shift the diffuse concept of 'tiger conservation' to a far more focused goal of 'securing a future for wild tigers'. Because of the shifting social and political Asian landscape this will never be a completed task; we see it as a search for road maps to the tiger's future. Tigers 2000 shifted the domain of the problem of saving tigers to the ground where wild tigers live. It shifted the focus from the alarm generated by declining tiger numbers to the opportunity presented by the substantial amount of tiger habitat remaining. It shifted thinking about saving the tiger from a 'bunker' mentality to one of reaching out to forge partnerships with neighbours. It shifted thinking about saving the tiger away from crisis management towards a recovery mode and towards understanding and encouraging landscape patterns and conditions where tigers can persist in all the bioregions over their vast geographical range; it also sought the vision needed to work towards encouraging these patterns. What the contributors and we as editors have done, in short, is to translate the many languages of tiger conservation into one language – the language of conservation biology.

Tigers 2000 highlighted the many good initiatives already in place to save the tiger. Going forward with this process is a matter of building on these efforts. We need to proceed with a consensus-driven process that moves beyond the domain of special interests – yet captures the power of those efforts – and seeks to improve the policy environment for the tiger, integrate conservation and development, and mobilise financial resources. Supportive partnerships are a central part of the vision for the future of wild tigers. Following Tigers 2000, we attended the TigerLink meeting in New Delhi in December 1997. At this meeting more than 60 NGOs and government officials sat together to discuss present threats to and needs of India's tigers. We believe we were watching the process that will secure a future for the tiger in India. Later that same month we attended The First International Symposium on Endangered Species used in Traditional East Asian Medicine: Substitutes for Tiger Bone and Musk, held in Hong Kong, and watched a remarkable effort by TCM practitioners and manufactures to interact and work with conservationists in an effort to remove tiger bone and musk from TCM. There is a long way yet to go on this road but the threat TCM poses to wild tigers is being communicated and heard.

There are good building blocks for realistic tiger conservation in place. Money, political will, key legislation, co-operation and integration are needed to start cementing these building blocks together into a future for the tiger. In the Chinese Year of the Tiger, the Save the Tiger Fund – a partnership between the US National Fish and Wildlife Foundation and the Exxon Corporation – convened a follow-on conference to Tigers 2000 entitled The Year of the Tiger Conference: Securing a Future for Wild Tigers. John Seidensticker and Peter Jackson represented the Save the Tiger Fund Council on the organising committee for the conference, along with Ron Tilson, Howard Quigley, Maurice Hornocker, Joshua Ginsberg, David Phemister and Nancy Sherman. Sarah Christie set up a network for, and served on, the conference advisory committee. Many of the authors contributing to this volume also served on the advisory committee. The Save the Tiger Fund invested in The Year of the Tiger Conference because there is simply no substitute for bringing so many of the critical players together in one place. There is no substitute for direct human contact in the evolution of co-operative relationships and strategies. It was a unique opportunity to pull together, and to build on the existing work in which so many people have invested so much of their energy and their lives to save the tiger.

Participants from 13 of the 14 tiger range countries (noone for North Korea was able to attend) met in Dallas, Texas, on February 10–13, 1998, to find what works and what has not worked, what fits and does not fit in seeking to secure a future for wild tigers. At this conference we sought to break down the problem of saving the tiger to technically practical and politically feasible scales. We wanted to be sure that we had identified the present context for action in order to support programmes that address present conditions, not problems in contexts that are no longer germane. Also we needed to be sure that we do not ignore the challenges of new problems arising from this shifting context that is the norm throughout tigerland. In more specific terms, The Year of the Tiger Conference asked how we are doing and appraised our current practices in tiger conservation. We wanted to be able to assess our tiger conservation efforts, and to clarify our standards of performance. We wanted to identify and build on the positive trends and conditions that are emerging in tiger conservation activities, especially efforts that make tiger conservation less combative and more co-operative for the people who live beside and in the midst of tigers. We wanted to build on the co-operation and goodwill that developed from the Tigers 2000 symposium, and on the power of coming and working together to foster and encourage innovation, diffusion of ideas and adaptation in tiger conservation activities. Also we wanted to ensure that the tiger's conservation needs were made apparent to the people living in tiger range countries and around the world.

The important difference between The Year of the Tiger Conference and Tigers 2000 is that the latter focused on using science to support tiger conservation. In London, and in these chapters, we have sought the ecological criteria that must be met in conservation strategies in order to sustain viable tiger populations and significant habitats. The Year of the Tiger Conference moved on to examine the political criteria that must be met if conservation actions that will maintain viable wild tiger populations are to proceed with confidence and success.

Cory Meacham, an astute observer of tiger conservation activities (1997), has noted that we must be able to articulate what we are about in a single sentence that cannot be misinterpreted. Thus, we selected as our theme 'Securing a future for wild tigers', a theme that resonates throughout this volume.

This was the largest international gathering on behalf of tigers ever held and an unprecedented multi-country collaboration to share ideas and to develop plans. This conference fostered cross-disciplinary and cross-regional communication, with participating government officials, representatives from NGOs, conservation biologists, forestry specialists, engineers, land managers and others. Lines of communication were opened that may help to secure tiger habitats that span international borders, including many of the largest TCUs. Small and large conservation groups found opportunities to form partnerships for more effective conservation action. Further consensus-building exercises, focused at the TCU and bio-regional levels through the tiger's range, will be an essential foundation for bringing people and institutions together in our ongoing efforts to secure a future for wild tigers.

The endangered tiger is an indicator of ecosystems in crisis. We must direct our attention to the tiger's long-term future and support sustainable ecosystems and landscapes. We must move from viewing tiger conservation as an isolated part of ecosystem conservation to viewing the maintenance of viable tiger populations as an essential component of an integrated system of sustainable ecosystem management. Protecting tigers means managing tiger habitats for long-term rather than short-term exploitation. This is good for people living in tigerland and for their economy in the long-term, and the tiger also benefits. It is not possible to separate the interests of tigers from those of humans on any temporal or spatial scale, yet many of our past conservation prescriptions have attempted to do just that. Instead, the tiger can be the star in our ongoing efforts to implement actions that enable people to live in balance with natural resources.

The future of wild tigers depends on the coordinated and thoughtful support of the citizens of the countries where tigers still live, and support of people everywhere.

Common and scientific names used in the text

Common name	Scientific name
Amur leopard	*Panthera pardus orientalis*
Amur tiger	*Panthera tigris altaica*
Asian elephant	*Elephas maximus*
Asian tapir	*Tapirus indicus*
Asiatic golden cat	*Catopuma temmincki*
Asiatic wild dog/Dhole	*Cuon alpinus*
Bactrian deer	*Cervus elaphus bactrianus*
Bali tiger	*Panthera tigris balica*
Banteng	*Bos javanicus*
Barasingha	*Cervus duvauceli*
Caspian tiger	*Panthera tigris virgata*
Chinkara	*Gazella bennetti*
Chital/Spotted deer	*Axis axis*
Chousingha	*Tetracerus quadricornis*
Clouded leopard	*Neofelis nebulosa*
Colugo	*Cynocephalus variegatus*
Cougar	*Puma concolor*
Flat-headed cat	*Prionailurus planiceps*
Florida panther	*Puma concolor coryi*
Gaur	*Bos frontalis*
Giant deer	*Megaloceros* spp.
Grey ghoral	*Nemorhaedus caudatus*
Hanuman langur	*Semnopithecus entellus*
Hog badger	*Arctonyx collaris*
Hog deer	*Axis porcinus*
Indian chevrotain	*Moschiola meminna*
Indian fox	*Vulpes bengalensis*
Indian porcupine	*Hystrix indica*
Indian tiger	*Panthera tigris tigris*
Indian/Black-naped hare	*Lepus nigricollis*
Indochinese tiger	*Panthera tigris corbetti*
Jaguar	*Panthera onca*
Javan tiger	*Panthera tigris sondaica*
Jungle cat	*Felis chaus*
Kouprey	*Bos sauveli*
Leopard	*Panthera pardus*
Lion	*Panthera leo*
Long-tailed macaque	*Macaca fascicularis*
Manchurian moose	*Alces alces cameloides*
Maral red deer	*Cervus elaphus maral*
Marbled cat	*Pardofelis marmorata*
Mongoose	*Herpestes* spp.
Moose	*Alces alces*
Mugger crocodile	*Crocodylus palustris*

Common name	Scientific name
Muntjac/Barking deer	*Muntiacus muntjak*
Nilgai	*Boselaphus tragocamelus*
Ocelot	*Leopardus pardalis*
Porcupine	*Hystrix brachyura*
Ratel	*Mellivora capensis*
Red deer	*Cervus elaphus*
Sambar deer	*Cervus unicolor*
Serow	*Nemorhaedus sumatraensis*
Siberian musk deer	*Mochus moschiferus*
Siberian roe deer	*Capreolus pygargus*
Sika deer	*Cervus nippon*
Sloth bear	*Melursus ursinus*
South China tiger	*Panthera tigris amoyensis*
Striped hyaena	*Hyaena hyaena*
Sumatran rhino	*Dicerorhinus sumatrensis*
Sumatran tiger	*Panthera tigris sumatrae*
Sunbear	*Helarctos malayanus*
Tibetan antelope	*Pantholops hodgsonii*
Tiger	*Panthera tigris*
Water buffalo	*Bubalus bubalis*
Wild pig/Wild boar	*Sus scrofa*
Wolf	*Canis lupus*
Woolly mammoth	*Mammuthus primigenius*

APPENDIX 2
The fossil tigers

Andrew C. Kitchener

Table A2.1. *Localities and ages of fossil tigers. Spellings of locality names are as given in original references and may well have changed since date of publication*

Geological epoch	Estimated age (millions of years ago, Mya)	Locality	Synonyms	Ref.
Late Pliocene to early Pleistocene	2	China: Anyan, Honan Province	*Felis palaeosinensis*	1,.4
		Java:		
	1.3–2.1	Gunung Butak (Jetis Beds);		3, 5
	1.66	Sangiran;		3, 30
		Punung		
Middle to late Pleistocene		Java:	*Felis groeneveldtii, Felis trinilensis, Panthera tigris soloensis, Felis palaeojavanica, Felis oxygnatha*	2, 5, 6
	0.7–1.3	Kendeng; Trinil; Kedung Brubus; Bangle; Jeruk; Kebon Duren; Teguan		
		Sumatra: Sangiran; Padang Highlands		8
		China:	*Felis acutidens Felis youngi*	
	0.23–0.46	Choukoutien;		7, 28, 29, 32
		Wanhsien, Szechuan Province;		14
		Chinkiang, W. Hupei Province;		27
		Fuminhsien, Yunnan Province;		18
		Kweilin, Kwangsi Province;		20
	0.65–0.8	Lantian, Shensi Province;		21, 32
		Gongwangling, Shensi Province;		26

Table A2.1. (*cont.*)

Geological epoch	Estimated age (millions of years ago, Mya)	Locality	Synonyms	Ref.
	0.24	Shihhung, N. Anhwei Province		25, 32
Late Pleistocene		Java:		
	0.27–0.053	Ngangdong		6, 31
		China:		
		Shangdong;		24
		Harbin		16
		Japan:		
		Mikkabi*;		4
		Akiyosi, Yamaguti Province;		19
		Isa, Yamaguchi Province;		22
		Gansuiji		23
		Russia:		
		Bolschoj Lyachow I., Siberia;		9
		Jana River Basin, Siberia;		
		Tscharyisch River, Altai		10
		Caucasus		12
		E. Beringia		13
Holocene		India:		
		Karnul Caves		11
		Java:		
		Sampung		14
		Borneo:		
		Niah Caves, Sarawak		15

*Also Kuzuu; Tadaki

Table A2.1 References:

1. Zdansky 1924. 2. von Koenigswald 1933. 3. Hemmer 1971, 1976, 1987. 4. Hemmer 1968. 5. Duboi 1908. 6. Brongersma 1935. 7. Pei 1934; Zdansky 1928. 8. Brongersma 1937. 9. Tscherski 1892. 10. Brandt 1871. 11. Lydekker 1886. 12. Vereshchagin 1959 in Herrington 1987. 13. Herrington 1987. 14. Hooijer 1947. 15. Hooijer 1963. 16. Loukashkin 1937, 1938. 17. Young 1939. 18. Young 1932. 19. Shikama & Okafuji 1958. 20. Pei 1935. 21. Minchen 1964. 22. Shikama & Okafuji 1963. 23. Takai & Hasegawa 1966. 24. Zhang Zulu 1994. 25. Young & Chow 1955. 26. Wu *et al.* 1966. 27. Chiu *et al.* 1960. 28. Teilhard de Chardin 1936. 29. Pei 1936. 30. Swisher *et al.* 1994. 31. Swisher *et al.* 1996. 32. Groves 1989.

Key to locations in Figs. 11.1–11.6

Alan Rabinowitz

Fig. 11.1 Cambodia

1 Virachey National Park (Samith *et al.* 1995)
2 Kirirom National Park (Lay Khim & Taylor-Hunt 1995)
3 Snoul Wildlife Sanctuary (Yem Sokhan;[1] Lic Vuthy[2] pers. comm)
4 Mondolkiri Survey Area (Desai & Vuthy 1996)
5 Rattanakiri Survey Area (Desai & Vuthy 1996)
6 Phnom Nam Lyr Wildlife Sanctuary (Lic Vuthy pers. comm.)
7 Phnom Bokor National Park (Yem Sokhan pers. comm.)
8 Ream National Park (Yem Sokhan pers. comm.)
9 Lomphat Wildlife Sanctuary (Yem Sokhan pers. comm.)

[1] Chief of the Office of Protected Areas, Dept. of Nature Conservation and Protection in the Ministry of Environment.
[2] Chief of Research Section, Wildlife Protection Office, Department of Forestry.

Fig. 11.2 Lao PDR

1a Xe Piane NBCA[1] (Duckworth *et al.* 1994)
1b Xe Piane NBCA (WCS 1995a)
2a Dong Hua Sao NBCA (Salter 1993; Evans *et al.* 1996b)
2b Dong Hua Soa NBCA (Duckworth *et al.* 1994)
3a Phou Xang He NBCA (Salter 1993)
3b Phou Xang He NBCA (Duckworth *et al.* 1994)
4 Nam Kading NBCA (WCS 1995b)
5 Phou Loeuy NBCA (WCS 1995c)
6 Nam Et NBCA (WCS 1995c)
7a Phou Khao Khouay NBCA (Salter 1993)
7b Phou Khao Khouay NBCA (Payne *et al.* 1995)
8 Khammouane Limstone NBCA (Rabinowitz 1996a)
9 Xe Bang Nouan NBCA (Timmins & Bleisch 1995)
10 Nakai Nam Theun NBCA (Timmins & Evans 1996)
11 Nam Poun NBCA (Salter 1993)
12 Phou Dene Dinh NBCA (Robichaud & Sounthala 1995)
13 Nam Xam NBCA (Salter 1993)
14 Dong Ampham NBCA (Salter 1993)
15 Hin Namno NBCA (Timmins & Khounboline 1996)
16 Phou Xiang Thong NBCA (Evans *et al.* 1996a)
17 Hongsa Special Zone (Bergmans 1995)
18 Nakai Nam Theun Extension (Tizard 1996)
19 Nam Theun Corridor (Rabinowitz 1996a)
20 Nam Pan and Nam Chat Valleys (Schaller 1995)
21 Dakchung Village Area (Schaller 1995)
22 Kaleum Village Area (Schaller 1995)
23 Nam Ao Forest/Nadee Limestone Area (WCS 1995b)
24 Dong Khanthung Area (Timmins & Vongkhamheng 1996)
25 Bolovens Northeast (IUCN 1995b)
26 Bolovens Southwest (WCS 1995a)
27 Nam Kan (IUCN 1995)
28 Phou Kathong (IUCN 1995)
29 Hou Theung (IUCN 1995)
30 Xe Kampho (IUCN 1995)
31 Nam Ha East (IUCN 1995)
32 Phou Xang He (Duckworth *et al.* 1993)

[1] NBCA = National Biodiversity Conservation Area.

Fig. 11.3 Peninsular Malaysia

1 Belum (Ratnam *et al.* 1995)
2 Endau-Rompin Area (Davison & Kiew Bong Heang 1987)
3 Taman Negara National Park (Rubeli 1979; Topani 1990)

4–6   Johor State (Topani 1990)

7–9   Perak State (Topani 1990)

10   Terangganu State (Topani 1990)

11   Kelantan State (Topani 1990)

12&13 Kedah State (Topani 1990)

14–17 Pahang State (Topani 1990)

18–20 Selangor State (Topani 1990)

21   Negeri Sembilan (Topani 1990)

22   Bukit Tarek (Laidlaw 1994)

23   Paya Pasir (Laidlaw 1994)

24   Terengun (Laidlaw 1994)

25   Jengka (Laidlaw 1994)

26   Kemasul (Laidlaw 1994)

27   Sungai Lalang (Laidlaw 1994)

28   Lesong (Laidlaw 1994)

29   Krau Wildlife Reserve (Topani 1990)

30   Pasoh Forest Reserve (C. Francis pers. comm.)[1]

[1]Charles Francis is a WCS Research Associate

Fig. 11.4 Myanmar

1   Pidaung Wildlife Sanctuary (Rabinowitz 1996b)

2   Kyatthin Wildlife Sanctuary (UNDP/FAO 1982a)

3   Shwesettaw Wildlife Sanctuary (UNDP/FAO 1982c)

4   Inle Lake Wetland Sanctuary (UNDP/FAO 1983a; Rabinowitz 1996b)

5   Irrawaddy Delta (UNDP/FAO 1983c; Saw Tun Khaing 1996)

6   Natma Taung (Mt. Victoria) National Park (UNDP/FAO 1983b; Saw Tun Khaing 1996)

7   Arakan Yomas (Sayer 1983; Saw Tun Khaing pers. comm.)[1]

8   Mt. Poppa National Park (Rabinowitz 1996b)

9   Tamanthi Wildlife Sanctuary (Rabinowitz et al. 1995)

10   Tamanthi Village Area (Rabinowitz et al. 1995)

11   Alaungdaw Kathapa National Park (UNDP/FAO 1984; Rabinowitz 1996b)

12   Pegu Yomas National Park (UNDP/FAO 1982b; Saw Tun Khaing pers. comm.)

13   Kaser Doo Wildlife Sanctuary (Latimer et al. 1993)

14   Lower Triangle (Rabinowitz 1996b)

15   Putao Township (Rabinowitz 1996b)

16   Naung Mung Township (Rabinowitz 1997a)

17   Hkakaborazi Protected Area (Rabinowitz 1997a)

[1]Saw Tun Khaing is the WCS Myanmar Program Coordinator.

Fig. 11.5 Thailand

1   Huai Kha Khaeng Wildlife Sanctuary (Rabinowitz 1991)

2   Phu Kieo Wildlife Sanctuary (Rabinowitz 1991)

3   Lum Nam Pai Wildlife Sanctuary (Rabinowitz 1991)

4   Thung Yai Naresuan Wildlife Sanctuary (Rabinowitz 1991)

5   Klong Saeng Wildlife Sanctuary (Rabinowitz 1991)

6   Phu Luang Wildlife Sanctuary (Rabinowitz 1991)

7   Phu Miang-Phu Tong Wildlife Sanctuary (Rabinowitz 1991)

8   Doi Chiang Dao Wildlife Sanctuary (Rabinowitz 1991)

9   Salawin Wildlife Sanctuary (Rabinowitz 1991)

10   Doi Pha Muang Wildlife Sanctuary (Rabinowitz 1991)

11   Om Koi Wildlife Sanctuary (Rabinowitz 1991)

12   Doi Luang Wildlife Sanctuary (Rabinowitz 1991)

13   Mae Yuam Wildlife Sanctuary (Rabinowitz 1991)

14   Khao Yai National Park (Rabinowitz 1991)

15   Phu Kradung National Park (Rabinowitz 1991)

16   Nam Nao National Park (Rabinowitz 1991)

17   Thung Salang Luang National Park (Boonratana 1988; Rabinowitz 1991)

18   Doi Khuntan National Park (Rabinowitz 1991; Gray et al. 1994)

19   Khao Luang National Park (Boonratana 1988; Rabinowitz 1991)

20   Doi Inthanon National Park (Rabinowitz 1991)

21   Phu Rua National Park (Rabinowitz 1991)

22   Lansang National Park (Rabinowitz 1991)

23   Ramkhamhaeng National Park (Rabinowitz 1991)

24   Ton Krabak Yai National Park (Rabinowitz 1991)

25   Kaeng Krachan National Park (Rabinowitz 1991)

26   Thap Lan National Park (Rabinowitz 1991)

27   Doi Suthep-Doi Pui National Park (Rabinowitz 1991)

28   Klong Lan National Park (Rabinowitz 1991)

29   Phu Hin Rong Gla National Park (Rabinowitz 1991)

30   Nam Tok Surin National Park (Rabinowitz 1991)

31   Mae Wong National Park (Rabinowitz 1991)

32   Mae Yom National Park. Rabinowitz (1991)

33   Nam Tok Chattrakan National Park (Rabinowitz 1991)

34   Phu Pha Man National Park (Rabinowitz 1991)

35   Huai Nam Dung National Park (Rabinowitz 1991)

36   Ob Luang National Park (Rabinowitz 1991)

37   Khao Sam Roi Yot National Park (Parr et al. 1993)

38   Phangnga National Park (Boonratana 1988)

39   Khao Lampi-Hat Thai Muang (Boonratana 1988)

40   Hat Laem Son National Park (Boonratana 1988)

41   Hat Noppharat Thara-Mu Ko Phi Phi National Park (Boonratana 1988)

42   Khao Phanom Bencha National Park (Boonratana 1988)

43   Hat Chao Mai National Park (Boonratana 1988)

44   Khao Sok National Park (Boonratana 1988; Gray et al. 1994)

45   Erawan National Park (Gray et al. 1994)

46   Sai Yok National Park (Gray et al. 1994)

47   Thale Ban National Park (Gray et al. 1994)

48   Khao Chamao-Khao Wong National Park (Gray et al. 1994)

49   Khao Laem National Park (Gray et al. 1994)

Fig. 11.6 Vietnam

1   Kon Ha Nung and Kon Cha Rang Forests (Le Xuan Canh 1995)

2   Yok Don National Park and Green Forest (Le Xuan Canh 1995)

3   Nui Bi Doup Area (Le Xuan Canh 1995)

4   Srepok River Basin (Dang Zui Khun & Le Suan Kan 1991)

5   Cuc Phuong National Park (Johnsingh 1995)

6   Tay Nguyen Plateau (Dang Zui Khun & Le Suan Kan 1991)

7   Bach Ma National Park (Vu Van Dung & Huynh Van Keo 1993)

8   Bu Gia Map Nature Reserve (Le Trong Trai 1996)

9   Vu Quang Nature Reserve (Vu Van Dung et al. 1995)

10   Ben En Nature Reserve (Hoang Hoe & Vo Quy 1991)

11   Nam Cat Tien Nature Reserve (Le Trong Trai 1996)

12   Ba Be National Park (Hoang Hoe & Vo Quy 1991).

13   Muong Nhe Nature Reserve (Le Trong Trai 1996)

14   Sop Cop Nature Reserve (Le Trong Trai 1996)

15   Xuan Son Nature Reserve (Le Trong Trai 1996)

16   Pu Huong Nature Reserve (Le Trong Trai 1996)

17   Pu Mat Nature Reserve (Kemp et al. 1995)

18   Phong Nha Nature Reserve (Le Trong Trai 1996)

19   Ba Na Nature Reserve (Le Trong Trai 1996)

20   Ngoc Linh Nature Reserve (Le Trong Trai 1996)

21   Mom Ray Nature Reserve (Le Trong Trai 1996)

22   Kon Ka Kinh Nature Reserve (Le Trong Trai 1996)

23   Kon Cha Rang Nature Reserve (Le Trong Trai 1996)

24   Chu Yang Sinh Nature Reserve (Le Trong Trai 1996)

25   Nam Ca Nature Reserve (Le Trong Trai 1996)

26   Nam Cat Tien Nature Reserve (Le Trong Trai 1996)

27   Sa Thay River Basin (Le Trong Trai 1996)

28   Ta Ke forest area (Boonratana & Le Xuan Canh 1994)

29   Nam Trang-Ban Bung forest area (Boonratana & Le Xuan Canh 1994)

30   Phong Nha Cultural Site (Eames *et al.* 1994)

Indices for ranking Tiger Conservation Units

Eric D. Wikramanayake, Eric Dinerstein, John G. Robinson,
K. Ullas Karanth, Alan Rabinowitz, David Olson, Thomas Mathew,
Prashant Hedao, Melissa Connor, Ginette Hemley and Dorene Bolze

A: Index for habitat integrity

The habitat integrity index takes into consideration the size and spatial configuration of habitat blocks containing tigers, the quality of the habitat within the forest blocks and intervening areas, and the extent to which a Tiger Conservation Unit (TCU) contains one or more protected areas that will provide effective refuge to tigers and prey.

When scoring note the following:

▸ For criteria 5 and 6, degraded habitat is defined as either: (1) forest in which the understory or the forest has been impacted by livestock grazing, firewood collection, swidden agriculture, or man-made fires; or (2) grasslands or savannahs in which the tall grass cover has been impacted by livestock grazing, collection of fodder/thatch, or man-made fires.

▸ TCUs positive for criteria 5b and 6b will be flagged for surveys to determine the status of habitat quality. No scores will be assigned.

1   TCU consists of a small ($\leq$200 km^2), isolated fragment or fragments with low potential for tiger dispersal – 1 point.

2   TCU consists of an isolated fragment or fragments, with at least one being >200 but ($\leq$500 km^2, but with low potential for tiger dispersal among them – 2 points.

3   TCU consists of several isolated fragments >200 but ($\leq$500 km^2, with potential for tiger dispersal among them, forming a network of tiger habitat that adds up to >1000 km^2 – 5 points.

4   TCU consists of one or more isolated, mid-sized fragments (>500 and (1000 km^2) of tiger habitat with low potential for tiger dispersal among the larger habitat blocks – 10 points.

5   TCU consists of one or more isolated mid-sized fragments (>500 and ($\leq$1000 km^2) of tiger habitat with potential for natural tiger dispersal (existing or potential for restoration) among the larger habitat blocks:

   5a. but with >50% of habitat known to be degraded (but not cleared) and/or not prime tiger habitat – 10 points.

   5b. but with habitat quality unknown across most of TCU (flag TCU for surveys)

   5c. and >50% of TCU is considered to be good quality habitat suitable for tigers – 16 points.

5.1   If >50% tiger habitat of TCU for category 5 consists of effectively protected areas, add 2 points to score.

6   TCU consists of one or more habitat blocks >1000 km^2 with potential for natural tiger dispersal among them (existing or potential for restoration):

   6a. but with >50% of habitat known to be degraded (but not cleared) and/or not prime tiger habitat – 14 points.

   6b. but with habitat quality unknown across most of TCU (flag TCU for surveys)

   6c. and >50% of TCU is considered to be good quality habitat suitable for tigers – 24 points.

6.1   If >25% tiger habitat of TCU for category 6 consists of effectively protected areas, add 4 points to score.

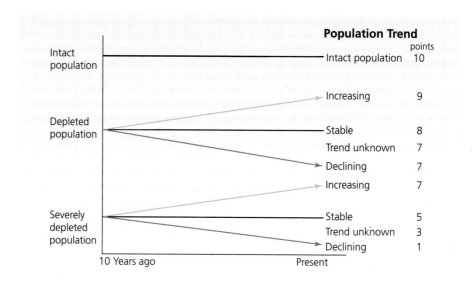

FIGURE A4.1

7 TCU consists of contiguous habitat throughout and exceeds 5000 km²; is relatively intact; contains the full range of habitat types necessary for tigers that is expected to occur in the Tiger Habitat Type and/or bioregion – 36 points.

7.1 If >20% tiger habitat of TCU for category 7 consists of effectively protected areas, add 4 points to score.

B: Index for poaching pressure

1 Low poaching; concentrated in a few areas and/or sporadic; prey-base relatively intact; effective anti-poaching programme and network in place – 20 points.

2 Low to medium poaching; concentrated in a few areas and/or sporadic; poaching on prey relatively high, but on tigers low; anti-poaching programme relatively effective. Potential for reversing poaching pressure – 19 points.

3 Medium poaching pressure; widespread, but low intensity; tigers and/or prey poached; potential for anti-poaching measures – 17 points.

4 Medium poaching pressure; widespread, but low intensity; tigers and/or prey poached; no potential for anti-poaching measures in near future, but tigers not severely threatened – 14 points.

5 Medium to high poaching pressure; poaching pressure on tigers and/or prey; but potential for anti-poaching measures – 11 points.

6 High poaching pressure; poaching on tigers and/or prey; but potential for effective anti-poaching measures – 8 points.

7 Medium to high poaching pressure; poaching pressure on tigers and/or prey; no potential for anti-poaching measures – 4 points.

8 High poaching pressure; poaching on tigers and/or prey; no potential for effective anti-poaching measures – 1 point.

9 Extent of poaching pressure unknown – 13 points.

C: Index for population status

The population status will be evaluated from the broad 10-year trends as depicted in Fig. A4.1.

A 10-year period was chosen because many park staff, local people and scientists are likely to remember the relative status of the present tiger population compared with that 10 years ago. Many park staff are also rotated periodically, and are thus unlikely to be familiar with the status of tiger populations before their arrival. However, any information that is available from a longer period can be so noted and evaluated in assigning scores.

An *intact population* is considered to be one that is 'minimally impacted' and will represent the best possible situation. An intact population will neither be in decline, nor increasing; therefore, only the stable trend is shown.

A *moderately depleted population* can be evaluated as a population that is known to be affected by habitat loss, loss of prey density, poaching, etc., but where tigers, or their signs, are still encountered 'rather frequently' for this habitat type.

A *severely depleted population* is one that is known to be *highly* impacted by poaching, habitat loss, lack of prey, etc., and tigers, or their signs, are very rarely encountered for this habitat type.

These are subjective and relative measures, but it is likely that any person who has been residing and/or working in the area of a TCU will be able to differentiate and make approximate judgements on whether a population is 'intact, moderately depleted, or severely depleted' on the basis of information, encounters, etc.

The population trends over the past 10 years are broad changes reflecting population increases, declines and stability. These are also relative and subjective; however, our field experience suggests that a person familiar with the area and/or tiger populations will be able to assign a trend to the status of tigers. (For example, 'there are fewer tigers in the area than before' suggests a declining population.)

Furthermore, population trends that do not fit into any one of these categories should be considered individually and assigned points on the basis of how they compare with the trends presented here, rather than 'forcing' a trend to conform to any of the above categories.

APPENDIX 5

Counting tigers, with confidence

K. Ullas Karanth

It may only be a slight exaggeration to say that there are as many methods of counting tigers as there are people out in the woods counting the beasts. These 'methods' range all the way from confident assertions, such as 'I *know* how to count tigers because of my experience and field-craft', to the rather more hesitant approaches of modern wildlife biology. The tools deployed to count tigers vary from notebooks, pugmark tracers and plaster casts to radio-tracking gear and camera traps. The practitioners of the craft are an equally diverse lot; officials, amateur naturalists, park guards, tribesmen, villagers and biologists. Because of this anarchic diversity of approaches, equipment and personnel involved, there has been no theoretical framework provided to the enterprise of counting tigers. Authoritative assertions – rather than proof and validation – sustain many methods that are quite demonstrably failure-prone (Karanth 1987).

Here, I have tried to look at the problem within the framework of formal population estimation theory. For conceptual details, the reader is referred to the excellent accounts in the literature (Nichols 1992; Lancia *et al.* 1994). In any given area (nature reserve, region, country) there is a population of N tigers that is annually turning over at the rate of 20% or so (Kenny *et al.* 1995; K. U. Karanth & B. M. Stith this volume). We need to estimate N 'reliably' (with a mean and its variance) at a given point in time. So we go out and get a 'count' (C) of tigers. The statistic C may result from photography, radiotelemetry, or counts of tiger sign (scats or tracks). The first two methods yield counts of individually identifiable tigers (McDougal 1977; Karanth 1995). Normally, although tiger sign (tracks and scats) can be identified in the field, identifying individual tigers from tracks poses many thorny problems (Karanth 1987).

In any case, assuming that the count C represents individually identified tigers, then:

$$N = C/p$$

where p is the sampling fraction (the proportion of tigers that were counted out of the total that are present). The count can be total *only* if $p=1$, that is, only if *each* tiger in the area is identified correctly and *every* tiger (or its sign) is detected during the count. The snow track counts of the Russian Far East (Miquelle *et al.* 1996), pugmark counts in Nepal (McDougal 1977), and the pugmark censuses of India (Panwar 1979a) are three field methods used for counting tigers from tracks. I would argue that in all these three cases, the basic requirements necessary to make the sampling fraction $p=1$ are not satisfied. Consequently, these counts are not censuses, as is sometimes claimed.

In the Russian track counts, distance between the geographical location of two sets of tracks, and their size differences, are used in combination to assign observed sets of tracks to different individuals, without claims of identifying individual tigers based on track shape.

In Nepal, an ability to recognise *some* individual tigers (mostly resident breeders) based on track shape is assumed, because of unique field conditions and high levels of skill. However, descriptions of these counting methods (Miquelle *et al.* 1996; C. McDougal this volume Box 3.1) suggest that C is an undercount of the true value of N ($C<N$). Although p is not estimated, it is consistently <1 in value. If the assigned individual tiger identities are correct (usually an invalid assumption), then these counts can at least be used as minimum bounds of the true tiger population size N.

The Indian pugmark counts do not even possess this virtue of consistency. While multiple counts

An adult male Indian tiger 'captured' with a camera-trap photo during a census in an Indian national park. Camera-traps are an efficient, non-invasive means of obtaining statistically reliable estimates of tiger abundance where tigers occur at high to moderate densities.

resulting from misidentification of tracks and from repeated counts in neighbouring census blocks tend to make $C>N$, inadequate effort and unsuitable substrate tend to make $C<N$ (Karanth 1987, 1988). Worse still, these two factors, which influence the magnitude of C in opposite directions, operate simultaneously, causing the value of C to be either lower or higher than N in a totally *unpredictable* manner.

Only formal methods permit the estimation of the sampling fraction p (Lancia *et al.* 1994). For instance, to derive p, line transect surveys estimate sighting probabilities and capture-recapture surveys estimate capture probabilities. If p cannot be estimated, how can tiger sign surveys be applied to

monitor populations? First, we have to accept that routine management needs only knowledge about tiger population trends (increase, decrease, or stability) rather than exact tiger numbers. Secondly, we must ensure that counts of tiger sign by field teams are based on replicated sample surveys. Then, the count statistic C can be considered as an index of the unknown population size N. Such indices are most useful when we are sure that the sampling fraction p remains constant between the two counts (constant proportion indices, Lancia *et al.* 1994), and consequently the magnitude of change (e.g. percentage decrease) can be assessed. Since most tiger sign counts are not based on formal population estimation methods, this assumption of constant p

cannot be tested from the field data. However, if the sampling is intensive and there are enough data points, perhaps C can be considered to at least reflect the direction of change of N (increase, decrease or stability).

In Nagarahole, India, I sampled a set of four 15–20 km forest road segments at four-day intervals 15 times, counting tiger scats, to generate a tiger abundance index (number of scats seen/10 km walked). I would argue that such sampling-based, quantitative indices are more valuable to wildlife managers than putative total counts generated from non-sampled surveys, based on uncertain individual tiger identities. If we cannot estimate the sampling fraction, then it is best to standardise sample surveys, so that, as far as possible, p remains consistent between successive counts (Lancia *et al.* 1994). Another approach is to make C a frequency index (e.g. percentage of road segments in which tiger sign was observed).

A fundamental point with such sample surveys of tiger sign is that it is not necessary, and may be even be disadvantageous, to try to identify individual tigers. By eliminating possible identification errors, and by treating each data point as a count or frequency of tiger sign observed/sampling unit (e.g. tiger sign/10 km; percentage of road segments with tiger sign) the data become more reliable. Secondly, if each data point is a 'generic' tiger sign/sampling unit, instead of 'sign from an individually identified tiger'/sampling unit, the overall increase in sample size increases the precision of the index. Conducting annual tiger sign surveys in the same season, on the same sampling routes, using similar sampling units, and investing the same amount of effort are other obvious ways to enhance the value of such indices of tiger population trend.

Can we use such sign surveys to compare tiger densities at different sites? I found that where tigers prefer to use roads as travel routes, road density (road length/unit area) strongly influenced my encounter rates with tiger sign, rendering the indices inadequate for comparing different sites. It may be difficult to standardise sampling procedures to overcome the influence of landscape features, such as road density, on tiger sign encounter rates.

There are many areas where tiger densities are so low that surveys can only record the presence or absence of tigers by their sign.

However, in the Russian Far East and other relatively unmodified tiger landscapes this may be feasible, enabling spatial comparison of tiger abundance among different sites.

Going beyond indices, if reliable estimates of tiger population parameters, such as size, density, survival, mortality or harvest rates, are needed, we have no escape from committing the skills and resources necessary for the task. You cannot go to the moon on a bullock cart. Early scientific estimates of tiger densities were based on home range mapping and social structure data obtained from radiotelemetry (Sunquist 1981; Smith 1993). However, the effort and costs involved in radio-tagging wild tigers, high population turnover rates and uncertainty about the sampling fraction p, constrain the application of radiotelemetry solely to count tigers. Home range mapping, based on individual identifications using DNA extracted from tiger scats or hair, is a new approach being investigated (K. U. Karanth & G. Amato unpubl. data).

Table A5.1. *Estimates of area sampled (A), population size (n) and density (D) for tiger populations derived from photographic capture-recapture surveys in India (Karanth & Nichols 1998)*

Study area	Sampling effort trap-nights	Area sampled A (SE [A]) km²	Population size n (SE[n])	Density/100 km² D (SE[D])
Pench	788	122 (17.7)	5 (1.41)	4.1 (1.31)
Nagarahole	936	243 (14.8)	28 (3.77)	11.5 (1.70
Kanha	803	282 (23.6)	33 (4.69)	11.7 (1.93)
Kaziranga	552	167 (12.1)	28 (4.51)	16.8 (2.96)

The means and their standard errors (SE) are reported.

Presently, camera trapping tigers (McDougal 1977; Karanth 1995) offers a reasonably efficient approach. However, investigators who camera trap on an *ad hoc* basis, hoping to get every tiger in the area, do not use their resources optimally. In the end, they may still be unsure of the sampled fraction (*p*). Tigers are naturally marked animals readily identifiable from photos. Therefore, we can apply formal capture-recapture models applicable to marked animal studies, and actually estimate *p*. The power of camera trapping can be enhanced manifold if the technique is used for sampling, rather than censusing, tiger populations. Recent capture-recapture models (Nichols 1992) can deal with biological factors such as differential trap access, trap response and temporal effects, which affect capture probabilities.

Photographic capture-recapture estimates of tiger abundance were first obtained in Nagarahole (Karanth 1995). Karanth & Nichols (1998) recently analysed camera-trap data from four sites in India using the program CAPTURE (White *et al.* 1982; Rexstad & Burnham 1991) to estimate tiger population sizes. Thereafter, using an approach developed by Wilson & Anderson (1985), they estimated the effectively sampled area, thus deriving the first rigorous estimates of population densities for wild tigers. Their results are summarised in Table A5.1, to illustrate the usefulness of this approach.

Because tiger cubs (age<1 year) had very low capture probabilities, the sampling effort required to estimate their numbers was too high. However, ancillary data on the number of females with cubs, or demographic structure data from past studies, can be used to estimate the numbers of cubs. It should also be noted that Karanth & Nichols (1998) found that camera-trapping gave adequate returns for the effort at sites with densities higher than 4.1 tigers/100 km². In the Russian Far East, where tigers occur at lower densities of 0.5–1.4 tigers/100 km² (D. G. Miquelle pers. comm.), the necessary sampling effort may be too high. There appears to be a threshold tiger density level somewhere between 1 and 4 tigers/100 km², below which camera-trapping may not be practical. At such sites, radiotelemetry (home range mapping) offers the best option for estimating tiger densities.

Where camera traps are prone to elephant damage or theft, they will need adequate protective structures. Clearly, in many areas where tiger densities are low, or other reasons preclude the use of camera traps, the best one can do is to obtain reliable indices of tiger abundance based on scats or tracks, as described earlier. Rough estimates of tiger numbers can also be derived, based on prey biomass, if reliable estimates of prey densities are available (K. U. Karanth this volume). Finally, there are many areas where tiger densities are so low that field surveys can reliably record only the presence or absence of tigers.

Literature cited

Abramov, K. G. (1960). Tiger conservation in the Far East. In *Nature Protection in the Reserves of USSR*, pp. 92–5. Moscow, Akademie Nauk Press. (In Russian.)

Abramov, K. G. (1961). On the method of registration of tigers. In *Organization and Methods of Registration of Ground Vertebrates*, pp. 53–4. Moscow, Akademie Nauk. (In Russian.)

Abramov, K. G. (1965). Amur tiger – relict fauna in the Far East. *Zapecki of the Primorski Affiliation of the USSR Geographic Society*, 1, 106-12. (In Russian.)

Abramov, V. (1962). A contribution to the biology of the Amur tiger, *Panthera tigris longipiiis* (Fitzinger 1868). *Vestník Ceskoslovenské Spolecnosti Zoologické Acta Societatis Zoologicae Bohemoslovenicae*, 26.

Abramov, V. K. (1977). On the reproductive potential and numbers of the Amur tiger. *Zoologicheskii Zhurnal*, 56, 268–75. (In Russian.)

Abramov, V. K., Pikunov, D. G., Bazylnikov, V. E. & Sablina, T. B. (1978). Ecological aspects of the winter distribution of tigers (*Panthera tigris* L.). In *International Tiger Studbook, 1st International Symposium on the Management and Breeding of the Tiger*, ed. P. Müller, pp. 10–12. Leipzig, Zoologischer Garten Leipzig. (In German.)

Ackerman, B. B., Lindzey, F. G. & Hemker, T. P. (1984). Cougar food habits in southern Utah. *Journal of Wildlife Management*, 48, 147–55.

Ahearn, S. C., Smith, J. L. D. & Wee, C. (1990). Framework for a geographically referenced tabular/visual conservation data base. *Photogrammetric Engineering and Remote Sensing*, 56, 1477–81.

Aiken, S. R. (1993). Struggling to save Malaysia's Endau-Rompin rain forest, 1972–1992. *Environmental Conservation*, 20(2), 157–62.

Aiken, S. R. (1994). Peninsular Malaysia's protected areas' coverage, 1903–92: Creation, rescission, excision, and intrusion. *Environmental Conservation*, 21(1), 49–56.

Aiken, S. R. & Leigh C. H. (1995). On the declining fauna of Peninsular Malaysia in the post-colonial period. *Ambio*, 14(1), 15–22.

Akcakaya, R. H. (1991). A method for simulating demographic stochasticity. *Ecological Modeling*, 54, 133–6.

Alldredge, J. R. & Ratti, J. T. (1986). Comparison of some statistical techniques for analysis of resource selection. *Journal of Wildlife Management*, 50, 157–65.

Anderson, S., Bankier, A. T., Barrell, B. G., de Bruijn, M. H., Coulson, A. R., Drouin, J., Eperon, I. C., Nierlich, D. P., Roe, B. A., Sanger, F., Schreier, P. H., Smith, A. J., Staden, R. & Young, I. G. (1981). Sequence and organization of the human mitochondrial genome. *Nature*, 290, 457–65.

Animal Traffic: 31 Tigers. A series of films by Ron Orders and Arpad Bondy. Cinecontact on Fuji.

Anon. (1994). *A Report to the CITES Secretariat on a Survey of Retail Availability of Rhinoceros Horn and Tiger Bone in the People's Republic of China*. Cambridge, UK, TRAFFIC International.

Anon. (1995a). Cambodia-Laos-Vietnam Workshop. *Cat News*, 22, 7. Bougy, Switzerland, IUCN Cat Specialist Group.

Anon. (1995b). International workshop to assess the status of tigers. *Cat News*, 22, 7. Bougy, Switzerland, IUCN Cat Specialist Group.

Anon. (1996a). Asian co-operation to save the tiger and its ecosystems. *Cat News*, 24, 2. Bougy, Switzerland, IUCN Cat Specialist Group.

Anon. (1996b). A sustainable land-use and allocation programme for the Ussuri/Wusuli River watershed and adjacent territories (northeastern China and the Russian Far East). A Co-operative Project of Ecologically Sustainable Development, Inc., FEB-RAS Institute of Aquatic and Ecological Problems, FEB-RAS Pacific Geographical Institute, Heilongjiang Province Territory Society, National Committee on United States-China Relations.

Anon. (1996c). Florida Panther reintroduction feasible, but issues complicated. *Cat News*, 26, 12. Bougy, Switzerland, IUCN Cat Specialist Group.

Anon. (1996d). *Gassinski Model Forest Workplan*. Gassinski, McGregor Model Forest Association.

Arseniev, V. K. (1941). *Dersu the Trapper* (translated by M. Burr). New York, E. P. Dutton.

Avise, J. (1994). *Molecular Markers, Natural History and Evolution*. New York, Chapman and Hall.

Avise, J. C. & Ball, R. M. (1990). Principles of genealogical concordance in species concepts and biological taxonomy. *Oxford Surveys in Evolutionary Biology*, **7**, 45–67.

Avise, J. C. & Hamrick, J. C. (1996). *Conservation Genetics Case Histories from Nature.* New York, Chapman and Hall.

Avise, J. C. & Lansman, R. A. (1983). Polymorphism of mitochondrial DNA in populations of higher animals. In *Evolution of Genes and Proteins*, ed. M. Nei & R. K. Koen, pp. 147–90. Sunderland, Massachusetts, Sinauer Associates.

Badan Pengawasan National, Republic of Indonesia (BAP-PENAS) (1993). Biodiversity Action Plan for Indonesia. Jakarta, BAPPENAS.

Badridze, J. (1992). *The Reintroduction of Captive-raised Large Mammals Into Their Natural Habitat: Problems and Methods.* Republic of Georgia, Institute of Zoology of the Academy of Sciences.

Badridze, J. (1994). Local participation in re-introduction research in Georgia: The case of the wolves. *Reintroduction News*, **12**, 16–17. Nairobi, Kenya, IUCN/SSC Reintroduction Specialist Group.

Baikov, N. A. (1925). *Manchurian Tiger.* Harbin, Society of the Study of Manchurian Krai. (In Russian.)

Bailey, T. N. (1993). *The African Leopard.* New York, Columbia University Press.

Bakels, J. (1994). But his stripes remain: On the symbolism of the tiger in the oral tradition of Kerinci, Sumatra. In *Text and Tales: Studies in Oral Tradition*, ed. J. Oosten, volume 22, pp. 33–51. Leiden, Netherlands, Research School CNWS.

Ballou, J. & Seidensticker, J. (1987). The genetic and demographic characteristics of the 1983 captive population of Sumatran tigers. In *Tigers of the World: The Biology, Biopolitics, Management, and Conservation of an Endangered Species.* New Jersey, Noyes Publications.

Balmford, A., Leader-Williams, N. & Green, M. (1995). Parks or arks: Where to conserve threatened mammals? *Biodiversity and Conservation*, **4**, 595–607.

Balmford, A., Mace, G. & Leader-Williams, N. (1996). Redesigning the Ark: Setting priorities for captive breeding. *Conservation Biology*, **10**(3), 719–27.

Bannikov, A. (1978). The present status of the Bactrian deer (*Cervus elaphus bactrianus*) in the USSR. In *Threatened Deer*, pp. 159–72. Morges, Switzerland, IUCN/SSC Deer Specialist Group.

Bart, J. (1995). Acceptance criteria for using individual-based models to make management decisions. *Ecological Applications*, **5**, 411–20.

Beier, P. (1993). Determining minimum habitat areas and habitat corridors for cougars. *Conservation Biology*, **7**, 94–108.

Beier, P. (1995). Dispersal of juvenile cougars in fragmented habitat. *Journal of Wildlife Management*, **59**, 228–37.

Belden, R. C. & McCown, J. W. (1995). *Florida Panther Reintroduction Feasibility Study: Final Report.* Florida Game and Fresh Water Fish Commission, Bureau of Wildlife Research.

Bensky, D. & Gamble, A. (1993). *Chinese Herbal Medicine Materia Medica.* Seattle, Eastland Press Inc.

Bergmans, W. (1995). On mammals from the People's Democratic Republic of Laos, mainly from Sekong Province and Hongsa Special Zone. *Zeitschrift Saugetierkunde*, **60**, 286–306.

Berkmuller, K., Evans, T., Timmins, R. & Vongphet, V. (1995). Recent advances in nature conservation in the Lao PDR. *Oryx*, **29**(4), 253–60.

Berwick, S. H. (1974). *The community of wild ruminants in the Gir Forest ecosystem, India.* Yale University, PhD dissertation.

Blanchard, R. F. (1977). *Preliminary Analysis of Tiger Mortality and Livestock Depredation in Trengganu and Kelantan, Peninsula Malaysia.* Report to the Department of Wildlife and National Parks, Peninsula Malaysia, Kuala Lumpur, Malaysia.

Bock, C. (1884). *Temples and Elephants.* Reprinted in 1995. Bangkok, Thailand, White Orchid Press.

Bodmer, J. G., Marsh, S. G. E., Albert, E. D., Bodmer, W. F., Dupont, B., Erlich, H. A., Mach, B., Mayr, W. R., Parham, P., Sasazuki, T., Schreuder, G. M. T., Strominger, J. L., Svejgaard, A. & Terasaki, P. I. (1994). Nomenclature for factors of the HLA system. *Tissue Antigens*, **44**, 1.

Bookbinder, M., Dinerstein, E., Rijal, A., Cauley, H. & Rajuria, A. (1998). Ecotourism's support of biodiversity conservation. *Conservation Biology*. (In press.)

Boonratana, R. (1988). Survey of mammals in South Thailand Parks. *Natural History Bulletin of the Siam Society*, **36**, 71–84.

Boonratana, R. & Le Xuan Canh (1994). *A Report on the Ecology, Status and Conservation of the Tonkin Snub-nosed Monkey (*Rhinopithecus avunculus*) in Northern Vietnam.* Report to the Wildlife Conservation Society, New York.

Bowcock, A., Ruiz-Linares, A., Tomfohrde, J., Minch, E., Kidd, J. R. & Cavalli-Sforza, L. L. (1994). High resolution of human evolutionary trees with polymorphic microsatellites. *Nature*, **368**, 455–7.

Boyce, M. S. (1992). Population viability analysis. *Annual Review of Ecology and Systematics*, **23**, 481–506.

Boyce, M. S. & Haney, A. (1997). *Ecosystem management.* New Haven, Yale University Press.

Brandon, K. (1997). Policy and practical considerations in land-use strategies for biodiversity conservation. In *Last Stand: Protected Areas and the Defense of Tropical Biodiversity*, ed. R. Kramer, C. van Schaik & J. Johnson, pp. 90–114. Oxford, Oxford University Press.

Brandt, F. (1871). Neue Untersuchungen über die in den altaischen Höhlen aufgefunden Säugethierreste, ein Beitrag zur quaternären Fauna des Russischen Reiches. *Bulletin de l'Académie Impériale de St. Petersbourg*, **15**, 147–202.

Bromley, G. F. (1964). *Ussuri Wild Boar*, Sus scrofa ussuricus *Heude, 1888.* Moscow, Nauka (In Russian.)

Bromley, G. F. & Kucherenko, S. P. (1983). *Ungulates of Southern Far East USSR.* Moscow, Nauka Press.

Brongersma, L. D. (1935). Notes on some recent and fossil cats, chiefly from the Malay Archipelago. *Zoologische Mededeelingen*, **18**, 1–89.

Brongersma, L. D. (1937). Notes on fossil and prehistoric remains of the 'Felidae' from Java and Sumatra. *Comptes Rendus 12th Congress Internationale de Zoologologie, Lisbonne, 1935*, pp. 1855–65.

Bruford, M., Cheesman, D. J., Coote, T., Green, H. A. A., Haines, S. A., O'Ryan, C. & Williams, T. R. (1996). Microsatellites and their application to conservation genetics. In *Molecular Genetic Approaches to Conservation*, ed. T. B. Smith & R. K. Wayne, pp. 278–97. Oxford, Oxford University Press.

Buckland, S. T., Anderson, D. R., Burnham, K. P. & Laake, J. L. (1993). *Distance Sampling; Estimating Abundance of Biological Populations.* New York, Chapman and Hall.

Budzan, D. V. (1996). Intended formation of cedar-broad-leaved forests under canopy of low value oak stands in the experimental forestry training enterprise of Primorski State Agriculture Academy. In *Cedar-broad-leaved Forests of the Far East.* Proceedings of an International Conference, 30 September to 6 October 1996, ed. D. F. Efremov, L. G. Kondrashov and A. P. Sapojnikov, pp. 146–7 (abstract). Khabarovsk, Khabarovski Krai, FEFRI (Far East Forestry Research Institute).

Burgman, M. A., Ferson, S. & Akcakaya, H. R. (1993). *Risk Assessment in Conservation Biology.* New York, Chapman and Hall.

Burnham, K. P., Anderson, D. R. & Laake, J. L. (1980). Estimation of density from line transect sampling of biological populations. *Wildlife Monographs*, **72**, 1–202.

Byers, C. R., Steinhorst, R. K. & Krausman, P. R. (1984). Clarification of a technique for analysis of utilization-availability data. *Journal of Wildlife Management*, **48**, 1050–3.

Callister, D. & Blythewood, T. (1995). *Of Tiger Treatments and Rhino Remedies: Trade in Endangered Species in Australia and New Zealand.* Sydney, TRAFFIC Oceania.

Caro, T. M. & Laurenson, K. (1994). Ecological and genetic factors in conservation: A cautionary tale. *Science*, **263**, 485–6.

Caswell, H. (1976). The validation problem. In *Systems Analysis and Simulation in Ecology*, ed. B. Patten, volume IV, pp. 6–16. New York, Academic Press.

Caswell, H. (1989). *Matrix Population Models: Construction, Analysis and Interpretation.* Sunderland, Massachusetts, Sinauer.

Caughley, G. (1977) *Analysis of Vertebrate Populations.* London, Wiley.

Caughley, G. (1994). Directions in conservation biology. *Journal of Animal Ecology*, **63**, 215–44.

Chalifour, N. (1996). *Canada's Role in the Tiger Trade – Recommendations for a Tiger Safe Nation.* Toronto, World Wildlife Fund-Canada.

Chazee, L. (1990). *The Mammals of Laos and the Hunting Practices.* Report to the Government of Laos, Vientiane, Lao PDR.

Chiu, C.-L., Chang, Y.-P. & Tung, Y.-S. (1960). Pleistocene mammalian fossils from Chinkiang District, W. Hupei. *Vertebrata Palasiatica*, **4**, 157–8.

Christensen, N. L., Bartuska, A., Brown, J., Franklin, S., MacMahon, J., Noss, R., Parsons, D., Peterson, C., Turner, M. & Woodmansee, R. (1996). The report of the Ecological Society of America Committee on the scientific basis for ecosystem management. *Ecological Applications*, **6**, 665–91.

Christie, S. (1997). *European Tiger Studbook*, 2nd edn. London, UK, Zoological Society of London.

CITES Secretariat (1993). *Decisions of the Standing Committee on trade in rhinoceros horn and tiger specimens. Notification to the Parties No. 774, 15 October 1993*. Geneva, Switzerland, CITES Secretariat.

CITES Secretariat (1997). Resolution Conf. 9.13 (Rev.): Conservation of and trade in tigers. In *CITES Resolutions: Resolutions of the Conference of the Parties to CITES that remain in effect after the 10th meeting*. Geneva, Switzerland, CITES Secretariat.

Clark, T. W., Curlee, A. P. & Reading R. P. (1996). Crafting effective solutions to the large carnivore problem. *Conservation Biology*, **10**, 940–8.

Collier, G. E. & O'Brien, S. J. (1985). A molecular phylogeny of the Felidae immunological distance. *Evolution*, **39**, 473–87.

Collins, M., Sayer, J. A. & Whitmore, T. C. (ed.) (1991). *The Conservation Atlas of Tropical Forests: Asia and the Pacific*. New York, Simon and Schuster.

Conover, D. (1980). *Practical Nonparametric Statistics*, 2nd edn. New York, John Wiley and Sons.

Conway, W. (1997). The changing role of zoos in international conservation and the WCS. *Society for Conservation Biology Newsletter*, **4**(2), 1–3.

Corbet, G. B. (1966). *The Terrestrial Mammals of Western Europe*. London, Foulis.

Corbet, G. B. (1970). Patterns of subspecific variation. In *Variation in Mammalian Populations*, Symposium of the Zoological Society of London, ed. R. J. Berry & H. N. Southern, volume 26, pp. 105–16. London, Academic Press.

Corbet, G. B. (1978). *The Mammals of the Palaearctic Region*. London, British Museum (Natural History) and New York, Cornell University Press.

Corbet, G. B. (1997). The species in mammals. In *The Units of Biodiversity*, ed. M. F. Claridge, H. A. Dawah & M. R. Wilson, pp. 341–56. London, Chapman and Hall.

Corbet, G. B. & Hill, J. E. (1992). *The Mammals of the Indomalayan Region*. Oxford, Oxford University Press. A Natural History Museum Publication.

Corbet, S. W., Grant, W. S. & Robinson, T. J. (1994). Genetic divergence in South African Wildebeest: Analysis of allozyme variability. *Journal of Heredity*, **85**, 479–83.

Corbett, J. (1944). *Man-eaters of Kumaon*. London, Oxford University Press.

Craighead, J. J., Varley, J. R. & Craighead, F. C. (1974). A population analysis of the Yellowstone grizzly bears. *Montana Forest Conservation Experiment Station Bulletin No. 40*.

Cronon, W., ed. (1996). *Uncommon Ground*. New York, W. W. Norton.

Crosby, M. (ed.) *Asian Important Bird Areas Briefing Book*. Cambridge UK, Birdlife International. (In press.)

Csuti, B. (1996). Mapping animal distribution areas for gap analysis. In *Gap Analysis: A Landscape Approach to Biodiversity Planning*, ed. J. M. Scott, T. H. Tear & F. W. Davis, pp. 135–45. Bethesda, Maryland, American Society for Photogrammetry and Remote Sensing.

d'Huart, J. P. (1991). Habitat utilisation of Old World wild pigs. In *Biology of Suidae*, ed. R. H. Barrett & F. Spitz, pp. 30–48. Briançon, France, Institut National de la Recherche Agronomique.

Dang Huy Huynh, Dao Van Tien, Cao Van Sung, Pham Trong Anh & Hoang Minh Khien (1994). *Checklist of Mammals in Vietnam*. Hanoi, Hanoi Science and Technic Publishing House. (In Vietnamese.)

Dang Zui Khun & Le Suan Kan (1991). Population of some mammals in the sclerophylous evergreen tropic forests near Konhkanyng. *Zoologicheskii Zhurnal*, **70**(2), 114–18. (In Russian with English abstract.)

Dang Zui Khun, Le Suan Kan & Puzachenko, Y. (1991). Large mammals of small-leafed deciduous forests in the Srepok River Basin (South Vietnam). *Zoologicheskii Zhurnal*, **70**(3), 154–7. (In Russian with English abstract.)

Danilkin, A. (1996). *Behavioural Ecology of Siberian and European Roe Deer*. London, Chapman and Hall.

Davis, F. W., Stoms, D. M., Estes, J. E., Scepan, J. & Scott, J. M. (1990). An information system approach to preservation of biological diversity. *International Journal of Geographic Information Systems*, **4**, 55–78.

Davison, G. & Kiew Bong Heang (1987). Mammals of Ulu Endau, Johore, Malaysia. *Malayan Nature Journal*, **41**, 435–9.

Dayan, T. & Simberloff, D. (1994). Character displacement, sexual dimorphism and morphological variation among British and Irish mustelids. *Ecology*, **75**, 1063–73.

Dayan, T. & Simberloff, D. (1996). Patterns of size separation in carnivore communities. In *Carnivore Behaviour, Ecology and Evolution*, ed. J. L. Gittleman, pp. 243–66. Ithaca, Cornell University Press.

Dayan, T., Simberloff, D., Tchernov, E. & Yom-Tov, Y. (1989). Inter- and intraspecific character displacement in mustelids. *Ecology*, **70**, 1526–39.

Dayan, T., Simberloff, D., Tchernov, E. & Yom-Tov, Y. (1990). Feline canines: Community-wide character displacement among the small cats of Israel. *American Naturalist*, **136**, 39–60.

Delacour, J. & Jabouille, P. (1931). *Les Oiseaux de l'Indochine Francaise*. 4 volumes. Paris, Exposition Coloniale Internationale.

Desai, A. & Vuthy, L. (1996). *Status and Distribution of Large Mammals in Eastern Cambodia*. Report of the IUCN/FFI/WWF Large Mammal Conservation Project, Phnom Penh.

Deuve, J. (1972). *Les Mammiferes du Laos*. Report to the Ministere de l'education nationale, Vientiane, Lao PDR.

Dillon, T. & Wikramanayake, E. D. (1997). Parks, peace, and progress: A forum for transboundary conservation in Indochina. *Parks*, **7**, 36–51.

Dinerstein, E. (1980). An ecological survey of the Royal Karnali Bardia Wildlife Reserve, Nepal. *Biological Conservation*, **18**, 5–38.

Dinerstein, E. & McCracken, G. F. (1990). Endangered greater one-horned rhinoceros carry high levels of genetic variation. *Conservation Biology*, **4**, 417–22.

Dinerstein, E. & Price, L. (1991). Demography and habitat use by greater one-horned rhinoceros in Nepal. *Journal of Wildlife Management*, **55**(3), 401–11.

Dinerstein, E. & Wikramanayake, E. D. (1993). Beyond 'hotspots': How to prioritise investments in biodiversity in the Indo-Pacific region. *Conservation Biology*, **7**, 53–65.

Dinerstein, E., Wikramanayake, E. D. & Forney, M. (1995). Conserving the reservoirs and remnants of tropical moist forest in the Indo-Pacific region. In *Ecology, Conservation and Management of Southeast Asian Rainforests*, ed. R. B. Primack & T. E. Lovejoy, pp. 140–75. New Haven, Washington DC, Yale University Press.

Dinerstein, E., Wikramanayake, E., Robinson, J., Karanth, U., Rabinowitz, A., Olson, D., Mathew, T., Hedao, P., Connor, M., Hemley, G. & Bolze, D. (1997). *A Framework for Identifying High Priority Areas and Actions for the Conservation of Tigers in the Wild*. Part I. Washington DC, World Wildlife Fund-US and Wildlife Conservation Society.

Dixon, K. R. & Chapman, J. A. (1980). Harmonic mean measure of animal activity areas. *Ecology*, **61**, 104–44.

Dorji, P., & Santiapillai, C. (1989). Status, distribution and conservation of the tiger in Bhutan. *Biological Conservation*, **48**, 311–19.

Dubois, E. (1908). Das geologische Alter der Kendeng- oder Trinil-Fauna. *Tijdschrift van het Koninklijke Nederlandsch Aardrijkskundig Genootschap, 2 Series*, **24**, 1235–71.

Duckworth, J. W., Timmins, R. J. & Cozza, K. (1993). *A Wildlife and Habitat Survey of Phou Xang He Proposed Protected Area*. Report to the Protected Areas and Wildlife Division of the National Office for Nature Conservation and Watershed Management, Vientiane, Lao PDR.

Duckworth, J. W., Timmins, R. J., Thewlis, R., Evans, T. & Anderson, G. (1994). Field Observations of Mammals in Laos, 1992–1993. *Natural History Bulletin of the Siam Society*, **42**, 177–205.

Eames, J. C., Lambert, F. R. & Nguyen Cu. (1994). A survey of the Annamese Lowlands, Vietnam, and its implications for the conservation of Vietnamese and Imperial Pheasants *Lophura hatinhensis* and *L. imperialis*. *Bird Conservation International*, **4**, 343–82.

Ecological Consulting, Inc. (1993). *CAMRIS: Computer Aided Mapping and Resource Inventory System*. Portland, Oregon.

Edwards, S. V. & Potts, W. K. (1996). Polymorphism of genes in the major histocompatibility complex (MHC): Implications for conservation genetics of vertebrates. In *Molecular Genetic Approaches to Conservation*, ed. T. B. Smith & R. K. Wayne, pp. 214–37. Oxford, Oxford University Press.

Eisenberg, J. F. (1980). The density and biomass of tropical mammals. In *Conservation Biology*, ed. M. Soulé & B. A. Wilcox, pp. 35–56. Sunderland, Massachusetts, Sinauer Associates.

Eisenberg, J. F. (1981). *The Mammalian Radiations: An Analysis of Trends in Evolution, Adaptations and Behaviour*. Chicago, University of Chicago Press.

Eisenberg, J. F. & Seidensticker, J. (1976). Ungulates in southern Asia: A consideration of biomass estimates for selected habitats. *Biological Conservation*, **10**, 293–308.

Evans, G. P. (1911). *Big Game Shooting in Upper Burma*. London, Longmans, Green and Co.

Evans, T., Stones, A. J. & Thewlis, R. (1996a). *A Wildlife and Habitat Survey of the Dong Hua Sao NBCA in 1996*. Report to the IUCN Biodiversity Conservation Project in Pakse. Vientiane, Lao PDR, IUCN.

Evans, T., Stones, A. J. & Thewlis, R. (1996b). *A Wildlife and Habitat Survey of the Phou Xiang Thong NBCA*. Report to the IUCN Biodiversity Conservation Project in Pakse. Vientiane, Lao PDR, IUCN.

Faust, T. & Tilson, R. (1994). Estimating How Many Tigers are in Sumatra: A Beginning. In *Sumatran Tiger Population and Habitat Viability Analysis Report*, ed. R. Tilson, K. Soemarna, W. Ramono, S. Lusli, K. Traylor-Holzer & U. S. Seal, pp. 11–38. Apple Valley, Minnesota, Indonesian Directorate of Forest Protection and Nature Conservation and IUCN/SSC Conservation Breeding Specialist Group.

Ferson, S. (1991). *RAMAS/STAGE. Generalized Stage-based Modeling for Population Dynamics*. Setauket, NY, Applied Biomathematics.

Flerov, K. K. (1960). *Fauna of USSR: Mammals; Musk deer and Deer*. Washington DC, Israel Program for Scientific Translation.

Forthman Quick, D. L., Gustavson, C. R. & Rusiniak, K. W. (1985). Coyote Control and Taste Aversion. *Appetite*, **6**, 253–64.

Frankel, O. H. & Soulé, M. E. (1981). *Conservation and Evolution*. New York, Cambridge University Press.

Frankham, R. (1995). Conservation Genetics. *Annual Review of Genetics*, **29**, 305–28.

Frankham, R. (1966). Relationship between genetic variation to population size in wildlife. *Conservation Biology*, **10**, 1500–8.

Franklin, N. & Wells, P. (1994). *Status of the Sumatran Rhino in Kerinci Seblat National Park*. Report to the Directorate of Forest Protection and Nature Conservation, Department of Forestry, Republic of Indonesia.

Gagliuso, R. A. (1991). Remote sensing and GIS technologies: An example of integration in the analysis of cougar habitat utilization in southwestern Oregon. In *GIS Applications in Natural Resources*, ed. M. Heit & A. Shortreid, pp. 323–29. Fort Collins, GIS World, Inc.

Gairdner, K. G. (1915). Notes on the fauna and flora of Ratburi and Petchaburi Districts. *Natural History Bulletin of the Siam Society*, **1**(3), 131–56.

Gargas, D. P. (1948). How far can a tiger swim? *Journal of the Bombay Natural History Society*, **47**, 545.

Garshelis, D. L. (1994). Density-dependent population regulation of black bears. In *Density-dependent Population Regulation of Black, Brown, and Polar Bears*, ed. M. Taylor, pp. 3–14. Ninth International Conference on Bear Research and Management Monograph Series 3. Montana, Missoula. Washington DC, Port City Press.

Gaski, A. & Johnson, K. A. (1994). *Prescription for Extinction: Endangered Species and Oriental Medicines in Trade*. Washington DC, TRAFFIC USA/World Wildlife Fund-US.

Gee, E. P. (1964). *The Wildlife of India*. London, Collins.

Geist, V. (1971). The relation of social evolution and dispersal in ungulates during the Pleistocene, with emphasis on the Old-World deer and the genus *Bison*. *Quaternary Research*, **1**, 283–315.

Geist, V. (1983). On the evolution of Ice Age mammals and its significance to an understanding of speciations. *Association of South-eastern Biologists Bulletin*, **30**, 109–33.

Geist, V. (1987a). Bergmann's Rule is invalid. *Canadian Journal of Zoology*, **65**, 1035–8.

Geist, V. (1987b). On the evolution and adaptations of *Alces*. *Swedish Wildlife Research Supplement*, **1**, 11–23.

Gersi, D. (1975). *Dans le Jungle de Borneo*. Paris, Editions GP.

Gervaise, N. (1688). *The Natural and Political History of the Kingdom of Siam*. Reprinted in 1989. Bangkok, White Lotus Press.

Gilbert, D. A., Packer, C., Pusey, A. E., Stephens, J. C. & O'Brien, S. J. (1991). Analytical DNA fingerprinting in lions: Parentage, genetic diversity, and kinship. *Journal of Heredity*, **82**, 378–86.

Gilpin, M & Hanski, I. (1991). *Metapopulation Dynamics: Empirical and Theoretical Investigations*. San Diego, Academic Press.

Gilpin, M. & Soulé, M. (1986). Minimum viable populations: Processes of species extinction. In *Conservation Biology*, ed. M. Soulé, pp. 19–34. Sunderland, Massachusetts, Sinauer Associates.

Ginzburg, L. R., Slobodkin, B., Johnson, K. & Bindman, A. G. (1982). Quasiextinction probabilities as a measure of impact on population growth. *Risk Analysis*, **21**, 171–81.

Girman, D. (1996). The use of PCR-based single-stranded conformation polymorphism analysis (PCR-SSCP) in conservation genetics. In *Molecular Genetic Approaches to Conservation*, ed. T. B. Smith & R. K. Wayne, pp. 167–82. Oxford, Oxford University Press.

Global Survival Network (1997). *CITES at Work – International Co-operation and the Comeback of the Siberian Tiger.* Washington DC, Global Survival Network.

Glowka, L., Burhenne-Guilmin, F. & Synge, H. (1994). *A Guide to the Convention on Biological Diversity. Environmental Policy and Law Centre Paper No. 30.* Gland, Switzerland, IUCN.

Goeble, A. & Whitmore, D. H. (1987). The use of electrophoretic data in the reevaluation of tiger systematics. In *Tigers of the World: The Biology, Biopolitics, Management, and Conservation of an Endangered Species*, ed. R. L. Tilson & U. S. Seal, pp. 36–50. Park Ridge, New Jersey, Noyes Publications.

Government of India (1994). *Report of the Committee for Prevention of Illegal Trade in Wildlife and Wildlife Products.* New Delhi, Ministry of Environment and Forests.

Goyal, S. P. & Johnsingh, A. J. T. (1996). Problems of identification of camera-trapped tigers. *TigerLink News*, 2(1), 27. New Delhi, TigerLink.

Gray, D., Piprell, C. & Graham, M. (1994). *National Parks of Thailand*, 2nd edn. Bangkok, Communication Resources Ltd.

Grebovoy, C. A., Karpenko, A. C., Kateninoy, G. D., Kurentsovoy, G. E., Peshkovoy, G. A. & Rosenberg, V. A. (1968). *Plant cover map of the Amur Basin.* Vladivostok, The USSR Academy of Sciences Botanical Institute and Far Eastern Affiliate Institute of Biology and Soils.

Griffin, J. G. (1994). An evaluation of protected area management: A case study of Khao Yai National Park, Thailand. *Tigerpaper*, 21(1), 15–23.

Griffiths, M. (1994). Population density of Sumatran tigers in Gunung Leuser National Park. In *Sumatran Tiger Population and Habitat Viability Analysis Report*, ed. R. Tilson, K. Soemarna, W. Ramono, S. Lusli, K. Traylor-Holzer & U. Seal, pp. 93–102. Apple Valley, Minnesota, Indonesian Directorate of Forest Protection and Nature Conservation and IUCN/SSC Conservation Breeding Specialist Group.

Griffiths, M. (1996) http:\\www.5tigers.org/griffiths.htm; http:\\www.5tigers.org/intrphva.htm

Groves, C. P. (1989). *A Theory of Human and Primate Evolution.* Oxford, Clarendon Press.

Grumbine, R. E. (1994). What is ecosystem management? *Conservation Biology*, 8, 27–38.

Guthrie, R. D. (1984). Mosaics, allelochemicals and nutrients. In *Pleistocene Extinctions*, ed. P. S. Martin & R. G. Klein, pp. 259–98. Tucson, Arizona, University of Arizona Press.

Hammond Gold Medallion World Atlas. (1988). Maplewood, New Jersey, Hammond Inc.

Hartl, G. B. & Pucek, Z. (1992). Genetic depletion in the European bison (*Bison bonasus*) and the significance of electrophoretic heterozygosity for conservation. *Conservation Biology*, 8, 167–83.

Hedges, S. (ed.) (1995). *Asian Wild Cattle and Buffaloes, Status Report and Conservation Action Plan, Part II.* Unpublished Draft of IUCN/SSC Action Plan for Asian Wild Cattle.

Hemley, G. & Bolze, D. (1997). *A Framework for Identifying High Priority Areas and Actions for the Conservation of Tigers in the Wild. Part II. Controlling Trade in and Reducing Demand for Tiger Products: A Preliminary Assessment of Priority Needs.* Washington DC, World Wildlife Fund.

Hemmer, H. (1967). Wohin gehört 'Felis' palaeosinensis (Zdansky 1924) in systematischer Hinsicht? *Neues Jahrbuch für Geologie und Paläontologie. Stuttgart. Abhandlungen*, 129, 83–96.

Hemmer, H. (1968). Der Tiger *Panthera tigris palaeosinensis* (Zdansky, 1924) im Jungpleistozän Japans. *Neues Jahrbuch für Geologie und Paläontologie Abhandlungen*, 10, 610–18.

Hemmer, H. (1969). Zur Stellung des Tigers (*Panthera tigris*) der Insel Bali. *Zeitschrift für Säugetierkunde*, 34, 216–23.

Hemmer, H. (1971). Fossil mammals of Java. II. Zur Fossilgeschichte des Tigers (*Panthera tigris* (L.)) in Java. *Koninklijke Nederlandse Akademie van Wetenschappen, Series B*, 74, 35–52.

Hemmer, H. (1976). Fossil history of the living Felidae. In *The World's Cats*, volume III (2), ed. R. L. Eaton, pp. 1–14. University of Washington, Seattle, The Carnivore Research Institute, Burke Museum.

Hemmer, H. (1981). Die Evolution der Pantherkatzen Modell zur Überprufung der Brauchbarkeit der Hennigschen Prinzipien der Phylogenetischen Systematik für wirbeltierpaläontoogische Studien. *Paläontologische Zeitschrift*, 55, 109–16.

Hemmer, H. (1987). The phylogeny of the tiger (*Panthera tigris*). In *Tigers of the World: The Biology, Biopolitics, Management and Conservation of an Endangered Species*, ed. R. L. Tilson & U. S. Seal, pp. 28–35. Park Ridge, New Jersey, Noyes Publications.

Hendrichs, H. (1975). The status of the tiger *Panthera tigris* (Linne, 1758) in the Sundarbans Mangrove Forest, (Bay of Bengal). *Saugetierkunaliche Mittelungen*, **3**, 161–99.

Heptner, V. G. & Sludskii, A. A. (1972). *Mammals of the Soviet Union – Carnivora, Hyenas and Cats*. Moscow, Vischnaya schkola. (In Russian.)

Heptner, V. G. & Sludskii, A. A. (1992a). *Mammals of the Soviet Union*, volume II, part 2, *Carnivora (Hyaenas and Cats)*. (English translation, ed. R. S. Hoffmann). Washington DC, Smithsonian Institution Libraries and the National Science Foundation.

Heptner, V. G. & Sludskii, A. A. (1992b). *Mammals of the Soviet Union*, volume II, part 2, *Carnivores (Feloidea)*. Leiden, E. J. Brill.

Heptner, V. G., Nasimovich, A. A. & Bannikov, A. G. (1989). *Mammals of the Soviet Union*, volume I, *Ungulates*. (English translation, ed. R. S. Hoffmann). Leiden, E. J. Brill.

Herrington, S. (1987). Subspecies and the conservation of *Panthera tigris*: Preserving genetic heterogeneity. In *Tigers of the World: The Biology, Biopolitics, Management and Conservation of an Endangered Species*, ed. R. L. Tilson & U. S. Seal, pp. 51–60. Park Ridge, New Jersey, Noyes Publications.

Hiby, L. & Lovell, P. (1991). DUNGSURV – A program for estimating elephant density from dung density without assuming 'steady state'. In *Proceedings of the Workshop on Censusing Elephants in Forests*, 2–10 January 1991, Bangalore, India, ed. U. Ramakrishnan, J. A. Santosh & R. Sukumar, pp. 73–80. Bangalore, India, Asian Elephant Conservation Centre.

Hill, G. (1994). Observations of Wildlife Trade in Mergui Tavoy District, Kawthoolei. *TRAFFIC Bulletin*, **14**.

Hilzheimer, (1905). Über einige Tigerschädel aus der Strassburger Zoologischen Sammlung. *Zoologischer Anzeiger*, **28**, 594–99.

His Majesty's Government of Nepal/National Planning Commission Secretariat. (1994). *Population of Nepal: Municipalities 1991 Population Census*. Ramshah Path, Kathmandu, Nepal, Central Bureau of Statistics.

Hoang Hoe & Vo Quy (1991). Nature conservation in Vietnam: An overview. *Tigerpaper*, **18**(4).

Hoelzel, A. R., Halley, J., O'Brien, S. J., Campagna, C., Arnbom, T., Le Boeuf, B., Ralls, K. & Dover, G. A. (1993). Elephant seal genetic variation and the use of simulation models to investigate historical population bottlenecks. *Journal of Heredity*, **84**, 443–9.

Hoogerwerf, A. (1970). *Ujung Kulon, the Land of the Last Javan Rhinoceros*. Leiden, E. J. Brill.

Hoogesteijn, R. & Mondolfi, E. (1996). Body mass and skull measurements in four jaguar populations and observations on their prey base. *Bulletin of the Florida Museum of Natural History*, **39**, 195–219.

Hooijer, D. A. (1947). Pleistocene remains of *Panthera tigris* (Linnaeus) subspecies from Wanhsien, Szechuan, China, compared with fossil and recent tigers from other localities. *American Museum Novitates*, **1346**, 1–17.

Hooijer, D. A. (1963). Further 'Hell' mammals from Niah. *Sarawak Museum Journal*, **11**, 196–200.

Hose, C. & McDougall, W. (1912). *The Pagan Tribes of Borneo*, volume II. London, Macmillan.

Hughes, A. L. (1991). MHC polymorphisms and the design of captive breeding programmes. *Conservation Biology*, **5**, 249–51.

Humphrey, S. R. & Bain, J. R. (1990). *Endangered Animals of Thailand. Flora and Fauna Handbook 6*. Gainesville, Florida, Sandhill Crane Press Inc.

Humphrey, S. R., & Stith, B. M. (1990). A balanced approach to conservation. *Conservation Biology*, **4**, 341–3.

Hutchins, M. & Wemmer, C. (1991). In defence of captive breeding. *Endangered Species UPDATE* **8**(9 & 10), 5–6.

Illiger, C. (1815). Ueberblick der Säugethiere nach ihrer Vertheilung über die Welttheile. *Abhandlungen Physikalischen Klasse der Königlich-Preussischen Akademie der Wissenschaften aus den Jahren 1804–1811*, pp. 39–159.

Iriarte, J. A., Franklin, W. L., Johnson, W. E. & Redford, K. H. (1990). Bioegographic variation of food habits and body size of the American puma. *Oecologia*, **85**, 185–90.

ISIS (1997). *ISIS Mammal Abstract Distribution Report, as of 31 December 1996*. Apple Valley, Minnesota, International Species Information System.

IUCN (1987). *The IUCN Policy Statement on Captive Breeding*. Gland, Switzerland, IUCN.

IUCN (1995). *Protected Area Fact Sheets, Annex 3. The 1995 Status Report on Protected Area System Planning and Management in Lao PDR*. Gland, Switzerland, IUCN.

IUCN (1996). *The 1996 IUCN Red List of Threatened Animals*. Gland, Switzerland, IUCN.

IUCN (1998). *IUCN Guidelines for Re-introductions*. Nairobi, Kenya, IUCN/SSC Reintroduction Specialist Group.

IUCN, UNEP & WWF. (1991). *Caring for the Earth: A Strategy for Sustainable Living*. Gland, Switzerland, IUCN.

IUDZG/CBSG. (1993). *The World Zoo Conservation Strategy: The Role of the Zoos and Aquaria of the World in Global Conservation*. Brookfield, Illinois, IUDZG – The World Zoo Organisation/Chicago Zoological Society.

Jackson, P. (1990). *Endangered Species: Tigers*. London, The Apple Press.

Jackson, P. (1993a). The status of the tiger in 1993. *Cat News*, **19**, 5–11. Bougy, Switzerland, IUCN Cat Specialist Group.

Jackson, P. (1993b). Tiger conservation moves again to centre stage. *Cat News*, **18**, 2–3. Bougy, Switzerland, IUCN Cat Specialist Group.

Jackson, P. (1995). Plight of the tiger. *Cat News*, **22**, 2–3. Bougy, Switzerland, IUCN Cat Specialist Group.

Jackson, P. (1996). The status of the tiger in 1996. *Cat News*, **25**, 6. Bougy, Switzerland, IUCN Cat Specialist Group.

Jackson, P. & Kemf, E. (1994, reprinted 1996). *Wanted Alive! Tigers in the Wild*. A WWF Status Report. Gland, Switzerland, WWF.

Janczewski, D. N., Modi, W. S., Stephens, J. C. & O'Brien, S. J. (1995). Molecular evolution of mitochondrial 12S RNA and cytochrome b sequences in the Pantherine lineage of Felidae. *Molecular Biology and Evolution*, **12**, 690–707.

Janhunen, J. (1996). *Manchuria – an ethnic history*. Vammalan Kirjapaino Oy, Vammala.

Ji, W. & Rabinowitz, A. (1995). *Proceedings for the Workshop of Trans-boundary Biodiversity Conservation in the Eastern Himalayas*. Yunnan, China, Kunming Institute of Zoology, Chinese Academy of Sciences.

Johnsingh, A. J. T. (1992). Prey selection in three large sympatric carnivores in Bandipur. *Mammalia*, **56**, 517–26.

Johnsingh, A. J. T. (1995). Vietnam venture: The primordial world of Sao La and Mang. *Frontline*, 94–7.

Johnsingh, A. J. T. (1996). A barren stage; efforts in China's Wuyishan Reserve. *Frontline*, April 5, 70–3.

Johnson, D. H. (1980). The comparison of usage and availability measurements for evaluating resource preference. *Ecology*, **61**(1), 65–71.

Johnson, P. (1991). *The Birth of the Modern World Society 1815–1830*. New York, Harper Collins.

Johnson, W. & O'Brien, S. J. (1997). Phylogenetic reconstruction of the Felidae using 16S and NADH-5 mitochondrial genes. *Journal of Molecular Evolution*, **44**, 598–616.

Johnson, W. E., Dratch, P. A., Martenson, J. S. & O'Brien, S. J. (1996). Resolution of recent radiations within three evolutionary lineages of Felidae using mitochondrial restriction fragment length polymorphism variation. *Journal of Molecular Evolution*, **3**, 97–120.

Kaplanov, L. G. (1948). *Tiger, Deer, Elk*. Moscow, Esdatelstvo Moskovskovo Obshestva Espitatelei Prerodi. (In Russian.)

Karanth, K. U. (1987). Tigers in India: A critical review of field censuses. In *Tigers of the World: The Biology, Biopolitics. Management and Conservation of an Endangered Species*, ed. R. L. Tilson & U. S. Seal, pp. 118–33. Park Ridge, New Jersey, Noyes Publications.

Karanth, K. U. (1988). Analysis of predator-prey balance in Bandipur Tiger Reserve with reference to census reports. *Journal of the Bombay Natural History Society*, **85**, 1–8.

Karanth, K. U. (1991). Ecology and management of the tiger in tropical Asia. In *Wildlife Conservation: Present Trends and Perspectives for the 21st Century*, ed. N. Maruyama, B. Bobek, Y. Ono, W. Reglin, L. Bartos & R. Ratcliffe, pp. 156–9. Tokyo, Japan Wildlife Research Centre.

Karanth, K. U. (1993). *Predator-prey relationships among large mammals in Nagarahole National Park, India*. Mangalore University, PhD Dissertation.

Karanth, K. U. (1995). Estimating tiger *Panthera tigris* populations from camera-trap data using capture-recapture models. *Biological Conservation*, **71**, 333–8.

Karanth, K. U. (1996). Identification of camera-trapped tigers. *TigerLink News*, **2**(2), 30. New Delhi.

Karanth, K. U. & Madhusudan, M. D. (1997). Avoiding paper tigers and saving real tigers: Response to Saberwal. *Conservation Biology*, **11**, 1–4.

Karanth, K. U. & Nichols, J. D. (1998). Estimation of tiger densities in India using photographic captures and recaptures. *Ecology*, **79**

Karanth, K. U. & Sunquist, M. E. (1992). Population structure, density and biomass of large herbivores in the tropical forests Nagarahole, India. *Journal of Tropical Ecology*, **8**, 21–35.

Karanth, K. U. & Sunquist, M. E. (1995). Prey selection by tiger, leopard and dhole in tropical forests. *Journal of Animal Ecology*, **64**, 439–50.

Keiter, R. B. (1995). Preserving Nepal's national parks: Law and conservation in the developing world. *Ecology Law Quarterly*, **22**, 591–675.

Kemp, N., Le Mong Chan & Dilger, M. (1995). *Site Description and Conservation Evaluation: Pu Mat Nature Reserve, Nghe An Province, Vietnam.* Frontier Vietnam Scientific Report 5. London, Society of Environmental Exploration.

Kenney, J. S., Smith, J. L. D., Starfield, A. M. & McDougal, C. W. (1994). Saving the tiger in the wild. *Nature*, **369**, 352.

Kenney, J. S., Smith, J. L. D., Starfield, A. M. & McDougal, C. (1995). The long-term effects of tiger poaching on population viability. *Conservation Biology*, **9**, 1127–33.

Khan, Mohammed Khan bin Momin (1987). Tigers in Malaysia: Prospects for the future. In *Tigers of the World: The Biology, Biopolitics, Management and Conservation of an Endangered Species*, ed. R. L. Tilson & U. S. Seal, pp. 75–85. Park Ridge, New Jersey, Noyes Publications.

Kirk, G. (1994). Insel-Tiger *Panthera tigris* (Linnaeus, 1758). *Säugetierkundliche Mitteilungen*, **35**, 151–76.

Kitchener, A. (1991). *The Natural History of the Wild Cats.* London, Helm.

Kitchener, A. C. (1993). A new look at subspecies in the Felidae. *Lifeline*, May 1993, pp. 6–13.

Kitchener, A. C. (1997). The role of museums and zoos in conservation biology. *International Zoo Yearbook*, **35**, 325–36.

Kleiman, D. (1974). The estrous cycle in the tiger (*Panthera tigris*). In *The World's Cats*, ed. R. L. Eaton, volume 2, pp. 60-75. Seattle, Woodland Park Zoo.

Klein, J. (1986). *Natural History of the Major Histocompatibility Complex.* New York, John Wiley & Sons, Inc.

Koch-Isenburg, L. (1963). *Through the Jungle Very Softly: Quest for Wild Animals in the Far East.* New York, Viking Press.

Kock, D. (1995). Zur Bennenung des Tigers *Panthera tigris* L. auf Sunda-Inseln. *Säugetierkundliche Mitteilungen*, **36**, 123–26.

Koehler, G. M. (1991). *Survey of remaining wild populations of south China tigers.* WWF Project 4512/China. Moscow, Idaho, Hornocker Wildlife Institute.

Koenigswald, G. H. R. von (1933). Beitrag zur Kenntnis der fossilen Wirbeltiere Javas. *Wetenschappelijke Mededeelingen: Dienst van de Mijnbouw in Nederlansch Oost-Indie*, **1**(23), 1–184.

Kolosova, T. I. & Kondrashov, L. G. (1996). Cedar-broad-leaved forest product exportation on the international market. In *Cedar-broad-leaved Forests of the Far East.* Proceedings of an International Conference, 30 September to 6 October 1996, ed. D. F. Efremov, L. G. Kondrashov and A. P. Sapojnikov, pp. 249–50 (abstract). Khabarovsk, Khabarovski Krai, FEFRI (Far East Forestry Research Institute).

Kotwal, P. C. (1984). Incidences of intraspecific fights and cannibalism among tigers in Kanha National Park. *Cheetal*, **27**, 28–33.

Kucherenko, S. P. (1985). *The Tiger.* Moscow, Argopromezdat. (In Russian.)

Kungurova, S. (undated). *Primorskii Krai.* Vladivostok, DalPress Printing and Publishing Centre.

Kurtén, B. (1967). *The Pleistocene Mammals of Europe.* London, Weidenfeld and Nicolson.

Kurtén, B. (1971). *The Age of Mammals.* New York, Columbia University Press.

Kurtén, B. (1976). Fossil puma (Mammalia: Felidae) in North America. *Netherlands Journal of Zoology*, **26**, 502–34.

Laake, J. L., Buckland, S. T., Anderson, D. R. & Burnham, K. P. (1993). *Distance User's Guide.* Fort Collins, Colorado, Colorado State University.

Lacy, R. (1997). Importance of genetic variation to the viability of mammalian populations. *Journal of Mammalogy*, **78**, 320–35.

Laidlaw, R. K. (1994). *The Virgin Jungle Reserves of Peninsular Malaysia: The Ecology and Dynamics of Small Protected Areas in Managed Forest.* Cambridge, Cambridge University, PhD Dissertation.

Laing, S. P. & Lindzey, F. G. (1993). Patterns of replacement of resident cougars in southern Utah. *Journal of Mammalogy*, **74**, 1056–8.

Lancia, R. A., Nichols, J. D. & Pollock, K. N. (1994). Estimation of number of animals in wildlife populations. In *Research and Management Techniques for Wildlife and Habitats*, ed. T. A. Bookhout, pp. 215–53. Bethesda, Maryland, The Wildlife Society.

Lande, R. (1988). Genetics and demography in biological conservation. *Science*, **241**, 1455–60.

Lande, R. & Barrowclough, G. F. (1987). Effective population size, genetic variation, and their use in population management. In *Viable Populations for Conservation*, ed. M. E. Soulé, 87–123. Cambridge, Cambridge University Press.

Latimer, W., Hill, G., Bhumpakkaphan, N. & Fehr, C. (1993). *Report on the Mergui District, Special Township Region and Kaser Doo Wildlife Sanctuary*. Report to the Regional Community Forestry Training Centre (RECOFTC), Bangkok.

Lay Khim (1995). *Protected Areas in the Kingdom of Cambodia*. Proceedings of the Workshop on Regional Trans-Boundary Protected Areas. Bangkok.

Lay Khim & Taylor-Hunt, D. (1995). *Kirirom General Survey*. Ministry of Environment/Department of Nature Conservation and Protection, Cambodia and International Development Research Center, Canada.

Le Trong Trai (1996). *The Status of the Tiger (*Panthera tigris*) in Vietnam*. Proceedings of the Second International Conference and Geographic Information System Workshop to Assess the Status of Tigers, 24–31 January 1996, Thailand.

Le Xuan Canh (1995). *A Report on the Survey for Large Carnivores in Tay Nguyen Plateau, South Vietnam with Emphasis on Tiger (*Panthera tigris*)*. Report to the Wildlife Conservation Society, New York.

Lefkovitch, L. P. (1965). The study of population growth in organisms grouped by stages. *Biometrics*, **21**, 1–18.

Lekagul, B. & McNeely, J. A. (1988). *Mammals of Thailand*, 2nd edn. Bangkok, Thailand, Saha Karn Bhaet Co.

Leng-EE, P. (1979). Status of the tiger in Thailand. *Tigerpaper*, **6**, 2–3.

Levins, R. (1969). Some demographic and genetic consequences of environmental heterogeneity for biological control. *Bulletin of the Entomology Society of America*, **15**, 237–40.

Levins, R. (1970). Extinction. In *Some Mathematical Questions in Biology. Lectures on Mathematics in the Life Sciences*, ed. M. Gerstenhaber, volume 2, pp 77–107. Province, Rhode Island, American Mathematical Society.

Leyhausen, P. (1979). *Cat Behavior*. New York, Garland STPM Press.

Lillesand, T. & Kiefer, R. (1994). *Remote Sensing and Image Interpretation*. New York, John Wiley & Sons, Inc.

Lindzey, F. G., Van Sickle, W. D., Laing, S. P. & Mecham, C. S. (1992). Cougar population response to manipulation in southern Utah. *Wildlife Society Bulletin*, **20**, 224–7.

Lindzey, F. G., Van Sickle, W. D., Ackerman, B. B., Barnhurst, D., Hemker, T. P. & Laing, S. P. (1994). Cougar population dynamics in southern Utah. *Journal of Wildlife Management*, **58**, 619–24.

Lister, A. & Bahn, P. (1994). *Mammoths*. London, Boxtree.

Locke, A. (1954). *The Tigers of Trengganu*. Republished 1993; Monograph 23. Kuala Lumpur, Malaysian Branch of the Royal Asiatic Society.

Lopez, J. V., Cevario, S. & O'Brien, S. J. (1996). Complete nucleotide sequence of the domestic cat (*Felis catus*) mitochondrial genome and a transposed mtDNA tandem repeat (Numt) in the nuclear genome. *Genomics*, **33**, 229–46.

Lopez, J. V., Culver, M., Stephens, J. C., Johnson, W. E. & O'Brien, S. J. (1997). Relative rate of nuclear versus cytoplasmic mitochondrial DNA sequence divergence in mammals. *Molecular Biology and Evolution*, **14**, 277–86.

Loukashkin, A. S. (1937). Some observations on the remains of a Pleistocene fauna and of the Palaeolithic age in northern Manchuria. In *Early Man*, ed. G. G. MacCurdy, 327–40. Freeport, Books for Libraries Press.

Loukashkin, A. S. (1938). The Manchurian tiger. *China Journal*, **28**, 127–33.

Lumpkin, S. (1991). Cats and Culture. In *Great Cats*, ed. J. Seidensticker & S. Lumpkin, pp. 190–203. Emmaus, Pennsylvania, Rodale Press.

Luscombe, B.W. (1986). *Spatial data handling in data poor environments*. Burnaby, British Columbia, Simon Fraser University, PhD Dissertation.

Lydekker, R. (1886). Preliminary notes on the Mammalia of the Karnul Caves. *Records of the Geological Survey of India*, **19**, 120–2.

Ma Yiqing & Li Xiaomin (1996). *The Status of Conservation of Tigers in China*. Presented at the 2nd International Symposium on Coexistence of Large Carnivores with Man, 19–23 November, 1996. Omiya, Saitama, Japan.

Mace, G. M., Smith, T. B., Bruford, M. W. & Wayne, R. (1996). An overview of the issues. In *Molecular Genetic Approaches to Conservation*, ed. T. B. Smith & R. K. Wayne, pp. 1–21. Oxford, Oxford University Press.

MacKinnon, J. & MacKinnon, K. (1986). *Review of the Protected Areas System in the Indo-Malayan Realm*. IUCN/UNEP Report. Gland, Switzerland, IUCN.

MacKinnon, J., MacKinnon, K., Child, G. & Thorsell, J. (1986). *Managing Protected Areas in the Tropics*. Gland, Switzerland, IUCN.

MacKinnon, J., Meng Sha, Cheung, C., Carey, G., Zhu Xiang & Melville, D. (1996). *A Biodiversity Review of China*. Hong Kong, WWF International.

MacKinnon, K. (1997). The ecological foundations of biodiversity protection. In *Last Stand: Protected Areas and the Defense of Tropical Biodiversity*, ed. R. Kramer, C. van Schaik & J. Johnson. Oxford, Oxford University Press.

Madhusudan, M. D. & Karanth, K. U. Hunting for an answer: Is local hunting compatible with large mammal conservation in India?. In *Hunting for Sustainability in Tropical Forests*, ed. J. G. Robinson & E. L. Bennet. New York, Columbia University Press. (In press.)

Mainka, S.A. (1997). *Tiger Progress? The Response to CITES Resolution Conf. 9.13*. Cambridge, TRAFFIC International.

Marks, R. B. (1998). *Tigers, Rice, Silt, & Silk*. Cambridge, Cambridge University Press.

Martin, E. B. (1992a). Observations on wildlife trade in Vietnam. *TRAFFIC Bulletin*, **13**(2), 61–7.

Martin, E. B. (1992b). The trade and uses of wildlife product in Laos. *TRAFFIC Bulletin*, **13**(1), 23–8.

Martin, E. B. & Phipps, M. (1996). A review of the wild animal trade in Cambodia. *TRAFFIC Bulletin*, **16**(2), 45–60.

Martin, R. B. & de Meulenaer, T. (1988). *Survey of the Status of the Leopard (Panthera pardus) in Sub-Saharan Africa*. Lausanne, Switzerland, CITES Secretariat.

Masuda, R., Lopez, J. V., Pecon Slattery, J., Yuhki, N. & O'Brien, S. J. (1996). Molecular phylogeny of mitochondrial cytochrome b and 12S rRNA sequences in the Felidae: Ocelot and domestic cat lineages. *Molecular and Phylogenetic Evolution*, **6**, 351–65.

Matyushkin, E. N. (1977). Choice of route and exploitation of territory by Amur tigers (based on winter tracking). In *Behavior of Mammals*, USSR Academy of Sciences All-Soviet Theriological Society, pp. 146–78. Moscow, Nauka Press.

Matyushkin, E. N. (1992). Tiger and elk on the coastal slopes of central Sikhote Alin. *Bulletin of the Moscow Organization for Investigation of Nature, Biology Branch*, **97**(1), 3–19. (In Russian.)

Matyushkin, E. N. & Yudakov, A. G. (1974). Traces of the Amur tiger. *Okhota and Okhot*, **5**, 12–17. (In Russian.)

Matyushkin, E. N., Zhivotchenko, V. I. & Smirnov, E. N. (1980). *The Amur tiger in the USSR*. Gland, Switzerland, IUCN.

Matyushkin. E. N., Astafiev, A. A., Zaitsev, V. A., Koctoglod, E. E., Palkin, B. A., Smirnov, E. N. & Yudt, R. G. (1981). The history, present situation, and perspectives of protection of the Amur tiger in Sikhote-Alin Zapovednik. In *Mammalian Predators*, ed. V. A. Zabroden, pp. 76–118. Moscow, TNEL Glavokhoti. (In Russian.)

Matyushkin, E. N., Pikunov, D. G., Dunishenko, Y. M., Miquelle, D. G., Nikolaev, I. G., Smirnov, E. N., Salkina, G. P., Abramov, V. K., Bazylnikov, V. I., Yudin, V. G. & Korkishko, V. G. (1996). *Numbers, Distribution, and Habitat Status of the Amur Tiger in the Russian Far East: 'Express-report'*. Final report to the USAID Russian Far East Environmental Policy and Technology Project.

Mauget, R. (1991). Reproductive biology of the wild Suidae. In *Biology of Suidae*, ed. R. H. Barrett & F. Spitz, pp. 49–64. Briançon, France, Institut National de la Recherche Agronomique.

Mayr, E. (ed.) (1963). *Animal Species and Evolution*. Cambridge, Massachusetts, Harvard University Press.

Mayr, E. (ed.) (1982). Processes of speciation in animals. In *Mechanisms of Speciation*, ed. C. Barigozzi, pp. 1–19. New York, Alan Liss.

Mayr, E. & Ashlock, P. D. (1991). *Principles of Systematic Zoology*, 2nd edn. New York, McGraw-Hill.

Mazák, V. (1967). Notes on Siberian long-haired tiger, *Panthera tigris altaica* (Temminck, 1844), with a remark on Temminck's mammal volume of the *Fauna Japonica*. *Mammalia*, **31**, 537–73.

Mazák, V. (1968). Nouvelle sous-espece de tigre provenant de l'Asie due Sud-Est. *Mammalia*, **32**, 104–12.

Mazák, V. (1976). On the Bali tiger, *Panthera tigris balica* (Schwarz, 1912). *Vestník Ceskoslovenské Spolecnosti Zoologické*, **40**, 179–95.

Mazák, V. J. (1981). *Panthera tigris*. *Mammalian Species*, **152**, 1–8.

Mazák, V. J. (1996). *Der Tiger*. Magdeburg, Westarp Wissenschaften. (Reprint of 1983 edn.).

Mazák, V., Groves, C. P. & van Bree, P. J. H. (1978). On a skin and skull of the Bali tiger, and a list of preserved specimens of *Panthera tigris balica* (Schwarz, 1912). *Zeitschrift für Säugetierkunde*, **43**, 108–13.

McCullough, D. R., ed. (1996). *Metapopulations and Wildlife Conservation*. Washington DC, Island Press.

McDougal, C. (1977). *The Face of the Tiger*. London, Rivington Books.

McDougal, C. & Tshering, K., (1998). *Tiger Conservation Strategy for the Kingdom of Bhutan*. Thimpu, Ministry of Agriculture/ WWF.

McNeely, J. A. & Wachtel, P.S. (1988). *Soul of the Tiger: Searching for Nature's Answers in Exotic Southeast Asia*. New York, Doubleday.

McNeely, J. A., Miller, K. R., Reid, W. V., Mittermeier, R. A. & Werner, T. B. (1990). *Conserving the World's Biological Diversity*. Gland, Switzerland and Washington DC, IUCN, World Resources Institute, Conservation International, WWF–US and the World Bank.

Meacham, C. J (1997). *How the Tiger Lost its Stripes*. Orlando, Florida, Harcourt Brace.

Mech, L. D. (1977). A recovery plan for the Eastern timber wolf. *National Parks and Conservation Magazine*, **51**, 17–21.

Medway, Lord (1977). *Mammals of Borneo*. Kuala Lumpur, Malaysian Branch of the Royal Asiatic Society.

Medway, Lord (1978). *The Wild Mammals of Malaya (Peninsular Malaysia) and Singapore*, 2nd edn. Kuala Lumpur, Oxford University Press.

Meffe, G. K. & Carroll, C. R. (1997). *Principles of Conservation Biology*, 2nd edn. Sunderland Massachusetts, Sinauer Associates.

Menotti-Raymond, M. & O'Brien, S. J. (1993). Dating the genetic bottleneck of the African cheetah. *Proceedings of the National Academy of Science, USA*, **90**, 3172–6.

Menotti-Raymond, M. A. & O'Brien, S. J. (1995). Evolutionary conservation of ten microsatellite loci in four species of Felidae. *Journal of Heredity*, **86**, 319–22.

Menotti-Raymond, M., David, V. A., Stephens, J. C. & O'Brien, S. J. (1997). Genetic individualization of domestic cats using feline STR loci for forensic analysis. *Journal of Forensic Science*, **42**, 1037–50.

Mescheryakov, V. S. & Kucherenko, S. P. (1990). *Number of Tigers and Ungulates in Primorsky Krai; Recommendations for Protection and Rational Use*. Final report. All Soviet Scientific Research Institute of Hunting Management, Far East Division of Primorski Krai Commercial Hunting, and Primorsky Co-operative for Commercial Hunting. (In Russian.)

Miller, R. I., Stuart, S. N. & Howell, K. M. (1989). A methodology for analyzing rare species distribution patterns utilizing GIS technology: The rare birds of Tanzania. *Landscape Ecology*, **2**, 173–89.

Mills, J. A. (1996). Unprecedented meeting gives TCM specialists a voice and could lead to co-operation. *TRAFFIC Dispatches*, January 1996, 8–9.

Mills, J. A. (ed.) (1997). *Rhinoceros Horn and Tiger Bone in China: An Investigation of Trade Since the 1993 Ban*. Cambridge, TRAFFIC International.

Mills, J. A. & Jackson, P. (1994). *Killed for a Cure: A Review of the World-Wide Trade in Tiger Bone*. Cambridge, TRAFFIC International.

Milton, O. & Estes, R. D. (1963). *Burma Wildlife Survey 1959–1960*. Special Publication. No. 15. New York, American Committee for International Wildlife Protection.

Minchen, C. (1964). Mammals of the 'Lantian man' locality at Lantian, Shensi. *Vertebrata Palasiatica*, **8**, 301–7.

Ministry of Forestry (1994). *Indonesian Sumatran Tiger Conservation Strategy*. Jakarta, Indonesian Ministry of Forestry.

Minta, S. & Karieva, P. M. (1994). A conservation science perspective: Conceptual and experimental improvements. In *Endangered Species Recovery*, ed. T. W. Clark, R. P. Reading & A. L. Clarke, pp. 275–304. Washington DC, Island Press.

Miquelle, D. G., Quigley, H., Hornocker, M. G., Smirnov, E. N., Nikolaev, I. G., Pikunov, D. G. & Quigley, K. (1993). *Present status of the Siberian tiger, and some threats to its conservation*. In *Proceedings of the International Union of Game Biologists*, ed. I. D. Thompson, volume XXI, pp. 274–8. Nova Scotia, Halifax.

Miquelle, D. G., Quigley, H. B. & Hornocker, M. G. (1995). *A Habitat Protection Plan for Amur Tiger Conservation: A Proposal Outlining Habitat Protection Measures for the Amur Tiger*. Idaho, Hornocker Wildlife Institute.

Miquelle, D. G., Arzhanova, T. D. & Solkin, V. A. (ed). (1996a). *A Recovery Plan for Conservation of the Far Eastern Leopard: Results of an International Conference Held in Vladivostock, Russia*. Final Report to the USAID Russian Far East Environmental Policy and Technology Project.

Miquelle, D. G., Smirnov, E. N., Quigley, H. G., Hornocker, M. G., Nikolaev, I. G. & Matyushkin, E. N. (1996b). Food habits of Amur tigers in Sikhote-Alin Zapovednik and the Russian Far East, and implications for conservation. *Journal of Wildlife Research*, **1**(2), 138–47.

Mishra, G. P. & Kotwal, P. C. (1990). Plant-herbivore-carnivore system in Kanha National Park. *Journal of Tropical Forestry*, **6**, 66–76.

Mishra, H. R. (1982). *The ecology and behaviour of chital (Axis axis) in Royal Chitwan National Park, Nepal.* University of Edinburgh, PhD Dissertation.

Mishra, H. R., Wemmer, C. & Smith, J. L. D. (1987). Tigers in Nepal: Management conflicts with human interests. In *Tigers of the World: The Biology, Biopolitics, Management and Conservation of an Endangered Species,* ed. R. L. Tilson & U. S. Seal, pp. 449–63. Park Ridge, New Jersey, Noyes Publications.

Miththapala, S., Seidensticker, J. & O'Brien, S. J. (1996). Phylogeographic subspecies recognition in leopards (*Panthera pardus*): Molecular genetic variation. *Conservation Biology,* 10, 1115–32.

Mohr, C. O. (1947a). Major fluctuations of some Illinois mammal populations. *Transactions of the Illinois Academy of Science,* 40, 197–204.

Mohr, C. O. (1947b). Table of equivalent populations of North American small mammals. *American Midland Naturalist,* 37, 223–49.

Morley, R. J. & Flenley, J. R. (1987). Late Cainozoic vegetational and environmental changes in the Malay Archipelago. In *Biogeographical Evolution of the Malay Archipelago,* ed. T. C. Whitemore, pp. 50–9. Oxford, Clarendon Press.

Mountfort, G. (1981). *Saving the Tiger.* London, Michael Joseph.

Mukherjee, S., Goyal, S. P. & Chellam, R. (1994). Refined techniques for the analysis of Asiatic lion *Panthera leo persica* scats. *Acta Theriologica,* 39(4), 425–30.

Müller, P. (1995). *International Tiger Studbook.* Leipzig, Germany, Leipzig Zoo.

Mulliken, T. & Haywood, M. (1994). Recent data on trade in rhino and tiger products, 1988–1992. *TRAFFIC Bulletin,* 14, 99–106.

Nepalese Department of National Parks & Wildlife Conservation (1996). *Report on a Tiger Census of Royal Chitwan, Parsa, and Bardia National Parks.* Kathmandu, Nepal.

Neu, C. W., Byers, C. R. & Peek, J. M. (1974). A technique for analysis of utilization-availability data. *Journal of Wildlife Management,* 38, 541–5.

Newell, J. & Wilson, E. (1996). *The Russian Far East: Forests, Biodiversity Hotspots, and Industrial Developments.* Tokyo, Japan, Friends of the Earth-Japan.

Newman, A., Bush, M., Wildt, D. E., van Dam, M., Frankehuis, M., Simmons, L., Phillips, L. & O'Brien, S. J. (1985). Biochemical genetic variation in eight endangered feline species. *Journal of Mammology,* 66, 256–7.

Ngui Siew Kong (1991). The management of protected areas in Malaysia. *Tigerpaper,* 18(2), 21–4.

Nichols, J. D. (1992). Capture-recapture models: Using marked animals to study population dynamics. *BioScience,* 42, 94–102.

Nikolaev, I. G. & Yudin, V. G. (1992). Tigers and people in conflict situations [Bulletin]. *Moscow Biology Society Search for Nature, Biology Dept.,* 98(3), 23–36. (In Russian.)

Norchi, D. & Bolze, D. (1995). *Saving the Tiger: A Conservation Strategy.* WCS Report No. 3. New York, Wildlife Conservation Society.

Noss, R. (1993). *The Wildlands Project Land Conservation Strategy. The Wildlands Project; plotting a North American Wilderness Recovery Strategy.* Wild Earth Special Issue no. 1. Richmond, VT, The Cenozoic Society Inc.

Noss, R. F., Quigley, H. B., Hornocker, M. G., Merrill, T. & Paquet, P. C. (1996). Conservation biology and carnivore conservation in the Rocky Mountains. *Conservation Biology,* 10, 949–63.

Nowak, R. M. (1991). *Walker's Mammals of the World,* volume II, 5th edn, pp. 643–1629. Baltimore, Maryland, John Hopkins University Press.

Nowell, K. & Jackson, P. (1996). *Wild Cats: Status Survey and Conservation Action Plan.* Gland, Switzerland, IUCN.

O'Brien, S. J. (1994a). *A Role for Molecular Genetics in Biological Conservation. Proceedings of the National Academy of Science, USA,* 91, 5748–55.

O'Brien, S. J. (1994b). Genetic and phylogenetic analyses of endangered species. *Annual Review of Genetics,* 28, 467–89.

O'Brien, S. J. & Mayr, E. (1991). Bureaucratic mischief: Recognizing endangered species and subspecies. *Science,* 251, 1187–8.

O'Brien, S. J., Roelke, M. E., Marker, L., Newman, A., Winkler, C. A., Meltzer, D., Colly, L., Evermann, J. F., Bush, M. & Wildt, D. E. (1985). Genetic basis for species vulnerability in the cheetah. *Science,* 227, 1428–34.

O'Brien, S., Collier, G., Benveniste, R., Nash, W., Newman, A., Simonson, J., Eichelberger, M., Seal, U., Janssen, D., Bush, M. & Wildt, D. (1987). Setting the molecular clock in Felidae: the great cats, Panthera. In *Tigers of the World: The Biology, Biopolitics, Management and Conservation of an Endangered Species*, ed. R. L. Tilson & U. S. Seal, pp. 36–50. Park Ridge, New Jersey, Noyes Publications.

O'Brien, S. J., Roelke, M. E., Yuhki, N., Richards, K. W., Johnson, W. E., Franklin, W. L., Anderson, A. E., Bass, O. L., Belden, R. C. & Martenson, J. S. (1990). Genetic introgression within the Florida panther (*Felis concolor coryi*). *National Geographic Research*, **6**, 485–94.

O'Brien, S. J., Martenson, J. S., Miththapala, S., Janczewski, D. N., Pecon Slattery, J., Johnson, W. E., Gilbert, D. A., Roelke, M. E., Packer, C., Bush, M. & Wildt, D. E. (1996). Conservation genetics of the Felidae. In *Conservation Genetics: Case Histories from Nature*, ed. J. C. Avise & J. L. Hamrick, pp. 50–74. New York, Chapman and Hall.

Ognev, S. I. (1962). *Mammals of the USSR and adjacent countries*, volume III, *Carnivora (Fissipedia and Pinnipedia)*. Jerusalem, Israel Program for Scientific Translations.

Orians, G. H. & Wittenberger, J. F. (1991). Spatial and temporal scales in habitat selection. *American Naturalist*, **137**, S29–S49.

Ortolani, A. & Caro, T. M. (1996). The adaptive significance of colour patterns in carnivores. In *Carnivore Behaviour, Ecology and Evolution*, ed. J. L. Gittleman, pp. 132–88. Ithaca, Cornell University Press.

Panwar, H. S. (1979a). A note on tiger census technique based on pugmark tracings. *Indian Forester*, special issue, 18–36.

Panwar, H. S. (1979b). Population Dynamics and Land Tenures of tigers in Kanha National Park. In *International Symposium on Tiger: Papers, Proceedings and Resolutions*, pp. 35–47. 22–24 February 1979, New Delhi. New Delhi, Project Tiger, Government of India.

Panwar, H.S. (1984). What to do when you've succeeded: Project Tiger, ten years later. In *National Parks, Conservation and Development: The role of protected areas in sustaining society*, ed. J. A. McNeely & K. R. Miller, pp. 183–9. Washington DC, Smithsonian Institution Press.

Panwar, H. S. (1987). Project Tiger: The reserves, the tigers and their future. In *Tigers of the World: The Biology, Biopolitics, Management and Conservation of an Endangered Species*, ed. R. L. Tilson & U. S. Seal, pp. 110–17. Park Ridge, New Jersey, Noyes Publications.

Parr, J. W. K., Mahannop, N. & Charoensiri, V. (1993). Khao Sam Roi Yot – one of the world's most threatened parks. *Oryx*, **27**(4), 245–9.

Pascal, J. P., Shyam Sunder, S. & Meher-Homji, V. M. (1982). *Forest Maps of South India: Mercara-Mysore*. Pondicherry, India, French Institute.

Payne, J., Bernazzani, P. & Duckworth, W. (1995). *Preliminary Wildlife and Habitat Survey of the Phou Khao Khouay National Biodiversity Conservation Area, Lao PDR*. Report for Protected Areas and Watershed Management, Vientiane, Lao PDR.

Peacock, E. H. (1933). *A Game Book for Burma and Adjoining Territories*. London, England, H. F. & G. Witherby.

Pedlar, J. H., Fahrig, L. & Merriam, H. G. (1997). Raccoon habitat use at 2 spatial scales. *Journal of Wildlife Management*, **61**, 102–12.

Pei, W. C. (1934). On the Carnivora from Locality 1 of Choukoutien. *Palaeontologia Sinica Series C*, **8**(1), 1–217.

Pei, W. C. (1935). Fossil mammals from the Kwangsi Caves. *Bulletin of the Geological Society of China*, **14**, 413–25.

Pei, W. C. (1936). On the mammalian remains from Locality 3 at Choukoutien. *Palaeontologia Sinica Series C*, **7**(5), 1–121.

Pennock, D. S. & Dimmick, W. W. (1997). Critique of the Evolutionary Significant Unit as a definition for 'distinct population segments' under the US Endangered Species Act. *Conservation Biology*, **11**, 611–19.

Pensen, I. D., Gnateshen, V. N., Aneichek, V. F., Veshnevskii, D. S., Gaifolina, T. N., Savina, S. V., Sapaev, V. M. & Semakov, V. I. (1995). *Khabarovski Krai: Atlas of Khabarovski Krai*. Moscow, RosKartographia.

Peranio, R. (1960). Animal teeth and oath taking among the Bisaya. *Sarawak Museum Journal*, **9**, 6–13.

Petropavlovski, B. S. (1996). *Pinus koraiensis* forests of Primorski Territory: Conditions, dynamics, geography, ecology, types of forest, and prospective research. In *Cedar-Broad-leaved Forests of the Far East*, Proceedings of an International Conference. Khabarovsk, 30 September to 6 October 1996, ed. D. F. Efremov, L. G. Kondrashov and A. P. Sapojnikov. Khabarovsk, Khabarovski Krai, FEFRI (Far East Forestry Research Institute).

Phillips, M. (1995). Conserving the red wolf. *Canid News*, **3**, 13–17. Oxford University, IUCN/SSC Canid Specialist Group.

Pikunov, D. G. (1988). Amur tiger (*Panthera tigris altaica*) present situation and perspectives for preservation of its population in the Soviet Far East. In *Proceedings of the 5th World Conference on Breeding Endangered Species in Captivity*, ed. B. L. Dresser, R. W. Reece & E. J. Maruska, pp. 175–84. Cincinnati, Ohio, USA.

Pikunov, D. G. (1990). Number of tigers in the Far East USSR. 5th Meeting of the All-Soviet Theriological Society, 29 January to 2 February 1990. Moscow, All-Soviet Theriological Society. (In Russian.)

Pikunov, D. G. (1991). Food intake of the Amur tiger. In *Rare and Endangered Land Animals of the Far East USSR*, pp. 71–5. Presented at a Conference at the USSR Academy of Sciences, Far Eastern Scientific Center, Institute of Biology and Soils, Vladivostok.

Pikunov, D. G. &. Bragin, A. P. (1987). Organization and census methods for the Amur tiger. In *Organization and methods of surveying commercial and rare species of mammals and birds in the Far East*. Vladivostok, Far Eastern Scientific Center, Academy of Sciences. (In Russian.)

Pikunov, D. G., Basilnikov, V. N. Rebachuk, V. N. & Abramov, V. K. (1983). Present area, number, structure, and distribution of tigers in Primorski Krai. In *Rare Mammals of USSR*, pp. 130–31. Moscow, Nauka Press. (In Russian.)

Pimm, S. (1996). Lessons from a kill. *Biodiversity and Conservation*, **5**, 1059–67.

Plowden, C. & Bowles, D. (1997). The illegal market in tiger parts in northern Sumatra, Indonesia. *Oryx*, **31** (1), 59–66.

Pocock, R. I. (1929). Tigers. *Journal of the Bombay Natural History Society*, **33**, 505–41.

Pocock, R. I. (1931a). The panthers and ounces of Asia. *Journal of the Bombay Natural History Society*, **34**, 64–82.

Pocock, R. I. (1931b). II. – Tiger or panther? Mr. Limouzin's specimen. *Journal of the Bombay Natural History Society*, **34**, 544–7.

Pocock, R. I. (1939). *The Fauna of British India, Including Ceylon and Burma. Mammalia*, volume I, *Primates and Carnivora (in part), Families Felidae and Viverridae.* London, Taylor and Francis.

Population Reference Bureau, Inc. (1995). *World Population Data Sheet: Demographic Data and Estimates for the Countries and Regions of the World.* Washington DC, Population Reference Bureau, Inc.

Prezhewalski, N. M. (1870). *Travels in the Ussuri Krai, 1867–1869.* St. Petersburg. (In Russian.)

Project Tiger (1993). *All India Tiger Census Report.* New Delhi, Ministry of Enviroment & Forests.

Rabinowitz, A. (1989). The density and behaviour of large cats in a dry tropical forest mosaic in Huai Kha Khaeng Wildlife Sanctuary, Thailand. *Natural History Bulletin of the Siam Society*, **37**(2), 235–51.

Rabinowitz, A. (1991). *Chasing the Dragon's Tail: the Struggle to Save Thailand's Wild Cats.* New York, Doubleday.

Rabinowitz, A. (1993). Estimating the Indochinese tiger (*Panthera tigris corbetti*) population in Thailand. *Biological Conservation*, **65**, 213–17.

Rabinowitz, A. (1995a). Asian nations meet in Thailand to discuss trans-boundary biodiversity conservation. *Natural History Bulletin of the Siam Society*, **43**, 23–6.

Rabinowitz, A. (1995b). Helping a species go extinct: the Sumatran rhino in Borneo. *Conservation Biology*, **9**(3), 482–8.

Rabinowitz, A. (1996a). *Additional Surveys and Recommendations on the Birds and Mammals for the Nam Theun 2 Hydropower Project.* Report to the Government of Lao PDR, Vientiane.

Rabinowitz, A. (1996b). *Setting Priorities for the Protected Area System of Myanmar.* Trip report to the Wildlife Conservation Society, New York.

Rabinowitz, A. (1997a). *A Biological Expedition to Hkakaborazi Protected Area in North Myanmar.* Trip report to the Wildlife Conservation Society, New York.

Rabinowitz, A. (1997b). Lost world of the Annamites. *Natural History Magazine*, **106**(3), 14–18.

Rabinowitz, A., Schaller, G. & Uga, U. (1995). A survey to assess the status of Sumatran rhinoceros and other large mammal species in Tamanthi Wildlife Sanctuary, Myanmar. *Oryx*, **28**(2), 123–8.

Rand, D. M. (1994). Thermal habit, metabolic rate and the evolution of mitochondrial DNA. *Trends in Ecology and Evolution*, **9**, 125–31.

Rangarajan, M. (1996). *Fencing the Forest: Conservation and Ecological Change in India's Central Provinces 1860–1914*. New Delhi, Oxford University Press.

Ratnam, L., Lim Boo Liat & Nor Azman Hussein (1995). Mammals of the Sungai Singgor Area in Temengor Forest Reserve, Hulu Perak, Malaysia. *Malayan Nature Journal*, **48**, 409–23.

Reading, R. & Clark, T. (1996). Carnivore reintroductions: An interdisciplinary examination. In *Carnivore Behaviour, Ecology and Evolution*, ed. J. Gittleman. Ithaca, Cornell University Press.

Rexstad, E. & Burnham, K. P. (1991). *User's Guide for Inter-active Program CAPTURE*. Fort Collins, Colorado State University.

Rice, C. G. (1986). Observations on predators and prey at Eravikulam National Park, Kerala. *Journal of the Bombay Natural History Society*, **83**, 283–305.

Riley, A. L. & Baril, L. L. (1976). Conditioned taste aversion: A bibliography. *Animal Learning Bahaviour*, **4**, 15–35.

Robichaud, W. & Sounthala, B. (1995). *A Preliminary Wildlife and Habitat Survey of Phou Dendin National Biodiversity Conservation Area, Phonsali*. Report to the Centre for Protected Areas and Watershed Management, Lao PDR, Vientiane.

Robinson, J. G. (1993). Limits to caring: sustainable living and the loss of biodiversity. *Conservation Biology*, **7**, 20–28.

Roelke, M. E., Martenson, J. S. & O'Brien, S. J. (1993). The consequences of demographic reduction and genetic depletion in the endangered Florida panther. *Current Biology* **3**, 340–50.

Rosenthal, M. (undated). *Siberian Tiger Museum Project, Panthera tigris altaica*. Chigaco, Lincoln Park Zoo.

Rubeli, K. (1979). Taman Negara. *Tigerpaper*, **6**(4), 6.

Rykiel, E. J. (1996). Testing ecological models: the meaning of validation. *Ecological Modelling*, **90**, 229–44.

Saberwal, V. K. (1997). Saving the tiger: More money or less power? *Conservation Biology*, **11**, 815–17.

Sadleir, R. (1966). Notes on reproduction in the larger Felidae. *International Zoo Yearbook*, **6**, 184–7.

Salkina, G. P. (1993). Tiger population present state in Sikhote-Alin South-east. In *Bulletin of the Moscow Society of Naturalists, Biological Series*, **98**, 45–53. (Summary in English.)

Salter, R. (1993). *Wildlife in Lao PDR. A Status Report*. Vientiane, Lao PDR, IUCN.

Salter, R. (1994). *Priorities for Further Development of the Protected Areas System in Myanmar*. Report for Conservation and Wildlife Sanctuary Mission MYA/91/015. Yangon, Government of Lao PDR.

Samith, C., Pisey, O. & Choir, R. (1995). *Current Status of the Virachey National Park, Ratanakiri Province*. ETAP/UNDP Report, Phnom Penh.

Samsudin, A. R. & Elagupillay, S. (1995). *A Review on the Protection of Tigers in Peninsular Malaysia*. Proceedings of the Second International Conference and Geographic Information System Workshop to Assess the Status of Tigers, 24–31 January 1996, Thailand.

Sangermano, V. (1833). *The Burmese Empire: A Hundred Years Ago*. Bangkok, White Orchid Press.

SAS Institute, Inc. (1985). *SAS Users' Guide: Statistics*. Version 5. Cary, North Carolina.

Saw Tun Khaing (1996). *Assessing the Protected Areas of Myanmar*. Trip report to the Wildlife Conservation Society, New York.

Sayer, J. A. (1982). How many national parks for Thailand? *Tigerpaper*, **9**(1), 2–5.

Sayer, J. A. (1983). *Wildlife in the Southern Arakan Yoma*. Field report for UNDP/FAO Nature Conservation and National Parks Project, Burma. Rangoon.

Schaller, G. B. (1967). *The Deer and the Tiger*. Chicago, University of Chicago Press.

Schaller, G. (1995). *A Wildlife Survey in the Annamite Mountains of Laos*. Trip report to the Wildlife Conservation Society, New York.

Schroeter, W. (1981). Über Färbung, Farbabweichungen, Streifenverminderungen und Farbaufhellungen beim Tiger, *Panthera tigris* (Linné, 1758). *Säugetierkundliche Mitteilungen*, **29**(4), 1–8.

Schwarz, E. (1912). Notes on Malay tigers, with description of a new form from Bali. *Annals and Magazine of Natural History, Series 8*, **10**, 324–6.

Scott, J. M., Davis, F., Csuti, B., Noss, R., Butterfield, B., Groves, C., Anderson, H., Caicco, S., D'Erchia, F., Edwards T. C. Jr., Ulliman, J. & Wright, R. G. (1993). Gap analysis: a geographic approach to protection of biological diversity. *Wildlife Monographs*, **123**, 1–41.

Seal, U. S. (1991). Global tiger plans underway. *Tiger Beat: The Newsletter of the Tiger Species Survival Plan*, **4**, 1&16. American Association of Zoological Parks and Zoos (AZA)/Species Survival Plan. Minnesota Zoo.

Seal, U. S., Jackson, P., & Tilson, R. (1987a). A global tiger conservation plan. In *Tigers of the World: The Biology, Biopolitics, Management, and Conservation of an Endangered Species*, ed. R. L. Tilson & U. S. Seal, pp. 487–98. Park Ridge, New Jersey, Noyes Publications.

Seal, U. S., Tilson, R. L., Plotka, E. D., Reindl, N. J. & Seal, M. F. (1987b). Behavioral indicators and endocrine correlates of estrus and anestrus in Siberian tigers. In *Tigers of the World: The Biology, Biopolitics, Management, and Conservation of an Endangered Species*, ed. R. L. Tilson & U. S. Seal, pp. 244–54. Park Ridge, New Jersey, Noyes Publications.

Seal, U., Soemarna, K. & Tilson, R. (1994). Population biology and analyses for Sumatran tigers. In *Sumatran Tiger Population and Habitat Viability Analysis Report*, ed. R. Tilson, K. Soemarna, W. Ramono, S. Lusli, K. Traylor-Holzer & U. Seal, pp. 45–70. Apple Valley, Minnesota, Indonesian Directorate of Forest Protection and Nature Conservation and IUCN/SSC Conservation Breeding Specialist Group.

Seidensticker, J. C. (1976). On the ecological separation between tigers and leopards. *Biotropica*, **8**, 225–34.

Seidensticker, J. (1986). Large carnivores and the consequences of habitat insularization: Ecology and conservation of tigers in Indonesia and Bangladesh. In *Cats of the World: Biology, Conservation, and Management*, ed. S. D. Miller & D. D. Everett, pp. 1–41. Washington DC, National Wildlife Federation.

Seidensticker, J. (1987). Bearing witness: Observations on the extinction of *Panthera tigris balica* and *Panthera tigris sondaica*. In *Tigers of the World: The Biology, Biopolitics, Management and Conservation of an Endangered Species*, ed. R. L. Tilson & U. S. Seal, pp. 1–8. Park Ridge, New Jersey, Noyes Publications.

Seidensticker, J. (1997). Saving the tiger. *Wildlife Society Bulletin*, **25**, 6–17.

Seidensticker, J. & McDougal, C. (1993). Tiger predatory behaviour, ecology and conservation. *Symposium of the Zoological Society of London*, **65**, 105–25.

Seidensticker, J. & Suyono. (1980). *The Javan Tiger and the Meru-Betiri Reserve: A Plan for Management*. Gland, Switzerland, IUCN.

Seidensticker, J., Hornocker, M. G., Wiles, W. V. & Messick, J. P. (1973). Mountain lion social organisation in the Idaho Primitive Area. *Wildlife Monographs*, **35**, 1–60.

Shackleton, N. J. & Opdyke, N. D. (1973). Oxygen isotope and palaeomagnetic stratigraphy of equatorial Pacific core V28 –238. *Quaternary Research*, **3**, 39–55.

Shafer, C. L. (1990). *Nature Reserves: Island Theory and Conservation Practice*. Washington DC, Smithsonian Press.

Shikama, T. & Okafuji, G. (1958). Quaternary cave and fissure deposits and their fossils in Akiyoshi District, Yamaguti Prefecture. *Scientific Report of the Yokohama National University, Section 2, Biological and Geological Sciences*, **7**, 43–103.

Shikama, T. & Okafuji, G. (1963). On some Choukoutien mammals from Isa, Yamaguchi Prefecture, Japan. *Scientific Report of the Yokohama National University, Section 2, Biological and Geological Sciences*, **9**, 51–8.

Siswomartono, D., Reddy, S., Ramono, W., Manansang, J., Tilson, R., Franklin, N. & Foose, T. (1996). The Sumatran rhino in Way Kambas National Park, Sumatra, Indonesia. *Pachyderm*, **21**, 13–14.

Smielowski, J. (1996). Reintroduction summary: European beaver, bison, elk and lynx in Poland. In *Reintroduction News*, **12**, pp. 17-19. Nairobi, Kenya, IUCN/SSC Reintroduction Specialist Group.

Smirnov, E. N. (1997). *One Hundred Encounters with Tigers*. Vladivostok, Zov Taiga. (In Russian.)

Smirnov, E. N. & Miquelle, D. G. (1995). *1995 Amur tigers Census of in Northeast Primorye*. Report to the Wildlife Conservation Society, New York.

Smirnov, E. N., Miquelle, D. G., Nikolaev, I. G., Hornocker, M. G., Quigley, H. B., Pikunov, D. G., Goodrich, J., Kerley, L., Schleyer, B., Reebin, N. N. & Kostira, A. V. (1997). Reproductive indicators in the Amur tiger [Abstract]. Rare mammalian species of Russia and neighboring territories. In Abstracts of presentations at an International Conference, 9–11 April, 1997. Moscow. [Hosted by the Therioloical Society of Russian Academy of Sciences, Institute of Ecological and Evolutionary Problems Russian Academy of Sciences, Scientific Soviet for the program 'Biological Diversity'.]

Smirnov, E. N., Nikolaev, I. G., Miquelle, D. G., Goodrich, J., Kerley, L., Quigley, H. B. & Hornocker, M. G. (1998) New data on Amur tiger reproduction. In *Endangered Species of Russia, 1997*. Proceedings of the Russian Academy of Sciences Meeting, Moscow.

Smith, H. (1993). *Report to Earth Island Institute*. Endangered Species Project Investigation, San Francisco.

Smith, J. L. D. (1978). *Smithsonian Tiger Ecology Project Report No. 13*. Washington DC, Smithsonian Institution.

Smith, J. L. D. (1984). *Dispersal, communication, and conservation stratgies for the tiger (*Panthera tigris*) in Royal Chitwan National Park, Nepal.* University of Minnesota, PhD Dissertation.

Smith, J. L. D. (1993). The role of dispersal in structuring the Chitwan tiger population. *Behaviour*, **124**, 165–95.

Smith, J. L. D. & McDougal, C. (1991). The contribution of variance in lifetime reproduction to effective population size in tigers. *Conservation Biology*, **5**, 484–90.

Smith, J. L. D., Sunquist, M. E., Tamang, K. M. & Rai, P. B. (1983). A technique for capturing and immobilizing tigers. *Journal of Wildlife Management*, **47**, 255–9.

Smith, J. L. D., Ahearn, S. C. & McDougal, C. (1998). A landscape analysis of tiger distribution and habitat quality in Nepal. *Conservation Biology.* **1**, **12**, 1338–46

Smith, J. L. D., McDougal, C. & Sunquist, M. E. (1987a). Female land tenure system in tigers. In *Tigers of the World: The Biology, Biopolitics, Management and Conservation of an Endangered Species,* ed. R. L. Tilson & U. S. Seal, pp. 97–109. Park Ridge, New Jersey, Noyes Publications.

Smith, J. L. D., Wemmer, C. & Mishra, H. R. (1987b). A tiger geographic information system: the first step in a global conservation strategy. In *Tigers of the World: The Biology, Biopolitics, Management and Conservation of an Endangered Species,* ed. R. L. Tilson & U. S. Seal, pp. 464–73. Park Ridge, New Jersey, Noyes Publications.

Smith, J. L. D., McDougal, C. & Miquelle, D. (1989). Scent marking in free-ranging tigers, *Panthera tigris. Animal Behaviour*, **37**, 1–10.

Smith, J. L. D., Mishra, H. R. & Wemmer, C. (1992). The role of land use planning in large mammal conservation. In *NORAGRIC Occasional Papers Series C. Proceedings of the International Theriological Congress,* August 1989, Rome, ed. Per Wegge, pp. 64–71. As, Norway, Norwegian Centre for Agricultural Development (NORAGRIC).

Smythies, E. A. (1942). *Big Game Shooting in Nepal.* Calcutta, Thacker, Spink & Co.

Soemarna, K., Ramona, W. & Sumardja, E. (1994). PHPA Sumatran Tiger Action Plan. In *Sumatran Tiger Population and Habitat Viability Analysis Report*, ed. R. Tilson, K. Soemarna, W. Ramono, S. Lusli, K. Traylor-Holzer & U. Seal, pp. 75–76. Apple Valley, Minnesota, Indonesian Directorate of Forest Protection and Nature Conservation and IUCN/SSC Conservation Breeding Specialist Group.

Somaiah, K. K. (1953). *Working Plan for the Eastern Deciduous Forests of Coorg.* Madikeri, India, Coorg Forest Department.

Soulé, M. E. (1980). Thresholds for survival: Criteria for maintenance of fitness and evolutionary potential. In *Conservation Biology: An Evolutionary-Ecological Perspective*, ed. M. E. Soulé & B. M. Wilcox, pp. 151–70. Sunderland, Massachusetts, Sinauer Associates.

Spillet, J. J. (1967). A report on wildlife surveys in North India and Southern Nepal: The Jaldapara WildLife Sanctuary, West Bengal. *Journal of the Bombay Natural History Society*, **63**, 534–56.

Srikosamatara, S. (1993). Density and biomass of large herbivores and other mammals in a dry tropical forest, western Thailand. *Journal of Tropical Ecology*, **9**, 33–43.

Srikosamatara, S., Siripholdej, B. & Suteethorn, V. (1992). Wildlife trade in Lao PDR and between Lao PDR and Thailand. *Natural History Bulletin of the Siam Society*, **40**, 1–47.

Stainton, J. D. A. (1972). Forests of Nepal. London, John Murray, Ltd.

Starfield, A. M. & Bleloch, A. L. (1986). *Models for Conservation and Wildlife Management*. New York, Macmillan.

State Committee of the Russian Federation for the Protection of the Environment. (1996). *Strategy for Conservation of the Amur tiger in Russia.* Moscow-Vladivostok, Text Publishers.

Stebbing, E. P. (1929). *The Forests of India*, volume 1. London, The Bodley Head Ltd.

Stephan, J. J. (1994). *The Russian Far East: A History*. Stanford, California, Stanford University Press.

Stiles, D. & Martin, E. B. (1994). New war in South East Asia. *Swara*, **17**(5), 19–21.

Støen, O. G. & Wegge, P. (1996). Prey selection and prey removal by tiger (*Panthera tigris*) during the dry season in lowland Nepal. *Mammalia*, **60**, 363–73.

Stringer, C. & Gamble, C. (1993). *In Search of the Neanderthals*. New York, Thames and Hudson.

Sunquist, M. E. (1981). The social organization of tigers (*Panthera tigris*) in Royal Chitawan National Park, Nepal. *Smithsonian Contributions to Zoology*, **336**, 1–98.

Sunquist, M. (1996).Tiger pause. *Wildlife Conservation*, **99**(3), 56–9, 64.

Sunquist, M. E. & Sunquist F. C. (1989). Ecological constraints on predation by large felids. In *Carnivore Behavior, Ecology, and Evolution*, ed. J. L. Gittleman, pp. 283–301. Ithaca, Cornell University Press.

Sutcliffe, A. J. (1985). *On the Trail of Ice Age Mammals*. London, British Museum (Natural History).

Swisher III, C. C., Curtis, G. H., Jacob, T., Getty, A. G., Suprijo, A. & Widiasmoro (1994). Age of the earliest known hominids in Java, Indonesia. *Science*, **263**, 1118–21.

Swisher III, C. C., Rink, R. J., Anton, S. C., Schwarcz, H. P., Curtis, G. H., Suprijo, A. & Widiasmoro (1996). Latest *Homo erectus* of Java: Potential contemporaneity with *Homo sapiens* in Southeast Asia. *Science*, **274**, 1870–72.

Ta-Kuan, C. (1993). *The Customs of Cambodia*, 3rd edn. Bangkok, The Siam Society.

Taberlet, P. (1996). The use of mitochondrial DNA control region sequencing in conservation genetics. In *Molecular Genetic Approaches to Conservation*, ed. T. B. Smith & R. K. Wayne, pp. 125–42. Oxford, Oxford University Press.

Takai, F. & Hasegawa, Y. (1966). Vertebrate fossils from the Gansuiji Formation. *Journal of the Anthropological Society of Nippon*, **74**, 155–67.

Tamang, K. M. (1982). *The status of the tiger (*Panthera tigris*) and its impact on principal prey populations in Royal Chitwan National Park, Nepal*. East Lansing, Michigan State University, PhD Dissertation.

Task Force (1972). *Project Tiger: A planning proposal for preservation of the tiger (*Panthera tigris tigris *Linn.) in India*. Government of India, New Delhi, Indian Board for Wild Life.

Teilhard de Chardin, P. (1936). Fossil mammals from locality 9 of Choukoutien. *Paleontologia Sinica, Series C*, **7**(4), 1–71.

Temminck, C. J. (1844). Aperçu général et spécifique sur les mammifères qui habitent le Japon et les Iles qui en dépendant. In *Fauna Japonica*, ed. P. T. von Siebold, pp. 25–59, plates 11–20. Amsterdam, Müller.

Thapar, V. (1989). *Tigers, The Secret Life*. Emmaus, Pennsylvania, Rodale Press.

Thapar, V. (1992). *The Tiger's Destiny*. London, Kyle-Cathie.

Thouless, C. (1987). Kampuchean wildlife – survival against the odds. *Oryx*, **21**(4), 223–8.

Tilson, R. (1991). Cats in zoos. In *Great Cats: Majestic Creatures of the World*, ed. J. Seidensticker & S. Lumpkin, pp. 214–19. Sydney, Weldon Owen Publishing.

Tilson, R. & Sriyanto, A. The conservation adoption of Ujung Kulon National Park, Indonesia. *AZA Field Conservation Resource Guide*. Bethesda, Maryland, American Zoo and Aquarium Association. (In press.)

Tilson, R. & Traylor–Holzer, K. (1994). Estimating poaching and removal rates of tigers in Sumatra. In *Sumatran Tiger Population and Habitat Viability Analysis Report*, ed. R. Tilson, K. Soemarna, W. Ramono, S. Lusli, K. Traylor-Holzer & U. Seal, pp. 75–6. Apple Valley, Minnesota, Indonesian Directorate of Forest Protection and Nature Conservation and IUCN/SSC Conservation Breeding Specialist Group.

Tilson, R., Foose, T., Princeé, F. & Traylor-Holzer, K. (1993). *Tiger Global Animal Survival Plan*. Apple Valley, Minnesota, IUCN/SSC Conservation Breeding Specialist Group.

Tilson, R. L., Soemarna, K., Ramono, W., Lusli, S., Traylor-Holzer, K. & Seal, U. S. (1994). *Sumatran Tiger Population and Habitat Viability Analysis Report*. Apple Valley, Minnesota, Indonesian Directorate of Forest Protection and Nature Conservation and IUCN/SSC Conservation Breeding Specialist Group.

Tilson, R. L., Dumnui, S., Traylor-Holzer, K., Armstrong, D., Kamolnorranarth, S., Wichasilpa, W. & Arsaithamkul, V. (ed.) (1995). *Indochinese Tiger Masterplan for Thailand*. Apple Valley, Minnesota, Minnesota Zoo.

Tilson, R. L., Franklin, N., Nyhus, P., Bastoni, Sriyanto, Siswomartono, D. & Manansang, J. (1996). *In situ* conservation of the Sumatran tiger in Indonesia. *International Zoo News*, **43**(5), 316–24. (Special edn. on zoos and *in-situ* conservation.)

Tilson, R. L., Traylor-Holzer, K. & Ming Jiang, Q. (1997). The decline and impending extinction of the South China tiger. *Oryx* **31**(4), 243–52.

Timmins, R. J. & Bleisch, W. (1995). *A Wildlife and Habitat Survey of Xe Bang Nouan National Biodiversity Conservation Area, Lao PDR*. Report to the Center for Protected Areas and Watershed Management, Vientiane, Lao PDR.

Timmins, R. J. & Evans, T. D. (1996). *A Wildlife and Habitat Survey of Nakai Nam Theun National Biodiversity Conservation Area, Lao PDR.* Report to the Center for Protected Areas and Watershed Management, Vientiane, Loa PDR.

Timmins, R. J. & Khounboline, K. (1996). *A Wildlife and Habitat Survey of Hin Namno National Biodiversity Conservation Area, Lao PDR.* Report to the Center for Protected Areas and Watershed Management, Vientiane, Lao PDR.

Timmins, R. J. & Vongkhamheng, C. (1996). *A Wildlife and Habitat Survey of the Dong Khanthung Area, Lao PDR.* Report to the Center for Protected Areas and Watershed Management, Vientiane, Loa PDR.

Tizard, R. J. (1996). *A Wildlife and Habitat Survey of the Proposed Northern Extension to the Nakai Nam Theun National Biodiversity Conservation Area, and the Adjacent Nam Gnouang Area, Lao PDR.* Report to the Center for Protected Areas and Watershed Management, Vientiane, Lao PDR.

Topani, R. (1990). Status and distribution of tiger in Peninsular Malaysia. *The Journal of Wildlife and Parks, Malaysia,* **IX**, 71–102.

TRAFFIC (1993). *Wildlife Trade Between the Southern Lao PDR provinces of Champasak, Sekong, and Attapeu, and Thailand, Cambodia and Vietnam.* TRAFFIC SE Asia Field Report No. 3.

Tscherski, J. D. (1892). Wissenschaftliche Resultate der von der Kaisertiche Akademie der Wissenschaften zur Erforschung des Janalandes und der neusibirischen Inseln in den Jahren 1885 und 1886 ausgesandten Expedition. Abt. IV. Beschreibung der Sammlung posttertiärer Säugethiere. *Mémoires de l'Académie Impériale des Sciences de St.-Petersbourg, VIIe Série,* **40**, 1–511.

Tun Yin (1993). *Wild Mammals of Myanmar.* Yangon, Myanmar Forest Department.

Umbgrove, J. H. F. (1949). *Structural History of the East Indies.* Cambridge, Cambridge University Press.

UNDP/FAO (1982a). *Kyatthin Wildlife Sanctuary: Report on a survey of the area and a preliminary census of the thamin.* Field Report for the Nature Conservation and National Parks Project, Burma, Rangoon.

UNDP/FAO (1982b). *Proposed Pegu Yomas National Park. Report on Preliminary Surveys of the Yenwe Chaung Area, 1981–1982.* Field Report to the Nature Conservation and National Parks Project, Burma. Rangoon.

UNDP/FAO (1982c). *Shwesettaw Wildlife Sanctuary. Report on a Reconnaissance Survey and Evaluation.* Field Report for the Nature Conservation and National Parks Project, Burma. Rangoon.

UNDP/FAO (1983a). *A Survey of Inle Lake and Adjacent Areas of the Shan Plateau.* Field Report to the Nature Conservation and National Parks Project, Burma. Rangoon.

UNDP/FAO (1983b). *A Survey of Natma Taung (Mount Victoria) in Southern Chin Hills.* Field Report to the Nature Conservation and National Parks Project, Burma. Rangoon.

UNDP/FAO (1983c). *Irrawaddy Delta: Potential for Nature Conservation and Recreation.* Field Report to the Nature Conservation and National Parks Project, Burma. Rangoon.

UNDP/FAO (1984). *Alaungdaw Kathapa National Park. Preliminary Master Plan.* UNDP/FAO Nature Conservation and National Parks Project, Field Document No. 8.

UNDP/FAO (1985). *Nature Conservation and National Parks. Burma. Survey Data and Conservation Priorities.* Technical Report 1. Rome, UNDP/FAO.

UNEP (United Nations Environment Program). (1996). *Global Biodiversity Assessment.* Cambridge, Cambridge University Press.

USAID Environmental Policy and Technology Project. (1995). Russian Workplan: EPT Russian Far East sustainable natural resources management project.

Varner, G. & Monroe, M. (1991). Ethical perspectives on captive breeding: Is it for the birds? *Endangered Species UPDATE,* **8**(1), 27–9.

Vu Van Dung & Huynh Van Keo (1993). Preliminary Results of Investigating the Flora and Fauna System of Bach Ma National Park in Thua Thien Hue Province. In *Proceedings of Regional Seminar-Workshop on Tropical Forest Ecosystem Research, Conservation and Repatriations.* Hanoi.

Vu Van Dung, Nguyen Ngoc Chinh, Ebregt, A. & Santiapillai, C. (1995). The status of the newly discovered large mammal, the long-horned bovid, in Vietnam. *Tigerpaper,* **22**(2), 3–19.

Wallace, A. R. (1869). *The Malay Archipelago.* Reprinted in 1989. Singapore, Oxford University Press.

Ward, G. (1997). Making room for wild tigers. *National Geographic,* **192**(6), 2–35.

Wayne, R. K. & Koepfli, K. P. (1996). Demographic and historical effects on genetic variation of carnivores. In *Carnivore Biology, Ecology and Evolution*, ed. J. L. Gittleman, pp. 453–84. Ithaca, Comstock Publishing Associates.

Wayne, R. K., Benveniste, R. E., Janczewski, N. & O'Brien, S. J. (1989). Molecular and biochemical evolution of the Carnivora. In *Carnivore Behaviour, Ecology and Evolution*, ed. J. L. Gittleman, pp. 465–94. London, Chapman and Hall.

WCS (1995a). *A Survey of Terrestrial Wildlife in the Area to be Affected by the Proposed Xe Nam oy-Xe Pian Hydroelectric Project*. Report to the Hydropower Office, Ministry of Industry and Handicrafts. Vientiane, Lao PDR.

WCS (1995b). *A Wildlife and Habitat Assessment of the Theun-Hinboun Hydropower Project Area*. Report to the Hydropower Office, Ministry of Industry and Handicrafts. Vientiane, Lao PDR.

WCS (1995c). *Reconnaissance Management Survey of Phou Loeuy and Nam Et National Biodiversity Conservation Areas*. Report to the Protected Areas and Watershed Management Office, Vientiane, Lao PDR.

WCS. (1995d). *Saving the Tiger: A Conservation Strategy*. Policy report number 3. New York, Wildlife Conservation Society.

Weaver, J. L., Paquet, P. C. & Ruggiero, L. F. (1996). Resilience and conservation of large carnivores in the Rocky Mountains. *Conservation Biology*, **10**, 964–76.

Weigel, I. (1961). Das Fellmuster der wildlebenden Katzenarten und der hauskatzen in vergleichender und stammesgeschichtlicher Hinsicht. *Säugetierkundliche Mitteilungen*, **9**, 1–120.

Wells, M. & Brandon, K. (1992). *People and Parks: Linking Protected Area Management with Local Communities*. Washington DC, The World Bank.

Wells, M., Khan, A., Wardojo, W. & Jepson, P. (1998). *A Study of Integrated Conservation Development Projects (ICDPs) in Indonesia*. Washington DC, The World Bank.

Wemmer, C. J., Smith, J. L. D. & Mishra, H. R. (1987). Tigers in the wild; the biopolitical challenges. In *Tigers of the World: The Biology, Biopolitics, Management and Conservation of an Endangered Species*, ed. R. L. Tilson & U. S. Seal, pp. 396–404. Park Ridge, New Jersey, Noyes Publications.

Wessing, R. (1986). *The Soul of Ambiguity: The Tiger in Southeast Asia*. Northern Illinois University, Center for Southeast Asian Studies.

Wharton, C. H. (1957). *An Ecological Study of the Kouprey, Novibos sauveli, (Urbain)*. Manila, Bureau of Printing.

Wharton, C. H. (1968). Man, Fire and Wild Cattle in Southeastern Asia. In *Proceedings of Tall Timbers Fire Ecology Conference, Series 8*, pp. 107–67.

White, G. C. & Garrott, R. A. (1990). *Analysis of wildlife radio-tracking data*. New York, Academic Press.

White, G. C., Anderson, D. R., Burnham, K. P. & Otis, D. L. (1982). *Capture-recapture Removal Methods for Sampling Closed Populations*. Los Alamos, New Mexico, Los Alamos National Laboratory. Publication LA-8787–NERP.

White, H. T. (1923). *Burma*. Provincial Geographies of India. Cambridge, Cambridge University Press.

Whitehead, G. K. (1972). *Deer of the World*. London, Constable.

Whitney, C. (1905). *Jungle Trails and Jungle People*. New York, Charles Scribners and Sons.

Whitten, A. J., Damanik, S. J., Anwar, J. & Hisyam, N. (1987). *The Ecology of Sumatra*. Yogyakarta, Indonesia, Gadjah Mada University Press.

Wielgus, R. B. & Bunnell, F. L. (1994). Dynamics of a small, hunted brown bear *Ursus arctos* population in southwestern Alberta, Canada. *Biological Conservation*, **67**, 161–6.

Wiese, R. & Hutchins, M. (1994). *Species Survival Plans: Strategies for Wildlife Conservation*. Bethesda, Maryland, American Zoo and Aquarium Association.

Wiese, R., Wildt, D., Byers, A. & Johnston, L. (1994). Tiger population management. In *Sumatran Tiger Population and Habitat Viability Analysis Report*, ed. R. L. Tilson, K. Soemarna, W. Ramono, S. Lusli, K. Traylor-Holzer & U. S. Seal, pp. 71–4. Apple Valley, Minnesota, Indonesian Directorate of Forest Protection and Nature Conservation and IUCN/SSC Conservation Breeding Specialist Group.

Wikramanayake, E. D., Dinerstein, E., Robinson, G., Karanth, K. U., Rabinowitz, A. R., Olson, D., Matthew, T., Hedao, P., Connor, M., Hemley, G. & Bolze, D. (1998) An ecology-based method of defining priorities for large mammal conservation: the tiger as case study. *Conservation Biology*, **12**, 865–78.

Wildt, D., Byers, A., Johnston, L., Howard, J., Willis, K., O'Brien, S., Tilson, R., Rall, W. & Seal, U. (1995). *Tiger Genome Resource Banking (GRB) Action Plan: Global Need and a Plan for North America*. Apple Valley, Minnesota, IUCN/SSC Conservation Breeding Specialist Group and American Zoo and Aquarium Association.

Wildt, D. E., Bush, M., Goodrowe, K. L., Packer, C., Pusey, A. E., Brown, J. L., Joslin, P. & O'Brien, S. J. (1987). Reproductive and genetic consequences of founding isolated lion populations. *Nature*, **329**, 328–31.

Wilson, E. O. (1993). *The Diversity of Life*. New York, W. W. Norton.

Wilson, K. R. & Anderson, D. R. (1985). Evaluation of two density estimators of small mammal population size. *Journal of Mammalogy*, **66**, 13–21.

World Bank (1995). *People's Republic of China Nature Reserve Management Project*. Washington DC, Global Environmental Coordination Division, Environmental Department, World Bank.

World Bank (1996a). *Cambodia: Forest Policy Assessment*. Report No. 15777-KH, Agriculture and Environment Division, East Asia and Pacific Region.

World Bank (1996b). *India Ecodevelopment Project*. Project Document, The World Bank, Washington DC.

Wright, B. (1997). *India's Tiger Poaching Crisis*. India, Wildlife Protection Society of India.

Wu Xin-zhi, Yuan Zhen-xin, Han De-fen, Qi Tao & Lu Qin-wu (1966). Report of the excavation at Lantian man locality of Gongwangling in 1965. *Vertebrata Palasiatica*, **10**, 23–9.

WWF. (1995). *WWF Technical Assistance and Conservation Fund for Conserving Biological Resources in the Russian Far East*. Workplan for FY 1996 through FY 1998. Report prepared by WWF-Russian Programme Office and WWF-US for United States Aid for International Development (USAID).

Yasuma, S. (1994). *An Invitation to the Mammals of East Kalimantan*. PUSREHUT Special Publication 3. Republic of Indonesia, Japan International Cooperation Agency and Directorate General of Higher Education.

Yelyakov, G. B., Aladin, V. E., Baklanov, P. Y., Bogatov, V. V., Zhuravlev, Y. N., Kachur, A. N., Kucanin, O. G., Lebedev, B. E., Ler, L. A., Medvedev, A. D., Rosenberg, V. A., Stometyuk, E. C. & Turyetskin, V. S. (1993). *Long-term Programme for Nature Protection and Rational Use of Natural Resources of Primorski Krai Through the Year 2005*, (Ecological programme). Part I. Vladivostok, Dalnauka. (In Russian.)

Yergin, D. & Gustafson, T. (1995). *Russia 2010 and What it Means for the World*. The Cera Report. New York, Vintage Books.

Yonzon, P. (1994). *Count Rhino '94*. Kathmandu, Nepal, Resources Nepal.

Young, C. C. (1932). On some fossil mammals from Yunnan. *Bulletin of the Geological Society of China*, **11**, 383–93.

Young, C. C. (1939). New fossils from Wanhsien (Szechuan). Bulletin of the Geological Society of China, 19, 317–31.

Young, C.-C. & Chow, M.C. (1955). Pleistocene stratigraphy and new fossil locations of Shihhung and Wuho, northern Anhwei. *Acta Palaeontological Sinica*, **3**, 43–53.

Yudakov, A. G. & Nikolaev, I. G. (1970). *Report on Population Census of Primorski Krai*. (In Russian.)

Yudakov, A. G. & Nikolaev I. G. (1973). Status of the Amur tiger population (*Panthera tigris altaica*) in Primorski Krai. *Journal of Zoology (Zoologicheski Djournal)*, **52**(6), 909–19. (Russian with English summary).

Yudakov, A. G. & Nikolaev, I. G. (1987). *Ecology of the Amur tiger. Winter observations during 1970–1973 in the western section of central Sikhote-Alin*. Moscow, Nauka Press. (In Russian.)

Yuhki, N. & O'Brien, S. J. (1990a). DNA recombination and natural selection pressure sustain genetic sequence diversity of the feline MHC class I genes. *Journal of Experimental Medicine*, **172**, 621–30.

Yuhki, N. & O'Brien, S. J. (1990b). DNA variation of the mammalian major histocompatibility complex reflects genomic diversity and population history. *Proceedings of the National Academy of Science, USA*, **87**, 836–40.

Yuhki, N. & O'Brien, S. J. (1997). Nature and origin of polymorphism in feline MHC class II *DRA* and *DRB* genes. *Journal of Immunology*, **158**, 2822–33.

Zar, J. H. (1984). *Biostatistical Analysis*. New Jersey, Prentice-Hall, Inc.

Zdansky, O. (1924). Jungtertiäre Carnivoren. *Palaeontologia Sinica, Series C*, **2**(1), 1–145.

Zdansky, O. (1928). Die Säugetiere der Quartärfauna von Chou -K'ou-Tien. *Palaeontologia Sinica, Series C*, **5**(4), 1–146.

Zhang Zulu (1994). The first discovery of *Panthera tigris* fossils in Shandong Province and its significance. *Marine Geology and Quaternary Geology*, **14**, 69–74.

Zhivotchenko, V. E. (1981). Food habits of the Amur tiger. In *Mammalian Predators*. ed. V. A. Zabroden, pp. 64–75. Moscow, TNEL Glavokhoti. (In Russian.)

Index